AAOS

Eighth Edition

Student Workbook to accompany
Emergency
Care and Transportation of the Sick and Injured

JONES AND BARTLETT PUBLISHERS

Sudbury, Massachusetts

BOSTON   TORONTO   LONDON   SINGAPORE

Emergency Care and Transportation of the Sick and Injured, Eighth Edition

Student Workbook

World Headquarters
40 Tall Pine Drive
Sudbury, MA 01776
978-443-5000
www.emtb.com
www.jbpub.com

Jones and Bartlett Publishers
Canada
2406 Nikanna Road
Mississauga, ON L5C 2W6
CANADA

Jones and Bartlett Publishers
International
Barb House, Barb Mews
London W6 7PA
UK

ISBN: 0-7637-1943-9
This Student Workbook is intended solely as a guide to the appropriate procedures to be employed when rendering emergency care to the sick and injured. It is not intended as a statement of the standards of care required in any particular situation, because circumstances and the patient's physical condition can vary widely from one emergency to another. Nor is it intended that this Student Workbook shall in any way advise emergency personnel concerning legal authority to perform the activities or procedures discussed. Such local determinations should be made only with the aid of legal counsel.

Note: The patients depicted in *Ambulance Calls* are fictitious.

Editorial Credits

Author: Rhonda J. Beck, NREMT-P

American Academy of Orthopaedic Surgeons
Vice President, Educational Programs: Mark W. Wieting
Director of Publications: Marilyn L. Fox, Ph.D.
Managing Editor: Lynne Roby Shindoll
Senior Editor: Barbara A. Scotese

Production Credits

Emergency Care Acquisitions Editor: Kimberly Brophy
Emergency Care Associate Editor: Carol Brewer
Manufacturing Buyer: Therese Bräuer
Senior Production Editor: Linda S. DeBruyn
Interior Design: Anne Spencer
Cover Design: Philip Regan
Composition: Nesbitt Graphics, Inc.
Printing and Binding: Courier Corporation

Printed in the United States of America

06 05 04 03 02   10 9 8 7 6 5 4 3 2 1

Table of Contents

Technology Resources

An important part of learning to become an EMT-B is by doing. You will find numerous types of interactivities and simulations at our web site: **www.emtb.com**.

www.emtb.com

Anatomy Review

allows interactive anatomical figure labeling.

Web Links

present current information, including trends in health care, the EMS community, and new equipment.

Online Outlook

activities further reinforce and expand on topics covered in each chapter.

Online Chapter Pretests

prepare you for training with instant results, feedback on incorrect answers, and page references to the Eighth Edition.

Online Review Manual

prepares you for regional, state, and national exams by providing instant results, feedback on incorrect answers, and page references to the Eighth Edition.

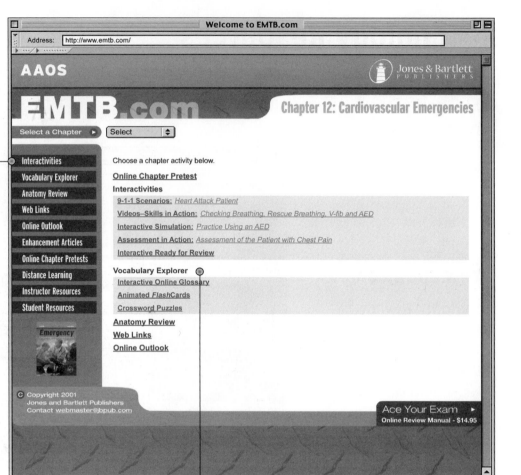

Interactivities

9-1-1 Scenarios

provide actual 9-1-1 calls and questions to prepare you for the field.

Videos: Skills in Action

show experienced providers performing critical skills.

Interactive Simulations

allow you to practice EMT-B skills in the safety of a virtual environment.

Assessment in Action

scenarios challenge your problem-solving abilities and give feedback.

Interactive Ready for Review

tests your comprehension of chapter material.

Vocabulary Explorer

Interactive Online Glossary

expand your medical vocabulary, complete with sound and images.

Animated FlashCards

review vital vocabulary and key concepts.

Student Resources

Review Manual

ISBN: 0-7637-1945-5 * Print
ISBN: 0-7637-1947-1 * CD-ROM
ISBN: 0-7637-1946-3 * Online at www.emtb.com

This Review Manual has been designed to prepare you to sit for exams by including the same type of scenario-based and multiple-choice questions that you are likely to see on classroom and national examinations. The manual is available in print, on CD-ROM, and online and provides answers to all questions with brief explanations and page references to the Eighth Edition.

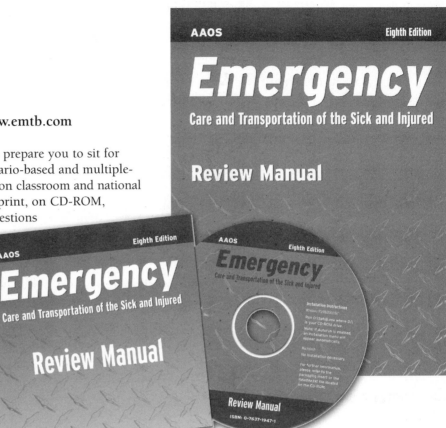

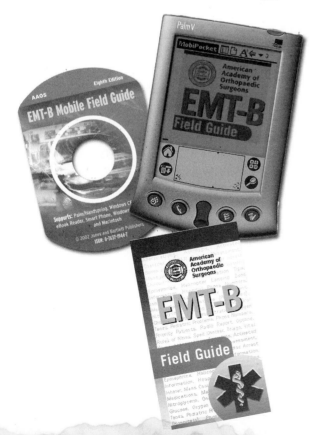

Mobile EMT-B Field Guide

ISBN: 0-7637-1944-7

This essential resource is packaged with each copy of *Emergency Care and Transportation of the Sick and Injured, Eighth Edition*. It provides instant, portable access to the EMT-B Field Guide on your PalmPilot™, eBook Reader, Windows CE™ -based handheld device, or Windows PC. The Field Guide provides the vital emergency information needed by prehospital personnel. It is a great resource for fast review of important step-by-step procedures, offering just the right amount of information to guide actions and reactions in the field.

The EMT-B Field Guide

ISBN: 0-7637-0880-1

This handy reference covers basic information from patient management tips to guidelines on helping a patient use medication. A few special features include a prescription medication reference and documentation tips. The EMT-B Field Guide is pocket-sized, spiral-bound, and water resistant for ready-reference in the field.

Workbook Activities

The following activities have been designed to help you. Your instructor may require you to complete some or all of these activities as a regular part of your EMT-B training program. You are encouraged to complete any activity that your instructor does not assign as a way to enhance your learning in the classroom.

Chapter Review

The following exercises provide an opportunity to refresh your knowledge of this chapter.

NOTES

Matching

Match each of the items in the left column to the appropriate definition in the right column.

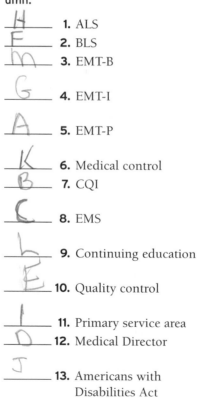

_H_ **1.** ALS

_F_ **2.** BLS

_M_ **3.** EMT-B

_G_ **4.** EMT-I

_A_ **5.** EMT-P

_K_ **6.** Medical control

_B_ **7.** CQI

_C_ **8.** EMS

_L_ **9.** Continuing education

_E_ **10.** Quality control

_I_ **11.** Primary service area

_D_ **12.** Medical Director

_J_ **13.** Americans with Disabilities Act

A. EMS professional trained in ALS

B. a system of internal reviews and audits

C. a system to provide prehospital care to the sick and injured

D. the physician who authorizes the EMT to perform in the field

E. responsibility of medical director to ensure appropriate care is delivered by EMT

F. basic lifesaving interventions, such as CPR

G. EMS professional trained in some ALS interventions

H. advanced procedures such as drug administration

I. designated area in which the EMS service is responsible

J. protects disabled individuals from discrimination

K. physician instruction to EMS team

L. a required amount of training to maintain skills

M. EMS professional trained in BLS

Introduction to Emergency Medical Care

Multiple Choice

Read each item carefully, then select the best response.

_B_ **1.** Control of external bleeding, provision of oxygen and CPR are included in the "scope of practice" of the:

A. EMT-P.

B. EMT-B.

C. EMT-I.

D. EMT-D.

_C_ **2.** The U.S. DOT 1994 EMT-Basic National Standard Curriculum requires a minimum of ____ hours of training.

A. 80

B. 100

C. 110

D. 160

_A_ **3.** All of the following are true of medical control except:

A. It is determined by the dispatcher.

B. It may be written or "standing orders."

C. It may require online radio or phone consultation.

D. It describes the care authorized by the medical director.

_D_ **4.** All of the following are components of continuous quality control except:

A. periodic run reviews.

B. remedial training.

C. internal reviews and audits.

D. public seminars and meetings.

_____C_____ **5.** The major goal of quality improvement is to ensure that:

    A. quarterly audits of the EMS System are done.

    B. EMTs have received BLS/CPR training.

    C. the public receives the highest standard of care.

    D. the proper information is received in the billing department.

_____A_____ **6.** Your main concern while responding to a call should be the:

    A. safety of the crew and yourself.

    B. number of potential patients.

    C. request for mutual assistance.

    D. type of call.

_____B_____ **7.** The first phase of the emergency care continuum consists of:

    A. recognition of the emergency by the public.

    B. patient assessment, stabilization, packaging, and transport.

    C. safe delivery of the patient to definitive care.

    D. accurate relay of information by the dispatcher.

_____B_____ **8.** Care for burns, delivery of a baby, and management of patients with behavioral problems are conditions covered in which category of the EMT-B's training?

    A. care of life-threatening problems

    B. care of problems not life-threatening

    C. important non-medical problems

    D. These are outside the EMT-B's scope of practice

_____C_____ **9.** Understanding legal and ethical issues, learning defensive driving, and stocking the ambulance are covered in which category of the EMT-B's training?

    A. care of life-threatening problems

    B. care of problems not life-threatening

    C. important non-medical problems

    D. These are not the concern of the EMT-B

_____B_____ **10.** Which of the following groups is responsible for the national standard curriculum for the EMT-B?

    A. American Academy of Orthopaedic Surgeons

    B. Department of Transportation

    C. American Heart Association

    D. National Association of Emergency Medical Technicians

Vocabulary emtb. vocab explorer

Define the following terms using the space provided.

1. Emergency medical services (EMS):

_____

_____

_____

2. First responder:

_____

3. Primary service area (PSA):

_____

_____

_____

4. Emergency medical technician (EMT):

_____

_____

_____

Fill-in

Read each item carefully, then complete the statement by filling in the missing word(s).

1. The training of the EMT-B should meet or exceed the guidelines of _____.

2. EMT-B training is divided into care of _____, _____, and important non-medical issues related to EMT abilities.

3. In some areas, EMT-Bs may provide selected ALS care such as _____, use of airway adjuncts, and assisting patients in taking prescription _____.

4. In most of the country, a communications center can be reached easily by the public by dialing _____.

5. The appropriate care for injury or illness as described by the medical director either by radio or in written form is _____.

True/False

If you believe the statement to be more true than false, write the letter "T" in the space provided. If you believe the statement to be more false than true, write the letter "F."

1. ___F___ EMT-B personnel are the highest qualified members of the prehospital care team.

2. ___T___ The EMT-B scope of practice may include the use of an automated or semi-automated defibrillator.

3. ___F___ The purpose of continuous quality improvement (CQI) is to support discipline of personnel.

4. ___T___ A professional appearance and manner by the EMT-B will help a patient build confidence.

5. ___T___ Essential keys to being a good EMT-B include compassion, commitment, and desire.

■ NOTES ■

Short Answer

Complete this section with short written answers using the space provided.

1. Describe the EMT-B's role in the EMS system.

_____

_____

_____

2. What role has the U.S. Department of Transportation played in the development of EMS?

_____

_____

_____

3. List five roles and/or responsibilities of being an EMT-B.

_____

_____

_____

_____

4. Describe the two basic types of medical direction that help the EMT-B provide care.

_____

_____

_____

■ CLUES ■

Across

2. System that provides pre-hospital emergency care

4. An EMT with training in BLS/CPR

7. Designed to protect individuals with disabilities

8. Section of the U.S. DOT that created programs to improve EMS

9. An individual trained to provide emergency care to the sick and injured

10. BLS intervention in cardiac arrest

Down

1. An EMT with extensive training in advanced life support

2. A type of 9-1-1 system that displays the caller's address

3. Simple lifesaving interventions, i.e., CPR

5. Advanced lifesaving procedures, i.e., defibrillation

6. Medical or Quality _____

10. A circular system of reviews and changes in an EMS system

Word Fun emtb.com vocab explorer

The following crossword puzzle is an activity provided to reinforce correct spelling and understanding of medical terminology associated with emergency care and the EMT-B. Use the clues in the column to complete the puzzle.

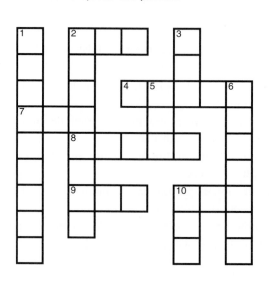

Ambulance Calls

The following real case scenarios provide an opportunity to explore the concerns associated with patient management. Read each scenario, then answer each question in detail.

1. While responding to a call for a "man down" at a local restaurant, you encounter a motor vehicle collision between a delivery van and a station wagon. There appear to be injuries among several of the occupants of the vehicles. The restaurant is several blocks away.

   How would you best manage this situation?

   _____

   _____

   _____

   _____

   _____

   _____

   _____

   _____

   _____

2. You are dispatched to a private residence. Several neighbors are gathered on the front lawn. There appears to be an argument taking place between two of them. A teenage boy is sitting on the doorstep, bleeding profusely from a cut above his left eye.

   How would you best manage this situation?

   _____

   _____

   _____

   _____

   _____

   _____

   _____

   _____

3. You have been called by the local fire service to transport a heart attack patient to an area hospital outside their primary service area. The paramedics have established an IV line and administered medication to the patient, who is now "pain-free." It will be a 15-minute transport to the hospital. The paramedics state that they will not be accompanying the patient but have contacted the hospital, which has accepted the patient.

How would you best manage this situation?

_____

_____

_____

_____

_____

_____

_____

_____

_____

Notes

I need to work on definitions of

Workbook Activities

The following activities have been designed to help you. Your instructor may require you to complete some or all of these activities as a regular part of your EMT-B training program. You are encouraged to complete any activity that your instructor does not assign as a way to enhance your learning in the classroom.

Chapter Review

The following exercises provide an opportunity to refresh your knowledge of this chapter.

Matching

Match each of the terms in the left column to the appropriate definition in the right column.

_D_ 1. Cover

_A_ 2. Burnout

_C_ 3. Occupational Safety and Health Administration

_F_ 4. Posttraumatic stress disorder

_B_ 5. Body substance isolation

_E_ 6. Pathogen

_N_ 7. Transmission

_H_ 8. Tuberculosis

_M_ 9. Virulence

_J_ 10. Hepatitis

_K_ 11. Exposure

_O_ 12. Infection control

_I_ 13. Meningitis

_G_ 14. Host

_L_ 15. Contamination

A. chronic fatigue and frustration

B. assuming all body fluids are potentially infected

C. regulatory compliance agency

D. concealment for protection

E. capable of causing disease in a susceptible host

F. delayed stress reaction

G. the organism or individual that is attacked by the infecting agent

H. chronic bacterial disease that usually affects the lungs

I. an inflammation of the meningeal coverings of the brain

J. an infection of the liver

K. contact with blood, body fluids, tissues, or airborne particles

L. the presence of infectious organisms in or on objects, or a patient's body

M. the strength or ability of a pathogen to produce disease

N. the way in which an infectious agent is spread

O. procedures to reduce transmission of infection among patients and health care personnel

The Well-Being of the EMT-B

Multiple Choice

Read each item carefully, then select the best response.

_____ **1.** Self-control is developed through all of the following, except:

    A. proper training.

    B. medication.

    C. a dedication to serve humanity.

    D. ongoing experience in dealing with all types of physical and mental distress.

_____ **2.** From the age of 1 to the age of 34, _____ is the leading cause of death.

    A. cardiac arrest

    B. congenital disease

    C. trauma

    D. AIDS

_____ **3.** Presumptive signs of death would not be adequate in cases of sudden death due to:

    A. hypothermia.

    B. acute poisoning.

    C. cardiac arrest.

    D. all of the above

_____ **4.** Definitive or conclusive signs of death that are obvious and clear to even nonmedical persons include all of the following, except:

    A. profound cyanosis.

    B. dependent lividity.

    C. rigor mortis.

    D. putrefaction.

_____ **5.** Medical examiner's cases include:

    A. violent death.

    B. suicide.

    C. suspicion of a criminal act.

    D. all of the above

_____C_____ **6.** The stage of the grieving process where an attempt is made to secure a prize for good behavior or promise to change lifestyle is known as:

   A. denial.

   B. acceptance.

   C. bargaining.

   D. depression.

_____A_____ **7.** The stage of the grieving process that involves refusal to accept diagnosis or care is known as:

   A. denial.

   B. acceptance.

   C. bargaining.

   D. depression.

_____D_____ **8.** The stage of the grieving process that involves an open expression of grief, internalized anger, hopelessness, and/or the desire to die is:

   A. denial.

   B. acceptance.

   C. bargaining.

   D. depression.

_____B_____ **9.** The stage of grieving where the person is ready to die is known as:

   A. denial.

   B. acceptance.

   C. bargaining.

   D. depression.

_____A_____ **10.** When providing support for a grieving person, it is okay to say:

   A. "I'm sorry."

   B. "Give it time."

   C. "I know how you feel."

   D. "You have to keep on going."

_____D_____ **11.** When grieving, family members may express:

   A. rage.

   B. anger.

   C. despair.

   D. all of the above

_____C_____ **12.** _____ is a response to the anticipation of danger.

   A. Rage

   B. Anger

   C. Anxiety

   D. Despair

_____A_(B)_ **13.** Signs of anxiety include all of the following, except:

   A. diaphoresis.

   B. fear.

   C. hyperventilation.

   D. tachycardia.

_____D_____ **14.** Fear may be expressed as:

   A. anger.

   B. bad dreams.

   C. restlessness.

   D. all of the above

_____ D **15.** If you find that you are the target of the patient's anger, make sure that you:

    A. are safe.

    B. do not take the anger or insults personally.

    C. are tolerant, and do not become defensive.

    D. all of the above

_____ B **16.** All of the following except _____ are common characteristics of mental health problems.

    A. confusion

    B. exhilaration

    C. distortion of perception

    D. abnormal mental content

_____ D **17.** When caring for critcally ill or injured patients, _____ will be decreased if you can keep the patient informed at the scene.

    A. confusion

    B. anxiety

    C. feelings of helplessness

    D. all of the above

_____ D **18.** When acknowledging the death of a child, reactions vary, but _____ is common.

    A. shock

    B. disbelief

    C. denial

    D. all of the above

_____ A **19.** Factors influencing how a patient reacts to the stress of an EMS incident include all of the following, except:

    A. family history.

    B. age.

    C. fear of medical personnel.

    D. socioeconomic background.

_____ B **20.** Negative forms of stress include all of the following, except:

    A. long hours.

    B. exercise.

    C. shift work.

    D. frustration of losing a patient.

_____ D **21.** Stressors include _____ situations or conditions that may cause a variety of physiologic, physical, and psychologic responses.

    A. emotional

    B. physical

    C. environmental

    D. all of the above

_____ C **22.** Physical symptoms of stress include all of the following, except:

    A. fatigue.

    B. changes in appetite.

    C. increased blood pressure.

    D. headaches.

_D_ **23.** Prolonged or excessive stress has been proven to be a strong contributor to:

    A. heart disease.

    B. hypertension.

    C. cancer.

    D. all of the above

_B_ **24.** _____ occur(s) when insignificant stressors accumulate to a larger stress-related problem.

    A. Negative stress

    B. Cumulative stress

    C. Psychological stress

    D. Severe stressors

_D_ **25.** Events that can trigger critical incident stress include:

    A. mass-casualty incidents.

    B. serious injury or traumatic death of a child.

    C. death or serious injury of a coworker in the line of duty.

    D. all of the above

_A_ **26.** The quickest source of energy is _____; however, this supply will last less than a day and is consumed in greater quantities during stress.

    A. glucose

    B. carbohydrate

    C. protein

    D. fat

_B_ **27.** The body conserves water during periods of stress through retaining _____ by exchanging and losing potassium from the kidneys.

    A. water-soluble B vitamins

    B. sodium

    C. calcium

    D. water-soluble C vitamins

_D_ **28.** Stress management strategies include:

    A. changing work hours.

    B. changing your attitude.

    C. changing partners.

    D. all of the above

_D_ **29.** _____ is a condition of chronic fatigue and frustration that results from mounting stress over time.

    A. Posttraumatic stress disorder

    B. Cumulative stress

    C. Critical incident stress

    D. Burnout

_B_ **30.** The safest, most reliable sources for long-term energy production are:

    A. sugars.

    B. carbohydrates.

    C. fats.

    D. proteins.

31. A _____ is any event that causes anxiety and mental stress to emergency workers.

    A. disaster

    B. mass-casualty incident

    C. critical incident

    D. stressor

32. A CISD meeting is an opportunity to discuss your:

    A. feelings.

    B. fears.

    C. reactions to the event.

    D. all of the above

33. Components of the CISM system include:

    A. preincident stress education.

    B. defusings.

    C. spouse and family support.

    D. all of the above

34. Sexual harrassment is defined as:

    A. any unwelcome sexual advance.

    B. unwelcome requests for sexual favors.

    C. unwelcome verbal or physical conduct of a sexual nature.

    D. all of the above

35. Drug and alcohol use in the workplace can result in all of the following, except:

    A. absence from work more often.

    B. enhanced treatment decisions.

    C. an increase in accidents and tension among workers.

    D. lessened ability to render emergency medical care because of mental or physical impairment.

36. You should begin protecting yourself:

    A. as soon as you arrive on the scene.

    B. before you leave the scene.

    C. as soon as you are dispatched.

    D. before any patient contact.

37. _____ is the way an infectious agent is spread.

    A. The route

    B. The mechanism

    C. Transmission

    D. Exposure

38. _____ is contact with blood, body fluids, tissues, or airborne droplets by direct or indirect contact.

    A. Transmission

    B. Exposure

    C. Handling

    D. all of the above

39. Modes of transmission for infectious diseases include:

    A. blood or fluid splash.

    B. surface contamination.

    C. needlestick exposure.

    D. all of the above

■ ■ NOTES ■ ■

_A_ **40.** The spread of HIV and hepatitis in the health care setting can usually be traced to:

A. careless handling of sharps.

B. improper use of BSI precautions.

C. not wearing PPE.

D. sexual interaction with infected persons.

_C_ **41.** _____ is equipment that blocks entry of an organism into the body.

A. Vaccination

B. Body substance isolation

C. Personal protective equipment

D. Immunization

_B_ **42.** _____ is a major factor in determining which hosts become ill from which germs.

A. Immunization

B. Immunity

C. Vaccination

D. A pathogen

_D_ **43.** Recommended immunizations include:

A. MMR vaccine.

B. hepatitis B vaccine.

C. influenza vaccine.

D. all of the above

_B_ **44.** _____ is the presence of an infectious organism on or in an object.

A. Virulence

B. Contamination

C. Immunity

D. Transmission

_D_ **45.** Why isn't tuberculosis more common?

A. Absolute protection from infection with the tubercle bacillus does not exist.

B. Everyone who breathes is at risk.

C. The vaccine for tuberculosis is only rarely used in the United States.

D. Infected air is easily diluted with uninfected air.

_B_ **47.** You can use a bleach and water solution at a _____ dilution to clean the unit.

A. 1:1

B. 1:10

C. 1:100

D. 1:1000

_D_ **48.** Hazardous materials are classified according to _____, which dictate(s) the level of protection required.

A. danger zones

B. flammability

C. toxicity levels

D. all of the above

B

___**49.** Breathing concentrations of carbon dioxide above _____ will result in
death within a few minutes.

    A. 5% to 8%

    B. 10% to 12%

    C. 13% to 15%

    D. 18% to 20%

D

___**50.** Factors to take into consideration for potential violence include:

    A. poor impulse control.

    B. substance abuse.

    C. depression.

    D. all of the above

Vocabulary

Define the following terms using the space provided.

    1. Critical incident stress management (CISM):

_____

_____

_____

    2. Posttraumatic stress disorder (PTSD):

_____

_____

_____

    3. Critical incident stress debriefing (CISD):

_____

_____

_____

Fill-in

Read each item carefully, then complete the statement by filling in the missing word.

    1. The personal health, safety, and _____ of all EMT-Bs are vital to an
EMS operation.

    2. The struggle to remain calm in the face of horrible circumstances contributes to

the _____ of the job.

    3. Sixty percent of all deaths today are attributed to _____ .

    4. Determination of the cause of death is the medical responsibility of a

_____ .

5. In cases of hypothermia, the patient should not be considered dead until the
patient is _____ and dead.

6. In most states, when trauma is a factor or the death involves suspected criminal
or unusual situations such as hanging or poisoning, the _____ must
be notified.

7. _____ is generally thought of in relation to the oncoming pain and
the outcome of the damage.

8. Almost all dying patients feel some degree of _____ because of
internalized anger and other factors.

9. You must also realize that the most _____ symptoms may be early
signs of severe illness or injury.

10. EMS is a _____ job.

True/False

If you believe the statement to be more true than false, write the letter "T" in the space
provided. If you believe the statement to be more false than true, write the letter "F."

1. _____ Developing self-control is aided by proper training.
2. _____ A low or decreased body temperature is sufficient evidence of death.
3. _____ Rigor mortis is a softening of body muscles shortly after death.
4. _____ Putrefaction is the decomposition of body tissues.
5. _____ Denial is generally the first step in the grieving process.
6. _____ Body fluids are generally not considered infectious substances.
7. _____ Most EMT-Bs never suffer from stress.
8. _____ Physical conditioning and nutrition are two factors the EMT-B can
control in helping reduce stress.

Short Answer

Complete this section with short written answers using the space provided.

1. Describe the basic concept of body substance isolation (BSI).

_____

_____

_____

2. List five of the presumptive signs of death.

_____

_____

_____

_____

3. What are the four definitive signs of death?

_____

_____

_____

_____

4. List the five stages of the grieving process.

_____

_____

_____

_____

_____

5. List five warning signs of stress.

_____

_____

_____

_____

_____

6. List two strategies for managing stress.

_____

_____

7. Describe the process for proper handwashing.

_____

_____

_____

8. Complete the following table on the toxicity of hazardous materials.

Level	Hazard	Protection Needed
0		
1		
2		
3		
4		

9. List the three layers of clothing recommended for cold weather.

_____

_____

_____

10. List the four principal determinants of violence.

_____

_____

_____

_____

Word Fun

The following crossword puzzle is an activity provided to reinforce correct spelling and understanding of medical terminology associated with emergency care and the EMT-B. Use the clues in the column to complete the puzzle.

Across

1. Impenetrable barrier for tactical use

2. Confronts responses and defuses them

5. Exposure by physical touching

7. Disease of the lungs

9. Delayed reaction to past incident

10. Infection of the liver

11. Capable of causing disease

Down

1. Disease spread from person to person

3. Chronic fatigue and frustration

4. Infection control process

6. Confidential discussion group

8. Federal workplace safety agency

9. Blocks entry of an organism into the body

12. May progress to AIDS

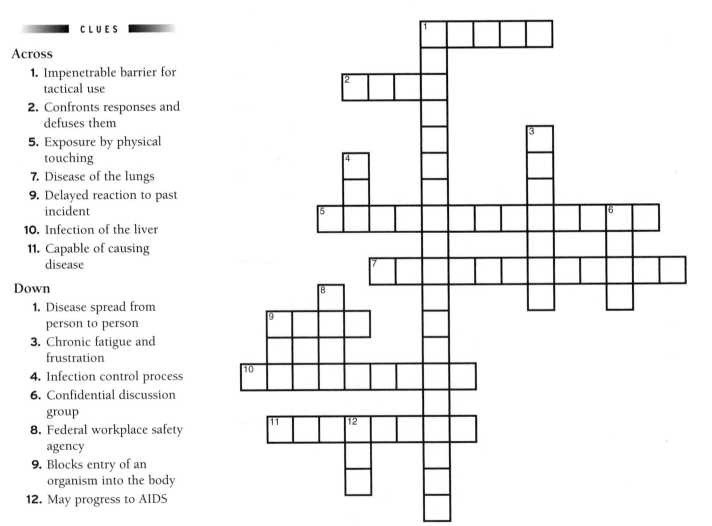

Ambulance Calls

The following real case scenarios provide an opportunity to explore the concerns associated with patient management. Read each scenario, then answer each question in detail.

1. You are dispatched to a residence for a "person not breathing." Upon arrival, you find a 78-year-old female with a cardiac history who is apneic and pulseless. She also shows signs of rigor mortis and dependent lividity. The family is extremely upset and asking you and your partner to do something.

   How would you best manage this situation?

   _____

   _____

   _____

   _____

   _____

   _____

   _____

   _____

   _____

   _____

2. You are on the scene of a motor vehicle crash where the driver, a 27-year-old female, is breathing shallowly and has blood in her airway. She is unresponsive and you can see obvious fractures of her lower legs. Her 3-year-old son is restrained in a car seat in the back and is screaming and covered in blood. There is no damage to the vehicle where he is sitting and from a quick check you determine that the blood probably belongs to his mother.

   How would you best manage these patients?

   _____

   _____

   _____

   _____

   _____

   _____

   _____

   _____

3. In the process of working a motor vehicle crash, your arm is gashed open and you are exposed to the blood of a patient who tells you that he is HIV positive. You have no water supply in which to wash. Your patient is stable and you are able to control his bleeding with direct pressure.

How would you best manage this situation?

_____

_____

_____

_____

_____

_____

_____

_____

Skill Drills

Skill Drill 2-1: Proper Glove Removal Technique
Test your knowledge of skill drills by placing the photos below in the correct order. Number the first step with a "1," the second step with a "2," etc.

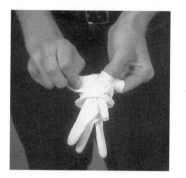

Grasp both gloves with your free hand touching only the clean, interior surfaces.

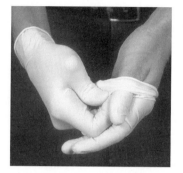

Partially remove the first glove by pinching at the wrist. Be careful to touch only the outside of the glove.

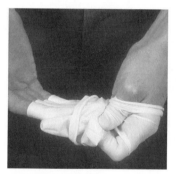

Pull the second glove inside out toward the fingertips.

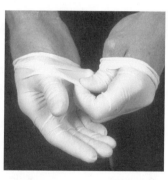

Remove the second glove by pinching the exterior with the partially gloved hand.

Workbook Activities

The following activities have been designed to help you. Your instructor may require you to complete some or all of these activities as a regular part of your EMT-B training program. You are encouraged to complete any activity that your instructor does not assign as a way to enhance your learning in the classroom.

Chapter Review

The following exercises provide an opportunity to refresh your knowledge of this chapter.

■ NOTES ■

Matching

Match each of the terms in the left column to the appropriate definition in the right column.

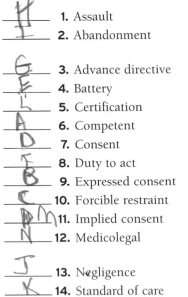

H	1. Assault	A. able to make decisions
I	2. Abandonment	B. specific authorization to provide care expressed by the patient
G	3. Advance directive	C. confining a person from mental or physical action
E	4. Battery	D. granted permission
L	5. Certification	E. touching without consent
A	6. Competent	F. legal responsibility to provide care
D	7. Consent	G. written documentation that specifies treatment
F	8. Duty to act	H. unlawfully placing a patient in fear of bodily harm
B	9. Expressed consent	I. unilateral termination of care
C	10. Forcible restraint	J. failure to provide standard of care
M	11. Implied consent	K. accepted level of care consistent with training
N	12. Medicolegal	L. process that recognizes that a person has met set standards
J	13. Negligence	M. legal assumption that treatment was desired
K	14. Standard of care	N. relating to law or forensic medicine

Multiple Choice

Read each item carefully, then select the best response.

   C   1. The care the EMT-B is able to provide, most commonly defined by state law, is:

    A. duty to act.

    B. competency.

    C. scope of practice.

    D. certification.

Medical, Legal, and Ethical Issues

_____A_____ **2.** How the EMT-B is required to act or behave is called the:

   A. standard of care.

   B. competency.

   C. scope of practice.

   D. certification.

_____D_____ **3.** The process by which an individual, institution, or program is evaluated and recognized as meeting certain standards is called:

   A. standard of care.

   B. competency.

   C. scope of practice.

   D. certification.

_____D_____ **4.** Negligence is based on the EMT-B's duty to act, cause, breach of duty, and:

   A. expressed consent.

   B. termination of care.

   C. mode of transport.

   D. real or perceived damages.

_____A_____ **5.** Which of the following forms of consent applies when the patient is considered mentally incompetent and the guardian is not readily available?

   A. protective custody

   B. informed

   C. expressed

   D. implied

_____C_____ **6.** While treating a patient with a suspected head injury, he becomes verbally abusive and tells you to "leave him alone." If you stop treating him you may be guilty of:

   A. neglect.

   B. battery.

   C. abandonment.

   D. slander.

_B_ **7.** Good Samaritan laws generally are designed to offer protection to persons who render care in good faith. They do not offer protection from:

A. properly performed CPR.

B. acts of negligence.

C. improvising splinting materials.

D. providing supportive BLS to a DNR patient.

_D_ **8.** Which of the following is generally NOT considered confidential?

A. assessment findings

B. patient's mental condition

C. patient's medical history

D. the location of the emergency

_D_ **9.** An important safeguard against legal implication is:

A. responding to every call with lights and siren.

B. checking ambulance equipment once a month.

C. transporting every patient to an emergency department.

D. a complete and accurate incident report.

Vocabulary emtb vocab explorer

Define the following terms using the space provided.

1. Abandonment:

_____

_____

_____

2. Advance directive:

_____

_____

_____

3. Assault:

_____

_____

_____

4. Battery:

_____

_____

_____

5. DNR order:

_____

_____

_____

6. Certification:

_____

_____

_____

7. Duty to act:

_____

_____

_____

8. Expressed consent:

_____

_____

_____

9. Good Samaritan laws:

_____

_____

_____

10. Implied consent:

_____

_____

_____

11. Negligence:

_____

_____

_____

NOTES

1

12. Standard of care:

_____

_____

_____

Fill-in

Read each item carefully, then complete the statement by filling in the missing word(s).

1. The _____ outlines the care you are able to provide.

2. The _____ is the manner in which the EMT must act when treating patients.

3. The legal responsibility to provide care is called the _____.

4. The determination of _____ is based on duty, breach of duty, damages, and cause.

5. Abandonment is the _____ of care without transfer to someone of equal or higher training.

6. _____ consent is given directly by an informed patient, where

_____ consent is assumed in the unconscious patient.

7. Unlawfully placing a person in fear of immediate harm is _____,

while _____ is unlawfully touching a person without their consent.

8. A(n) _____ is a written document that specifies authorized treatment in case a patient becomes unable to make decisions. A written document that authorizes the EMT not to attempt resuscitation efforts is a(n)

_____.

9. Mentally competent patients have the right to _____.

10. Incidents involving child abuse, animal bites, childbirth, and assault have

_____ requirements in many states.

True/False

If you believe the statement to be more true than false, write the letter "T" in the space provided. If you believe the statement to be more false than true, write the letter "F."

1. __T__ Failure to provide care to a patient once you have been called to the scene is considered negligence.

2. __F__ For expressed consent to be valid, the patient must be a minor.

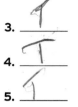

3. _____ If a patient is unconscious and a true emergency exists, the doctrine of implied consent applies.

4. _____ The EMT can legally restrain a patient against their will if the patient poses a threat to themselves or others.

5. _____ The best defense against legal action for the EMT is to always provide care in a manner consistent with ethical and moral standards.

NOTES

Short Answer

Complete this section with short written answers using the space provided.

1. In many states, certain conditions allow a minor to be treated as an adult for the purpose of consenting to medical treatment. List three of these conditions.

_____

_____

_____

2. When does your responsibility for patient care end?

_____

_____

3. There will be some instances when you will not be able to persuade the patient, guardian, conservator, or parent of a minor child or mentally incompetent patient to proceed with treatment. List five steps you should take to protect all parties involved.

_____

_____

_____

4. List the two rules of thumb courts consider regarding reports and records.

_____

_____

_____

5. List four steps to take when you are called to the scene involving a potential organ donor.

_____

_____

_____

_____

■■■■ CLUES ■■■■

Across

1. Care that the EMT-B is authorized to provide

3. Relating to medical law

8. Failure to provide standard of care

10. Touching another person without their expressed consent

11. Direct permission to provide care

12. Responsibility to provide care

Down

2. Evaluation and recognition of meeting standards

4. Able to make rational decisions

5. Placing one in fear of bodily harm

6. Assumed permission to provide care

7. Unilateral termination of care

9. A serious situation, such as an injury or illness

Word Fun

The following crossword puzzle is an activity provided to reinforce correct spelling and understanding of medical terminology associated with emergency care and the EMT-B. Use the clues in the column to complete the puzzle.

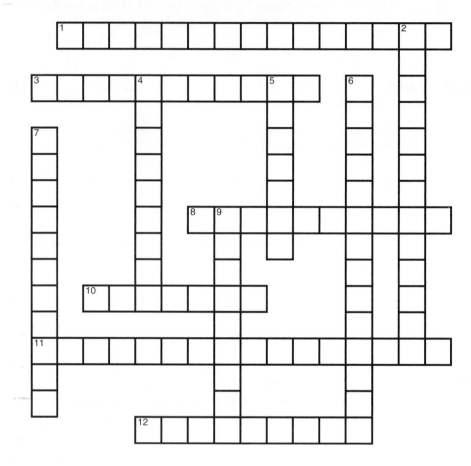

Ambulance Calls

The following real case scenarios provide an opportunity to explore the concerns associated with patient management. Read each scenario, then answer each question in detail.

1. You and your partner have arrived on the scene of a domestic dispute where the wife has stabbed the husband while "defending herself." The husband has a minor cut to his left forearm. Law enforcement officers have not yet arrived.

What actions are necessary in the management of this situation?

_____

_____

_____

_____

_____

_____

_____

_____

_____

_____

2. You are "flagged down" by a teenager at a local playground. She tells you there is a small boy injured on one of the baseball diamonds. You arrive to find a 10-year-old boy complaining of a "twisted ankle." There is obvious deformity to his right ankle with a noted loss of function. He says he came to the park to play with his friends while his mother went shopping. There are no adults immediately available.

What actions are necessary in the management of this situation?

_____

_____

_____

_____

_____

_____

_____

_____

_____

_____

3. It is late in the night when police summon you to an auto crash. On arrival the officer directs you to the back of his patrol car. Sitting on the seat is your patient, snoring loudly with blood covering his face. The officer states that the patient was involved in a drunk driving accident in which he hit his head on the rear-view mirror. The patient initially refused care at the scene. You were called because his wound continues to bleed. Assessment reveals a sleeping 56-year-old male with a deep, gaping wound over the right eye with moderate venous bleeding. During assessment the patient wakes suddenly and pushes you away. He tells you to "leave him alone."

What actions are necessary in the management of this situation?

_____

_____

_____

_____

_____

NOTES

Notes

Workbook Activities

The following activities have been designed to help you. Your instructor may require you to complete some or all of these activities as a regular part of your EMT-B training program. You are encouraged to complete any activity that your instructor does not assign as a way to enhance your learning in the classroom.

Chapter Review

The following exercises provide an opportunity to refresh your knowledge of this chapter.

■ NOTES ■

Matching

Match each of the terms in the left column to the appropriate definition in the right column.

___F___ **1.** Anterior

___K___ **2.** Capillary

___D___ **3.** Anatomic position

___G___ **4.** Superior

___E___ **5.** Midline

___L___ **6.** Carotid

___A___ **7.** Medial

___C___ **8.** Inferior

___M___ **9.** Femoral

___J___ **10.** Proximal

___O___ **11.** Brachial

___B___ **12.** Distal

___H___ **13.** Midaxillary

___N___ **14.** Radial

___I___ **15.** Posterior

A. closer to the midline

B. farther from the midline

C. farther from the head; lower

D. standing, facing forward, palms facing forward

E. imaginary vertical line descending from the middle of the forehead to the floor

F. front surface of the body

G. closer to the head; higher

H. imaginary vertical line descending from the middle of the armpit to the ankle

I. back or dorsal surface of the body

J. closer to the midline

K. connects arterioles to venules

L. major artery that supplies blood to the head and brain

M. major artery that supplies blood to the lower extremities

N. major artery of the lower arm

O. major artery of the upper arm

For each of the bones listed in the left column, indicate whether it is an upper extremity bone (A) or a lower extremity bone (B).

___B___ **16.** Acetabulum

___B___ **17.** Patella

___A___ **18.** Clavicle

___B___ **19.** Fibula

A. upper extremity bone

B. lower extremity bone

CHAPTER 4

The Human Body

_____ **20.** Calcaneus

_____ **21.** Ulna

_____ **22.** Acromion

For each of the muscle characteristics described in the left column, select the type of muscle from the right column.

_____ **23.** Attaches to the bone   **A.** skeletal

_____ **24.** Found in the walls of   **B.** smooth

_____   the gastrointestinal tract   **C.** cardiac

_____ **25.** Carries out much of the automatic work of the body

_____ **26.** Forms the major muscle mass of the body

_____ **27.** Under the direct control of the brain

_____ **28.** Found only in the heart

_____ **29.** Responds to primitive stimulus, such as heat

_____ **30.** Can tolerate blood supply interruption for only a very short period

_____ **31.** Responsible for all bodily movement

_____ **32.** Has its own blood supply and electrical system

For each of the parts of the nervous system in the left column, select the phrase in the right column with which it is associated.

_____ **33.** Spinal cord   **A.** exits the brain through an opening at the base of the skull

_____ **34.** Central nervous system   **B.** transmits electrical impulses to the muscles, causing them to contract

_____ **35.** Sensory nerves   **C.** brain and spinal cord

_____ **36.** Motor nerves   **D.** links the central nervous system to various organs in the body

_____ **37.** Brain   **E.** carries sensations of taste and touch to the brain

_____ **38.** Peripheral nervous system   **F.** controlling organ of the body

Multiple Choice

Read each item carefully, then select the best response.

_B_ **1.** The topographic term used to describe the location of an injury that is toward the midline center of the body is:

   A. lateral.

   B. medial.

   C. midaxillary.

   D. midclavicular.

_C_ **2.** Topographically, the term distal means:

   A. near the trunk.

   B. near a point of reference.

   C. below a point of reference.

   D. toward the center of the body.

_B_ **3.** The firm cartilaginous ring that forms the inferior portion of the larynx is called the:

   A. costal cartilage.

   B. cricoid cartilage.

   C. thyroid cartilage.

   D. laryngo cartilage.

_C_ **4.** Which of the following types of muscle carries out the automatic work of the body?

   A. striated

   B. skeletal

   C. smooth

   D. sensory

_D_ **5.** Which of the following is the main supporting structure of the skeleton?

   A. thorax

   B. upper extremities

   C. lower extremities

   D. spinal column

_C_ **6.** The ilium, ischium, and pubis are fused together, forming the:

   A. clavicle.

   B. sternum.

   C. pelvis.

   D. olecranon.

_B_ **7.** The leaf-shaped flap of tissue that prevents food and liquid from entering the trachea is called the:

   A. uvula.

   B. epiglottis.

   C. laryngopharynx.

   D. cricothyroid membrane.

_B_ **8.** Which of the following systems is responsible for releasing chemicals that regulate body activities?

   A. nervous

   B. endocrine

   C. cardiovascular

   D. skeletal

_____ **9.** Which of the following vessels does NOT carry blood to the heart?

    A. inferior venae cavae

    B. superior venae cavae

    C. pulmonary vein

    D. pulmonary artery

_____ **10.** The peripheral nervous system is composed of:

    A. brain, spinal cord, and motor nerves.

    B. brain, sensory, and connecting nerves.

    C. motor, sensory, and connecting nerves.

    D. spinal cord, sensory, and motor nerves.

Labeling

Label the following diagrams with the correct terms.

    1. Directional Terms

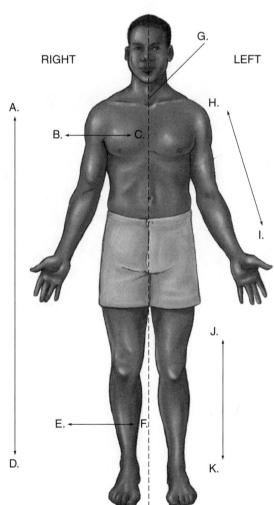

RIGHT   LEFT

A. _Superior_

B. _latteral_

C. _Medial_

D. _Inferior_

E. _latteral_

F. _medial_

G. _Midline_

H. _Proximal_

I. _Distal_

J. _Proximal_

K. _Distal_

2. Anatomic Positions

Fill in the names of the anatomic positions shown.

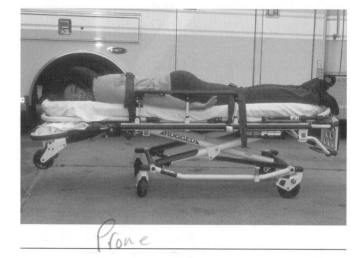

Prone

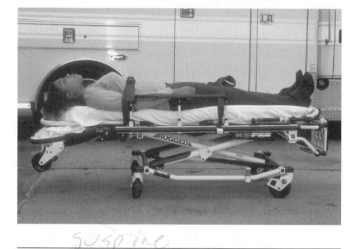

Supine

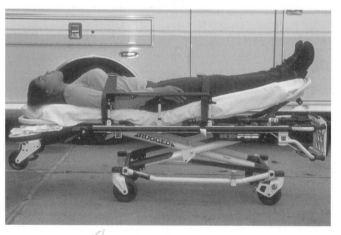

Shock Position

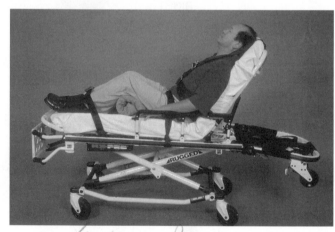

Fowlers Position

3. The Skeletal System

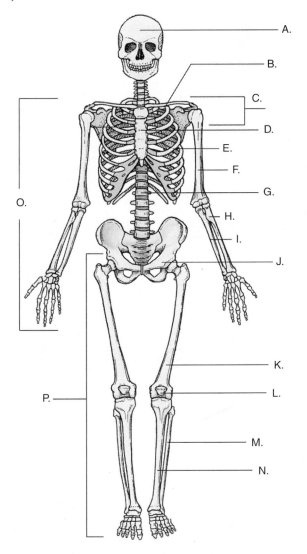

A. —— Skull
B. Clavical
C. Shoulder Girdle
D. Sternum
E. ~~Rib~~ Thorax
F. Humurus
G. Spinal Collum
H. Radius
I. Ulna
J. Pelvis
K. Femer
L. Patella
M. Fibia
N. Tibia
O. ~~Spral~~ Upper extremity
P. ~~Inferior~~ Lower Extremity

A. ——————————
B. ——————————
C. ——————————
D. ——————————
E. ——————————
F. ——————————
G. ——————————
H. ——————————
I. ——————————
J. ——————————
K. ——————————
L. ——————————
M. ——————————
N. ——————————
O. ——————————
P. ——————————
Q. ——————————

4. The Skull

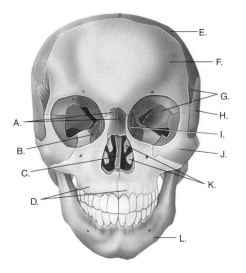

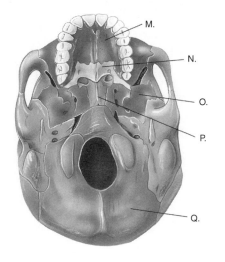

A. _____ C-7 Cervic

B. _____ T-1-S Thoraxiz

C. _____ Lumbar

D. _____ Sacrum

E. _____ Coccyx

5. The Spinal Column

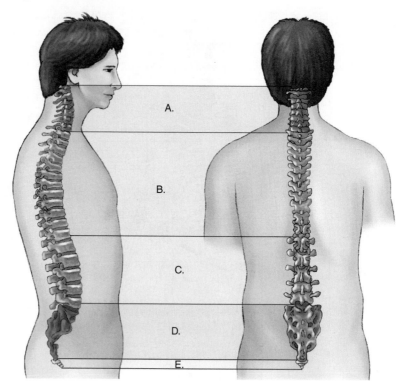

6. The Thorax

A. _____

B. _____

C. _____

D. _____

E. _____

F. _____

G. _____

H. _____

I. _____

J. _____

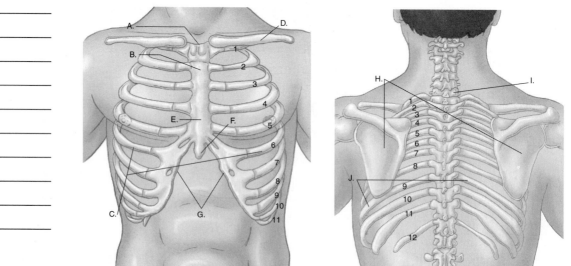

7. The Pelvis

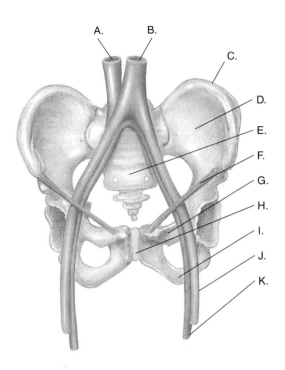

A. _____

B. _____

C. _____

D. _____

E. _____

F. _____

G. _____

H. _____

I. _____

J. _____

K. _____

8. The Lower Extremity

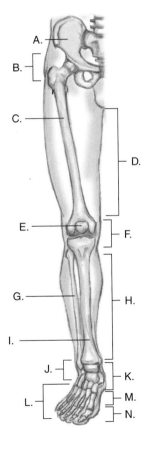

A. _____

B. _____

C. _____

D. _____

E. _____

F. _____

G. _____

H. _____

I. _____

J. _____

K. _____

L. _____

M. _____

N. _____

9. The Shoulder Girdle

A. _____

B. _____

C. _____

D. _____

E. _____

F. _____

G. _____

H. _____

I. _____

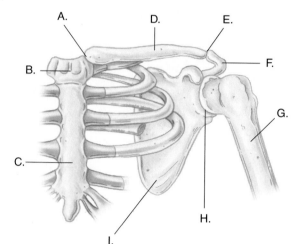

10. The Upper Extremity

A. _____

B. _____

C. _____

D. _____

E. _____

F. _____

G. _____

H. _____

I. _____

J. _____

K. _____

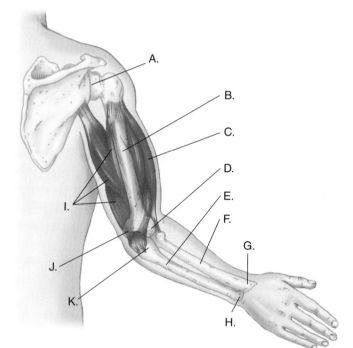

11. Wrist and Hand

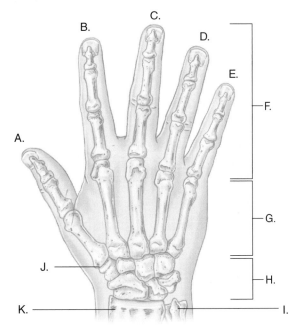

A. _____
B. _____
C. _____
D. _____
E. _____
F. _____
G. _____
H. _____
I. _____
J. _____
K. _____

12. The Respiratory System

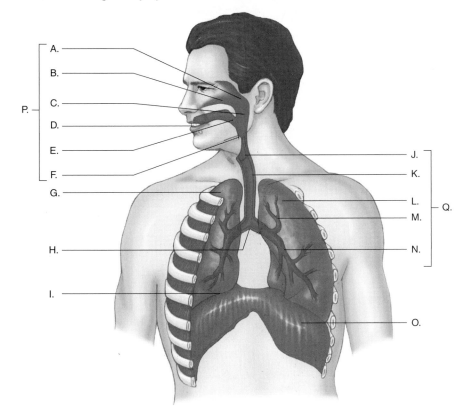

A. _____
B. _____
C. _____
D. _____
E. _____
F. _____
G. _____
H. _____
I. _____
J. _____
K. _____
L. _____
M. _____
N. _____
O. _____
P. _____
Q. _____

13. The Circulatory System

A. _____

B. _____

C. _____

D. _____

E. _____

F. _____

G. _____

H. _____

I. _____

J. _____

K. _____

L. _____

M. _____

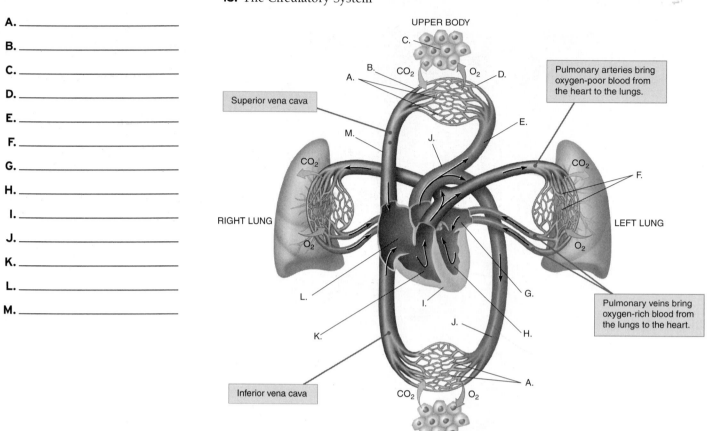

NOTES

14. Electrical Conduction

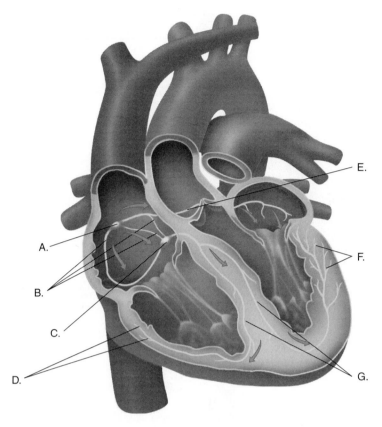

A. _____

B. _____

C. _____

D. _____

E. _____

F. _____

G. _____

1

NOTES

15. Central and Peripheral Pulses

A. _____

B. _____

C. _____

D. _____

E. _____

F. _____

G. _____

H. _____

I. _____

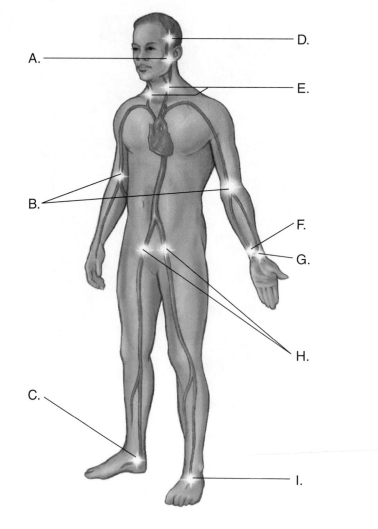

A.

B.

C.

D.

E.

F.

G.

H.

I.

NOTES

16. The Brain

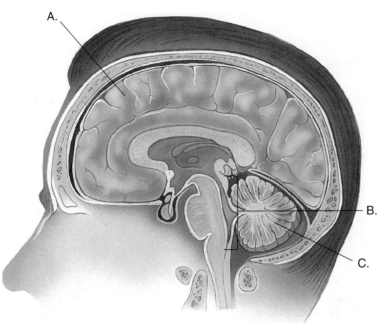

A. _____

B. _____

C. _____

17. Anatomy of the Skin

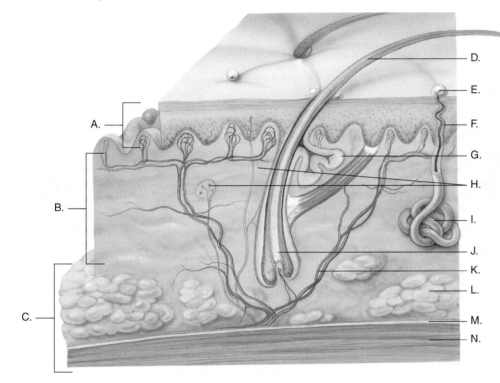

A. _____

B. _____

C. _____

D. _____

E. _____

F. _____

G. _____

H. _____

I. _____

J. _____

K. _____

L. _____

M. _____

N. _____

18. The Male Reproductive System

A. _____

B. _____

C. _____

D. _____

E. _____

F. _____

G. _____

H. _____

I. _____

J. _____

K. _____

L. _____

M. _____

N. _____

O. _____

P. _____

Q. _____

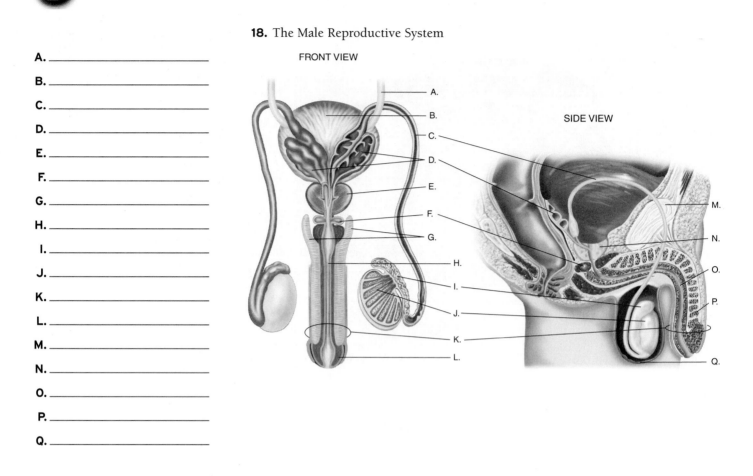

FRONT VIEW

SIDE VIEW

19. The Female Reproductive System

FRONT VIEW   SIDE VIEW

A. _____

B. _____

C. _____

D. _____

E. _____

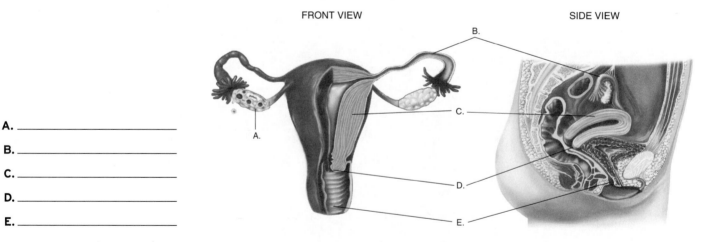

Vocabulary

Define the following terms using the space provided.

1. Perfusion:

_____

_____

_____

2. Agonal respirations:

_____

_____

_____

3. Autonomic nervous system:

_____

_____

_____

4. Pleural space:

_____

_____

_____

5. Trendelenburg's position:

_____

_____

_____

6. Fowler's position:

_____

_____

_____

7. Somatic nervous system:

_____

_____

_____

8. Endocrine system:

_____

_____

_____

9. Peripheral nervous system:

_____

_____

_____

10. Epiglottis:

_____

_____

_____

11. Metabolism:

_____

_____

_____

12. Brain stem:

_____

_____

_____

Fill-in

Read each item carefully, then complete the statement by filling in the missing word.

1. There are _____ cervical vertebrae.

2. The movable bone in the skull is the _____.

3. There is a total of _____ lobes in the right and left lungs.

4. There are _____ pairs of ribs that attach posteriorly to the thoracic vertebrae.

5. The spinal column has _____ vertebrae.

6. The vocal cords are located in the _____.

7. The ankle bone is known as _____.

8. The 11th and 12th rib are called _____.

True/False

If you believe the statement to be more true than false, write the letter "T" in the space provided. If you believe the statement to be more false than true, write the letter "F."

1. _____ The aorta is the major artery that supplies the groin and lower extremities with blood.

2. _____ The largest joint in the body is the knee.

3. _____ The phalanges are the bones of the finger and toes.

4. _____ The right atrium receives blood from the pulmonary veins.

5. _____ There are 12 ribs that attach to the sternum.

Short Answer

Complete this section with short written answers using the space provided.

1. List the four components of blood and each of their functions.

_____

_____

_____

_____

2. List the five sections of the spinal column and indicate the number of vertebrae in each.

_____

_____

_____

_____

_____

3. What organs are in each of the quadrants of the abdomen?

RUQ_____

_____

LUQ_____

_____

RLQ_____

_____

LLQ_____

_____

4. List in the proper order the parts of the heart that blood flows through.

1. _____   6. _____

2. _____   7. _____

3. _____   8. _____

4. _____   9. _____

5. _____

Word Fun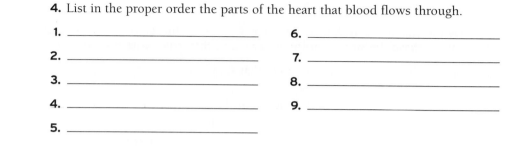

The following crossword puzzle is an activity provided to reinforce correct spelling and understanding of medical terminology associated with emergency care and the EMT-B. Use the clues in the column to complete the puzzle.

Across

1. Inner layer of skin

8. Nearer to the feet

10. Lower chamber of heart

13. Front surface of the body

14. Away from the midline, sides

15. Slow, dying respirations or pulse

16. Bone on thumb side of forearm

Down

1. Nearer the end

2. Behind the abdomen

3. Lower jawbone

4. Adequate circulation of blood

5. Upper chamber of the heart

6. Back surface of the body

7. Appears on both sides

9. Windpipe

11. Large solid organ in RUQ

12. Sitting up with knees bent

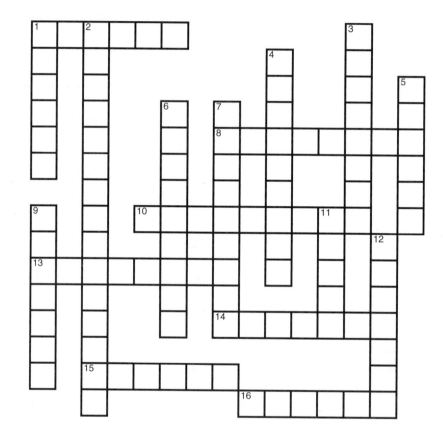

Ambulance Calls

The following real case scenarios provide an opportunity to explore the concerns associated with patient management. Read each scenario, then answer each question in detail.

1. You are dispatched to the scene of a bar fight. A 34-year-old male has been stabbed in the right upper quadrant of the abdomen with a knife.

Using medical terminology, indicate which organ(s) might be affected. How would you describe this patient's injuries?

_____

_____

_____

1

_____

_____

_____

_____

_____

_____

2. You are dispatched to a school playground to find an 8-year-old who has fallen from the monkey bars and is complaining of pain to his left forearm just above his wrist. You see it sticking out at an odd angle.
Using medical terminology, how would you describe this patient's injuries?

_____

_____

_____

_____

_____

_____

_____

_____

3. You are dispatched to a scene where a 14-year-old male was hit by a car. Upon arrival, you find the patient's left lower leg to be deformed and swollen.
Using medical terminology, how would you describe this patient's injuries?

_____

_____

_____

_____

_____

_____

_____

Workbook Activities

The following activities have been designed to help you. Your instructor may require you to complete some or all of these activities as a regular part of your EMT-B training program. You are encouraged to complete any activity that your instructor does not assign as a way to enhance your learning in the classroom.

Chapter Review

The following exercises provide an opportunity to refresh your knowledge of this chapter.

■ NOTES ■

Matching

Match each of the terms in the left column to the appropriate definition in the right column.

N **1.** Pulse

E **2.** Auscultation

D **3.** Palpation

L **4.** Perfusion

A **5.** Blood pressure

H **6.** SAMPLE history

K **7.** Sphygmomanometer

F **8.** Capillary refill

M **9.** Bradycardia

O **10.** Cyanosis

B **11.** Diaphoretic

G **12.** Hypertension

I **13.** Hypotension

C **14.** Sclera

J **15.** Tachycardia

A. the pressure of the circulating blood against the walls of the arteries

B. characterized by profuse sweating

C. white portion of the eye

D. examination by touch

E. listening to sounds within the organs, usually with a stethoscope

F. the ability of the circulatory system to restore blood to the capillary blood vessels after it is squeezed out

G. blood pressure that is higher than normal range

H. a patient's history consisting of signs/symptoms, allergies, medications, pertinent past history, last oral intake, and events leading to the illness/injury

I. blood pressure that is lower than normal range

J. rapid heart rate

K. a blood pressure cuff

L. the process whereby blood enters an organ or tissue through its arteries and leaves through its veins, providing nutrients and oxygen and removing waste

M. slow heart rate

N. the pressure wave that is felt with the expansion and contraction of an artery

O. bluish-gray skin color caused by reduced blood-oxygen levels

Baseline Vital Signs and SAMPLE History

Multiple Choice

Read each item carefully, then select the best response.

_D_ **1.** When assessing the patient, you will have to:
   A. visualize.
   B. auscultate.
   C. palpate.
   D. all of the above

_C_ **2.** The reason that a patient or other individual calls 9-1-1 is called the:
   A. signs.
   B. symptoms.
   C. chief complaint.
   D. primary problem.

_A_ **3.** Signs include all of the following, except:
   A. dizziness.
   B. marked deformities.
   C. external bleeding.
   D. wounds.

_D_ **4.** A critical problem or deficit in any of the body's vital systems or functions will affect and be reflected by changes in the:
   A. respiratory system.
   B. circulatory system.
   C. central nervous system.
   D. all of the above

_B_ **5.** The first set of vital signs that you obtain is called the:
   A. original vital signs.
   B. baseline vital signs.
   C. actual vital signs.
   D. real vital signs.

_____ D _____ **6.** Besides pulse, respirations, and blood pressure, you should also include evaluation of:

   A. level of consciousness.

   B. pupillary reaction.

   C. capillary refill in children.

   D. all of the above

_____ A _____ **7.** During inspiration:

   A. the chest rises up and out.

   B. the phase is passive.

   C. carbon dioxide is released.

   D. all of the above

_____ D _____ **8.** Assess breathing by:

   A. watching for chest rise and fall.

   B. feeling for air through the mouth and nose during exhalation.

   C. listening for breath sounds with a stethoscope.

   D. all of the above

_____ D _____ **9.** The normal range for adult respirations is _____ breaths/min.

   A. 8 to 20

   B. 15 to 30

   C. 25 to 50

   D. none of the above

_____ D _____ **10.** Normal breathing does not affect a patient's:

   A. speech.

   B. posture.

   C. positioning.

   D. all of the above

_____ B _____ **11.** In the _____ position, the patient sits leaning forward on outstretched arms with the head and chin thrust slightly forward.

   A. Fowler's

   B. tripod

   C. sniffing

   D. lithotomy

_____ C _____ **12.** In the _____ position, the patient sits upright with the head and chin thrust slightly forward.

   A. Fowler's

   B. tripod

   C. sniffing

   D. lithotomy

_____ B _____ **13.** Signs of labored breathing include all of the following, except:

   A. accessory muscle use.

   B. dyspnea.

   C. retractions.

   D. gasping.

_____ C _____ **14.** The _____ is the pressure wave that occurs as each heartbeat causes a surge in the blood circulating through the arteries.

   A. systolic pressure

   B. diastolic pressure

   C. pulse

   D. ventricular pressure

_____ **15.** In responsive patients who are older than 1 year, you should palpate a pulse at the _____ artery.

    A. carotid

    B. femoral

    C. radial

    D. brachial

_____ **16.** In unresponsive patients who are older than 1 year, you should palpate a pulse at the _____ artery.

    A. carotid

    B. femoral

    C. radial

    D. brachial

_____ **17.** A pulse that is weak and _____ should be palpated and counted for a full minute.

    A. difficult to palpate

    B. irregular

    C. extremely slow

    D. all of the above

_____ **18.** When the interval between each ventricular contraction of the heart is short, the pulse is:

    A. slow.

    B. rapid.

    C. regular.

    D. irregular.

_____ **19.** The rhythm of cardiac contractions is considered _____ if the heart periodically has a premature or late beat or if a pulse beat is missed.

    A. slow

    B. rapid

    C. regular

    D. irregular

_____ **20.** When assessing the skin, you should evaluate:

    A. color.

    B. temperature.

    C. moisture.

    D. all of the above

_____ **21.** Perfusion may be assessed in the:

    A. fingernail beds.

    B. lips.

    C. conjunctiva.

    D. all of the above

_____ **22.** Poor peripheral circulation will cause the skin to appear:

    A. pale.

    B. ashen.

    C. gray.

    D. all of the above

_B_ 23. Liver disease or dysfunction may cause _____ , resulting in the patient's skin and sclera turning yellow.

   A. cyanosis

   B. jaundice

   C. diaphoresis

   D. lack of perfusion

_D_ 24. The skin will feel cool when the patient:

   A. is in early shock.

   B. has mild hypothermia.

   C. has inadequate perfusion.

   D. all of the above

_D_ 25. Capillary refill reflects the patient's perfusion and is often affected by the patient's:

   A. body temperature.

   B. position.

   C. medications.

   D. all of the above

_B_ 26. With adequate perfusion, the color in the nail bed should be restored to its normal pink within _____ seconds when checking capillary refill.

   A. 1½

   B. 2

   C. 2½

   D. 3

_A_ 27. Adequate _____ is necessary to maintain proper circulation and perfusion of the vital organ cells.

   A. blood pressure

   B. pulse

   C. capillary refill

   D. body temperature

_D_ 28. A decrease in blood pressure may indicate:

   A. loss of blood.

   B. loss of vascular tone.

   C. a cardiac pumping problem.

   D. all of the above

_B_ 29. When blood pressure drops, the body compensates to maintain perfusion to the vital organs by:

   A. decreasing pulse rate.

   B. decreasing the blood flow to the skin and extremities.

   C. decreasing respiratory rate.

   D. dilating the arteries.

_A_ 30. Blood pressure is usually measured through:

   A. auscultation.

   B. palpation.

   C. visualization.

   D. rationalization.

31. When obtaining a blood pressure by palpation, you should place your fingertips on the _____ artery.

   A. carotid

   B. brachial

   C. radial

   D. posterior tibial

32. You must assume that a patient who has a critically _____ can no longer compensate sufficiently to maintain adequate perfusion.

   A. low blood pressure

   B. high blood pressure

   C. low pulse rate

   D. high pulse rate

33. The patient's level of consciousness reflects the status of the:

   A. peripheral nervous system.

   B. central nervous system.

   C. peripheral perfusion.

   D. distal perfusion.

34. The Glasgow Coma Scale evaluates all of the following, except:

   A. eye opening.

   B. verbal response.

   C. distal circulation.

   D. motor response.

35. The diameter and reactivity to light of the patient's pupils reflect the status of the brain's:

   A. perfusion.

   B. oxygenation.

   C. condition.

   D. all of the above

36. You should reassess vital signs:

   A. every 15 minutes in a stable patient.

   B. every 5 minutes in an unstable patient.

   C. after every medical intervention.

   D. all of the above

37. In the mnemonic "SAMPLE," the "P" stands for:

   A. pupillary response.

   B. pulse rate.

   C. pertinent past history.

   D. pain level.

Vocabulary emtb.com vocab explorer

Define the following terms using the space provided.

   1. Glasgow Coma Scale:

   _____

   _____

   _____

2. AVPU scale:

_____

_____

_____

3. Chief complaint:

_____

_____

_____

4. Stridor:

_____

_____

_____

Fill-in

Read each item carefully, then complete the statement by filling in the missing word.

1. _____ is the amount of air that is exchanged with each breath.

2. The _____ is the delicate membrane lining the eyelids and covering the exposed surface of the eye.

3. By using your _____ powers, you will be able to interpret the meaning and implications of your findings and the information that you have gathered while assessing the patient.

4. The severity of a _____ is subjective because it is based on the patient's interpretation and tolerance.

5. A patient who is breathing without assistance is said to have _____.

6. _____ are the key signs that are used to evaluate the patient's initial general condition.

7. When assessing respirations, you must determine the rate, _____, and depth of the patient's breathing.

8. When you can actually see the effort of the patient's breathing, it is described as

_____.

9. If you can hear bubbling or gurgling, the patient probably has _____ in the airway.

10. The condition of the patient's skin can tell you a lot about the patient's peripheral

circulation and _____, blood oxygen levels, and body temperature.

True/False

If you believe the statement to be more true than false, write the letter "T" in the space provided. If you believe the statement to be more false than true, write the letter "F."

1. _____ A pulse is an indicator of the condition of the heart.

2. _____ A blood pressure determined by palpation is less accurate than if determined by auscultation.

3. _____ Only the diastolic pressure can be measured by the palpation method.

4. _____ Labored breathing can be described as increased breathing effort, grunting, and use of accessory muscles.

5. _____ A conscious patient is likely to alter his or her breathing if he or she is aware that you are evaluating it.

6. _____ The pulse is most commonly palpated at the femoral artery.

7. _____ The normal pulse range for a newborn is 140 to 160 beats/min.

8. _____ To assess skin color in an infant, you should look at the palms of the hands and soles of the feet.

9. _____ Cyanosis indicates a need for oxygen.

10. _____ BSI precautions should be followed when a patient is jaundiced.

11. _____ The skin will feel hot when the patient is in profound shock or has hypothermia.

12. _____ Normal reaction to a bright light shone in one eye is pupil constriction in only that eye.

13. _____ Normal respirations in an adult are 15 to 30 breaths/min.

Short Answer

Complete this section with short written answers using the space provided.

1. Name the seven basic vital signs.

_____

_____

_____

_____

_____

_____

_____

2. List four abnormal skin colors.

_____

_____

_____

_____

3. Name four factors to be considered when assessing adequate or inadequate breathing.

_____

_____

_____

_____

4. Define systolic blood pressure.

_____

_____

_____

5. Define diastolic blood pressure.

_____

_____

_____

6. Name the three factors to consider when assessing a patient's pulse.

_____

_____

_____

7. Name the three factors to consider when assessing a patient's skin.

_____

_____

_____

8. Describe the process for assessing capillary refill.

_____

_____

_____

9. Define the acronym PEARRL.

_____

_____

_____

10. Explain the difference between a sign and a symptom.

_____

_____

_____

Word Fun

The following crossword puzzle is an activity to reinforce correct spelling and understanding of medical terminology associated with emergency care and the EMT-B. Use the clues in the column to complete the puzzle.

CLUES

Across

1. Heartbeat, respirations, BP, etc.

3. Circulation within the organs

5. Acronym used to assess LOC

9. BP lower than normal

11. Pressure against the wall of arteries

13. Harsh, high-pitched inspiratory sound

14. Coma scale

Down

2. Listening

4. Yellow

6. Objective finding

7. Bluish

8. BP higher than normal

10. Cardiac pressure wave

12. Subjective finding

1

Ambulance Calls

The following real case scenarios provide an opportunity to explore the concerns associated with patient management. Read each scenario, then answer each question in detail.

1. You are dispatched to the residence of a patient who is not feeling well. You arrive on the scene to find an 82-year-old male confined to a wheelchair. He is very warm to the touch. He tells you he had some nausea and vomiting this morning and feels dizzy.

How would you best manage this patient?

_____

_____

_____

_____

_____

_____

_____

_____

_____

2. You are called to a school nurse's office for a sick child. Upon arrival, you find a 9-year-old female lying on the bed in the office. She is flushed and appears weak. Closer inspection reveals a yellow tint to her sclera. The nurse tells you she has a history of liver disease.

How would you best manage this patient?

_____

_____

_____

_____

_____

_____

_____

_____

_____

3. You are called to a residence for a possible cardiac arrest. Family members tell you that the patient, a 57-year-old male, stopped breathing and had no pulse. They also tell you he has a cardiac history. They stopped CPR when he resumed breathing and regained a pulse, approximately 2 minutes before your arrival. How would you best manage this patient?

_____

_____

_____

_____

_____

_____

_____

_____

Skill Drills

Skill Drill 5-1: Obtaining a Blood Pressure by Auscultation or Palpation
Test your knowledge of this skill drill by placing the photos below in the correct order.
Number the first step with a "1," the second step with a "2," etc.

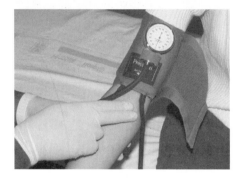

Palpate the brachial artery.

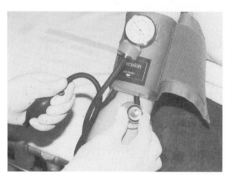

Close the valve and pump to 20 mm Hg above the point at which you stop hearing pulse sounds. Note the systolic and diastolic pressures as you let air escape slowly.

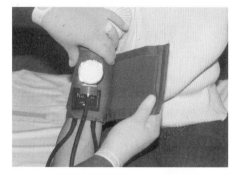

Apply the cuff snugly.

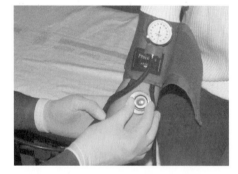

Place the stethoscope and grasp the ball-pump and turn-valve.

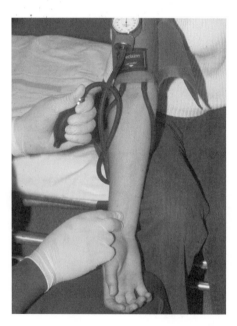

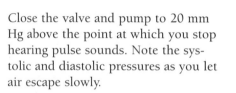

When using the palpatation method, you should place your fingertips on the radial artery so that you feel the radial pulse.

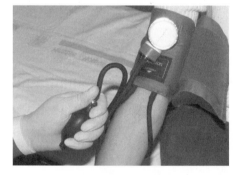

Open the valve and quickly release remaining air.

Notes

Workbook Activities

The following activities have been designed to help you. Your instructor may require you to complete some or all of these activities as a regular part of your EMT-B training program. You are encouraged to complete any activity that your instructor does not assign as a way to enhance your learning in the classroom.

Chapter Review

The following exercises provide an opportunity to refresh your knowledge of this chapter.

■ NOTES ■

Matching

Match each of the terms in the left column to the appropriate definition in the right column.

_____ **1.** Extremity lift

_____ **2.** Flexible stretcher

_____ **3.** Stair chair

_____ **4.** Basket stretcher

_____ **5.** Scoop stretcher

_____ **6.** Backboard

_____ **7.** Direct ground lift

_____ **8.** Portable stretcher

_____ **9.** Wheeled ambulance stretcher

A. separates into two or four pieces

B. tubular framed stretcher with rigid fabric stretched across it

C. used for patients without a spinal injury who are supine

D. specifically designed stretcher that can be rolled along the ground

E. used to carry patients across uneven terrain

F. used for patients who are found lying supine with no suspected spinal injury

G. can be folded or rolled up

H. used to carry patients up and down stairs

I. spine board or longboard

Multiple Choice

Read each item carefully, then select the best response.

_____ **1.** _____ safety depends on the use of proper lifting techniques and maintaining a proper hold when lifting or carrying a patient.

A. Your

B. Your team's

C. The patient's

D. all of the above

Lifting and Moving Patients

_____ **2.** You should perform an urgent move:

   A. if a patient has an altered level of consciousness.

   B. if a patient has inadequate ventilation or shock.

   C. in extreme weather conditions.

   D. all of the above

_____ **3.** The _____ is the mechanical weight-bearing base of the spinal column and the fused central posterior section of the pelvic girdle.

   A. coccyx

   B. sacrum

   C. lumbar region

   D. thorax

_____ **4.** You may injure your back if you lift:

   A. with your back curved.

   B. with your back straight, but bent significantly forward at the hips.

   C. with the shoulder girdle anterior to the pelvis.

   D. all of the above

_____ **5.** When lifting you should:

   A. spread your legs shoulder width apart.

   B. never lift a patient while reaching any significant distance in front of your torso.

   C. keep the weight that you are lifting as close to your body as possible.

   D. all of the above

_____ **6.** When carrying a patient on a cot, be sure:

   A. to flex at the hips.

   B. to bend at the knees.

   C. that you do not hyperextend your back.

   D. all of the above

_____ **7.** In lifting with the palm down, the weight is supported by the _____ rather than the palm.

  A. fingers

  B. forearm

  C. lower back

  D. wrist

_____ **8.** When you must carry a patient up or down a flight of stairs or other significant incline, use a _____ if possible.

  A. backboard

  B. stair chair

  C. stretcher

  D. short spine board

_____ **9.** If you need to lean to either side to compensate for a weight imbalance, you have probably _____ your weight limitation.

  A. met

  B. exceeded

  C. increased

  D. countered

_____ **10.** A backboard is a device that provides support to patients who you suspect have:

  A. hip injuries.

  B. pelvic injuries.

  C. spinal injuries.

  D. all of the above

_____ **11.** The team leader should do all of the following, except _____, before any lifting is initiated.

  A. give a command of execution

  B. indicate where each team member is to be located

  C. rapidly describe the sequence of steps that will be performed

  D. give a brief overview of the stages

_____ **12.** Special _____ are usually required to move any patient who weighs more than 300 lb to an ambulance.

  A. techniques

  B. equipment

  C. resources

  D. all of the above

_____ **13.** When carrying a patient in a stair chair, always remember to:

  A. keep your back in a locked-in position.

  B. flex at the hips, not at the waist.

  C. keep the patient's weight and your arms as close to your body as possible.

  D. all of the above

_____ **14.** When you use a body drag to move a patient:

  A. your back should always be locked and straight.

  B. you should avoid any twisting so that the vertebrae remain in normal alignment.

  C. avoid hyperextending.

  D. all of above

_____ **15.** When pulling a patient, you should do all of the following, except:

    A. extend your arms no more than about 15 to 20 inches.

    B. reposition your feet so that the force of pull will be balanced equally.

    C. when you can pull no farther, lean forward another 15 to 20 inches.

    D. pull the patient by slowly flexing your arms.

_____ **16.** When log rolling a patient:

    A. kneel as close to the patient's side as possible.

    B. lean solely from the hips.

    C. use your shoulder muscles to help with the roll.

    D. all of the above

_____ **17.** If the weight you are pushing is lower than your waist, you should push from:

    A. the waist.

    B. a kneeling position.

    C. the shoulder.

    D. a squatting position.

_____ **18.** To protect your _____ from injury, never push an object with your arms fully extended in a straight line and elbows locked.

    A. elbows

    B. shoulders

    C. neck

    D. arms

_____ **19.** If you are alone and must remove an unconscious patient from a car, you should first move the patient's:

    A. legs.

    B. head.

    C. torso.

    D. pelvis.

_____ **20.** Situations in which you should use an emergency move include those where:

    A. there is the presence of fire, explosives, or hazardous materials.

    B. you are unable to protect the patient from other hazards.

    C. you are unable to gain access to others in a vehicle who need lifesaving care.

    D. all of the above

_____ **21.** You should use a one-person technique to move a patient only:

    A. if a potentially life-threatening danger exists and you are alone.

    B. because of the pressing nature of the danger.

    C. if your partner is moving a second patient.

    D. all of the above

_____ **22.** You can move a patient on his or her back along the floor or ground by using all of the following methods, except:

    A. pulling on the patient's clothing in the neck and shoulder area.

    B. placing the patient on a blanket, coat, or other item that can be pulled.

    C. pulling the patient by the legs if they are the most accessible part.

    D. placing your arms under the patient's shoulders and through the armpits, while grasping the patient's arms, drag the patient backward.

_____23. An urgent move may be necessary for moving a patient with:

    A. an altered level of consciousness.

    B. inadequate ventilation.

    C. shock.

    D. all of the above

_____24. Use the rapid extrication technique in the following situation(s):

    A. The vehicle on the scene is unsafe.

    B. The patient's condition cannot be properly assessed before being removed from the car.

    C. The patient blocks access to another seriously injured patient.

    D. all of the above

_____25. Before you attempt any move, the team leader must be sure:

    A. that there are enough personnel and that the proper equipment is available.

    B. that any obstacles have been identified or removed.

    C. that the procedure and path to be followed have been clearly identified and discussed.

    D. all of the above

_____26. To avoid the strain of unnecessary lifting and carrying, you should use _____ or assist an able patient to the cot whenever possible.

    A. the direct ground lift

    B. the extremity lift

    C. the draw sheet method

    D. a scoop stretcher

_____27. To move a patient from the ground or the floor onto the cot, you should:

    A. lift and carry the patient to the nearby prepared cot using a direct body carry.

    B. use a scoop stretcher.

    C. use a log roll or long-axis drag to place the patient onto a backboard, and then lift and carry the backboard to the cot.

    D. all of the above

_____28. The _____ is the most uncomfortable of all the various devices; however, it provides excellent support and immobilization.

    A. portable stretcher

    B. flexible stretcher

    C. wooden backboard

    D. scoop stretcher

_____29. If _____ are used, you must follow infection control procedures before you can reuse the backboards.

    A. plastic backboards

    B. wooden backboards

    C. metal backboards

    D. all of the above

_____30. You should use a rigid _____, often called a Stokes litter, to carry a patient across uneven terrain from a remote location that is inaccessible by ambulance or other vehicle.

    A. basket stretcher

    B. scoop stretcher

    C. molded backboard

    D. flotation device

_____ **31.** Basket stretchers can be used:

    A. for technical rope rescues and some water rescues.

    B. to carry a patient across fields on an all-terrain vehicle.

    C. to carry a patient on a toboggan.

    D. all of the above

_____ **32.** Every time you have to move a patient, you must take special care that _____ are (is) not injured.

    A. you

    B. your team

    C. the patient

    D. all of the above

_____ **33.** Certain patient conditions, such as _____ , call for special lifting and moving techniques.

    A. head or spinal injury

    B. shock

    C. pregnancy

    D. all of the above

Vocabulary emtb.com vocab explorer

Define the following terms using the space provided.

1. Diamond carry:

_____

_____

_____

2. Rapid extrication technique:

_____

_____

_____

3. Power grip:

_____

_____

_____

4. Power lift:

_____

_____

_____

5. Emergency move:

_____

_____

_____

Fill-in

Read each item carefully, then complete the statement by filling in the missing word.

1. To avoid injury to you, the patient, or your partners, you will have to learn how to lift and carry the patient properly, using proper _____ and a power grip.

2. The key rule of lifting is to always keep the back in a straight, _____ position and to lift without twisting.

3. The safest and most powerful way to lift, lifting by extending the properly placed flexed legs, is called a _____ .

4. The arm and hand have their greatest lifting strength when facing _____ up.

5. Be sure to pick up and carry the backboard with your back in the _____ position.

6. You should not attempt to lift a patient who weighs more than _____ pounds with fewer than four rescuers, regardless of individual strength.

7. During a body drag where you and your partner are on each side of the patient, you will have to alter the usual pulling technique to prevent pulling _____ and producing adverse lateral leverage against your lower back.

8. When you are rolling the wheeled ambulance stretcher, your back should be _____ , straight, and untwisted.

9. Be careful that you do not push or pull from a(n) _____ position.

10. Remember to always consider whether there is an option that will cause _____ _____ to you and the other EMT-Bs.

11. The manual support and immobilization that you provide when using the rapid extrication technique produce a greater risk of _____ .

12. The _____ is used for patients with no suspected spinal injury who are found lying supine on the ground.

13. The _____ may be especially helpful when the patient is in a very narrow space or there is not enough room for the patient and a team of EMTs to stand side by side.

14. The mattress on a stretcher must be _____ so that it does not absorb any type of potentially infectious material, including water, blood, or other body fluid.

15. A _____ may be used for patients who have been struck by a motor vehicle.

True/False

If you believe the statement to be more true than false, write the letter "T" in the space provided. If you believe the statement to be more false than true, write the letter "F."

1. _____ Patient packaging and handling are technical skills you will learn and perfect through practice and training.

2. _____ A portable stretcher is typically a lightweight folding device that does not have the undercarriage and wheels of a true ambulance stretcher.

3. _____ The term "power lift" refers to a posture that is safe and helpful for EMT-Bs when they are lifting.

4. _____ If you find that lifting a patient is a strain, try to move to the ambulance as quickly as possible to minimize the possibility of back injury.

5. _____ The use of adjunct devices and equipment, such as sheets and blankets, may make the job of lifting and moving a patient more difficult.

6. _____ One-person techniques for moving patients should only be used when immediate patient movement is necessary due to a life-threatening hazard and only one EMT-B is available.

7. _____ A scoop stretcher may be used alone for a standard immobilization of a patient with a spinal injury.

8. _____ When carrying a patient down stairs or on an incline, make sure the stretcher is carried with the head end first.

9. _____ The rapid extrication technique is the preferred technique to use on all sitting patients with possible spinal injuries.

10. _____ It is unprofessional for you to discuss and plan a lift at the scene in front of the patient.

Short Answer

Complete this section with short written answers using the space provided.

1. List the one-rescuer drags, carries, and lifts.

_____

_____

_____

2. List the situations where the rapid extrication technique is used.

_____

_____

_____

3. List the three guidelines for loading the cot into the ambulance.

_____

_____

_____

4. List the five guidelines for carrying a patient on a cot.

_____

_____

_____

_____

_____

5. Describe the key rule of lifting.

_____

_____

_____

Word Fun

The following crossword puzzle is an activity provided to reinforce correct spelling and understanding of medical terminology associated with emergency care and the EMT-B. Use the clues in the column to complete the puzzle.

CLUES

Across

3. Stretcher used with technical rescues, particularly when water is involved

5. Safest way to lift

7. Stretcher that becomes rigid when secured around patient

10. Tubular framed stretcher with fabric across

11. Four-rescuer carry with one at the head, one at the foot, and one on each side

Down

1. A ground lift used with no suspected spinal injury

2. Used to support patient with hip, pelvic, or spine injury

4. Folding device for moving seated patients up or down floors

6. Stretcher designed to roll along ground

8. Using the patient's limbs to lift

9. Stretcher that can be split into two or four sections

Ambulance Calls

The following real case scenarios provide an opportunity to explore the concerns associated with patient management. Read each scenario, then answer each question in detail.

1. You are dispatched to a construction site for a 26-year-old male who fell into a ravine. He is approximately 35 feet down a rocky ledge. He is alert with an unstable pelvis and weak radial pulses. You have all the help you need from the construction crew and the volunteer fire department.

   How would you best manage this patient?

   _____

   _____

   _____

   _____

   _____

   _____

   _____

   _____

   _____

   _____

2. You are working a head-on motor vehicle crash where you have patients in critical condition who were driving each vehicle. As your partner works with the driver of one vehicle, you assess the driver of the second car. He is a 58-year-old male with weak radial pulses and a respiratory rate of 4 breaths/min. No other help has arrived on the scene and your partner's patient is equally critical. You cannot effectively ventilate the patient where he is seated. The fire department is en route to your location.

   How would you best manage this patient?

   _____

   _____

   _____

   _____

   _____

   _____

   _____

   _____

   _____

3. You are working a motor vehicle crash and note that the patient, a 34-year-old female, responds to pain, and has absent radial pulses. She is breathing shallowly at a rate of 10 breaths/min.

How would you best manage this patient?

_____

_____

_____

_____

_____

_____

_____

_____

NOTES

1

Skill Drills emt-b video clips

Skill Drill 6-1: Performing the Power Lift
Test your knowledge of this skill drill by filling in the correct words in the photo captions.

1. Lock your back into a(n) _____, inward curve. _____ and bend your legs. Grasp the backboard, palms up and just in front of you. _____ and _____ the weight between your arms.

2. Position your feet, _____ the object, and _____ weight.

3. _____ your legs and lift, keeping your back locked in.

Skill Drill 6-2: Performing the Diamond Carry
Test your knowledge of this skill drill by placing the photos below in the correct order. Number the first step with a "1," the second step with a "2," etc.

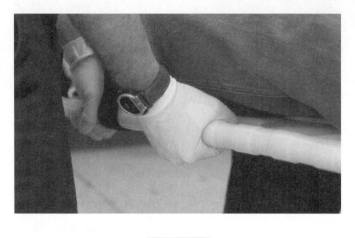

_____

EMT-Bs at the side each turn the head-end hand palm down and release the other hand.

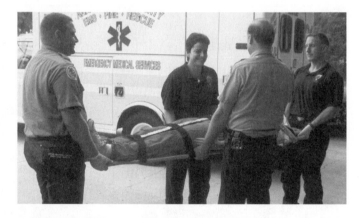

_____

After the patient has been lifted, the EMT-B at the foot turns to face forward.

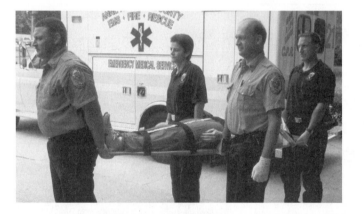

_____

EMT-Bs at the side turn toward the foot end.

_____

Position yourselves facing the patient.

Skill Drill 6-3: Performing the One-Handed Carrying Technique

Test your knowledge of this skill drill by filling in the correct words in the photo captions.

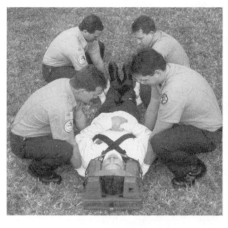

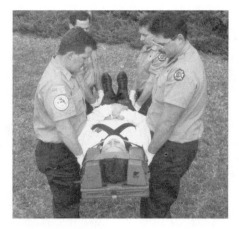

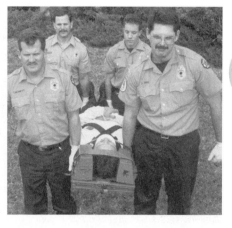

1. _____ each other and use both _____ .

2. Lift the backboard to _____ _____ .

3. _____ in the direction you will walk and _____ to using one hand.

Skill Drill 6-4: Carrying a Patient on Stairs

Test your knowledge of this skill drill by filling in the correct words in the photo captions.

1. _____ the patient securely.

2. Carry a patient down stairs with the _____ end first, _____ elevated.

3. Carry the _____ end first going up stairs.

Skill Drill 6-5: Using a Stair Chair
Test your knowledge of this skill drill by filling in the correct words in the photo captions.

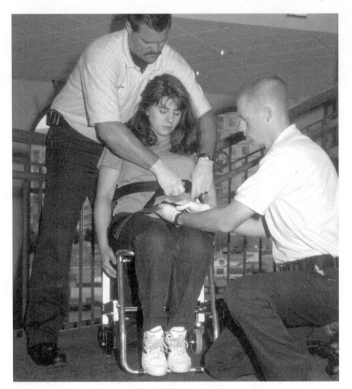

1. Position and secure the patient on the chair, _____ strapped down.

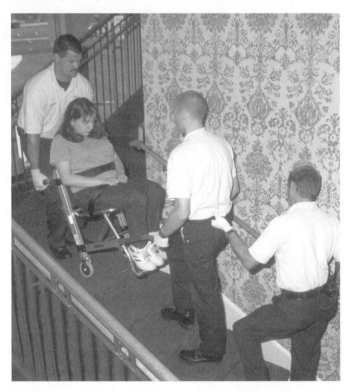

2. Take your places at the _____ and _____ of the chair.

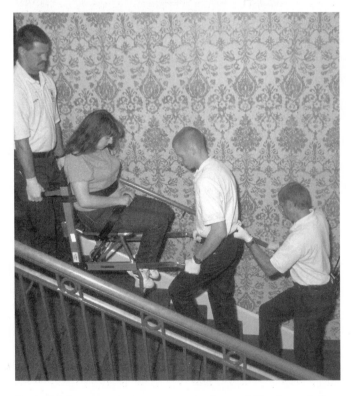

3. A third _____ precedes and "backs up" the rescuer carrying the _____.

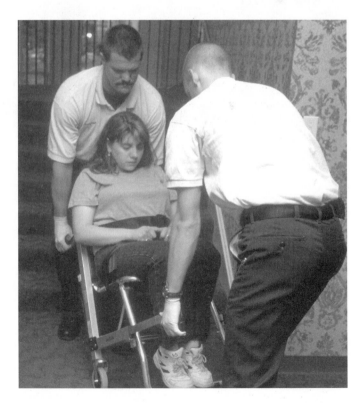

4. _____ the chair to roll on landings, or for transfer to the cot.

Skill Drill 6-6: Performing Rapid Extrication Technique

Test your knowledge of this skill drill by placing the photos below in the correct order.
Number the first step with a "1," the second step with a "2," etc.

Second EMT-B supports the torso.

Third EMT-B frees the patient's legs from the pedals and moves the legs together, without moving pelvis or spine.

First EMT-B provides in-line manual support of the head and cervical spine.

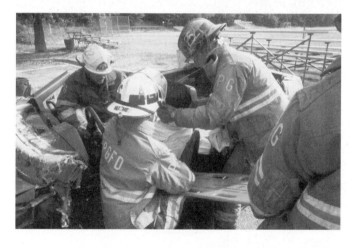

Third EMT-B exits the vehicle, moves to the backboard opposite Second EMT-B, and they continue to slide the patient until patient is fully on the board.

First (or Fourth) EMT-B places the backboard on the seat against patient's buttocks.

Second and Third EMT-Bs lower the patient onto the long spine board.

(continued)

Third EMT-B moves to an effective position for sliding the patient.

Second and Third EMT-Bs slide the patient along the backboard in coordinated, 8" to 12" moves until the hips rest on the backboard.

First (or Fourth) EMT-B continues to stabilize the head and neck while Second and Third EMT-Bs carry the patient away from the vehicle.

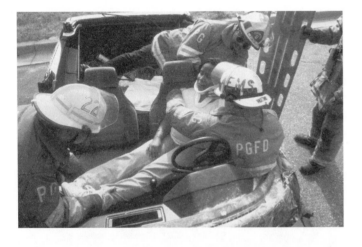

Second and Third EMT-Bs rotate the patient as a unit in several short, coordinated moves.

First EMT-B (relieved by Fourth EMT-B or bystander as needed) supports the head and neck during rotation (and later steps).

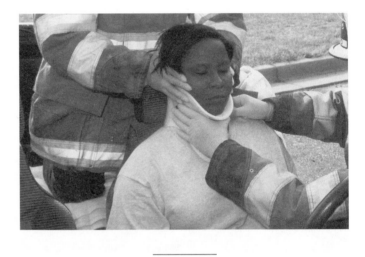

Second EMT-B gives commands, applies a cervical collar, and performs the initial assessment.

Skill Drill 6-7: Extremity Lift
Test your knowledge of this skill drill by filling in the correct words in the photo captions.

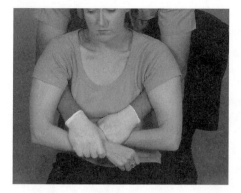

1. Patient's hands are
_____ over the chest.
First EMT-B grasps patient's wrists
or _____ and pulls
patient to a _____
position.

2. When the patient is sitting, First
EMT-B passes his or her arms
through patient's _____
and grasps the patient's opposite
(or his or her own)
_____ or
_____.
Second EMT-B kneels between the
_____, facing the feet,
and places his or her hands under
the _____.

3. Both EMT-Bs rise to _____.
On _____, both lift and
begin to move.

Skill Drill 6-8: Using a Scoop Stretcher
Test your knowledge of this skill drill by filling in the correct words in the photo captions.

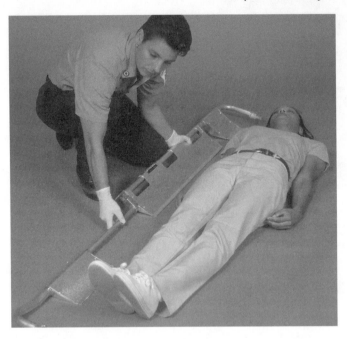

1. Adjust stretcher _____.

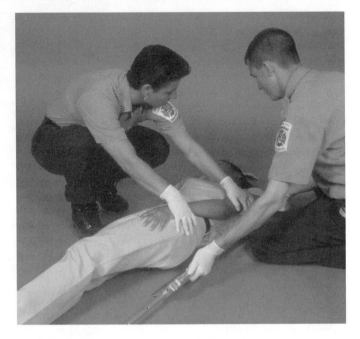

2. _____ patient slightly and _____ stretcher into place, one side at a time.

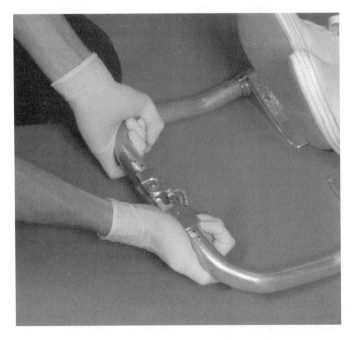

3. _____ the stretcher ends together, avoiding _____.

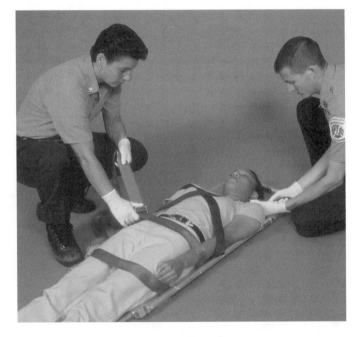

4. _____ the patient and _____ to the cot.

Notes

Workbook Activities

The following activities have been designed to help you. Your instructor may require you to complete some or all of these activities as a regular part of your EMT-B training program. You are encouraged to complete any activity that your instructor does not assign as a way to enhance your learning in the classroom.

Chapter Review

The following exercises provide an opportunity to refresh your knowledge of this chapter.

■ NOTES ■

Matching

Match each of the terms in the left column to the appropriate definition in the right column.

_____ **1.** Inhalation	**A.**	moves down slightly when it contracts
_____ **2.** Exhalation	**B.**	irregular breathing pattern with increased rate and depth, followed by apnea
__H___ **3.** Alveoli	**C.**	active part of breathing
_____ **4.** Automatic function	**D.**	voice box
_____ **5.** Hypoxic drive	**E.**	amount of air moved during one breath
_____ **6.** Tidal volume	**F.**	raises ribs when it contracts
__F___ **7.** Diaphragm	**G.**	controls breathing when we sleep
_____ **8.** Intercostal muscle	**H.**	site of oxygen diffusion
_____ **9.** Ventilation	**I.**	thorax size decreases
__D___ **10.** Larynx	**J.**	insufficient oxygen for cells and tissues
__A___ **11.** Hypoxia	**K.**	backup system to control respiration
__D___ **12.** Cheyne-Stokes respirations	**L.**	exchange of air between lungs and environment

Multiple Choice

Read each item carefully, then select the best response.

 1. What percentage of the air we breathe is made up of oxygen?

- **A.** 78%
- **B.** 12%
- **C.** 16 %
- **D.** 21%

Airway

___*D*___ **2.** Regarding the maintenance of the airway in an unconscious adult, which of the following is false?

    A. Insertion of an oropharyngeal airway helps keep the airway open.

    B. The head tilt–chin lift maneuver should always be used to open the airway.

    C. Secretions should be suctioned from the mouth as necessary.

    D. Inserting a rigid suction catheter beyond the tongue may cause gagging.

___*B*___ **3.** The normal respiratory rate for an adult is:

    A. about equal to the person's heart rate.

    B. 12 to 20 breaths per minute.

    C. faster when the person is sleeping.

    D. the same as in infants and children.

___*C*___ **4.** All of the following conditions are associated with hypoxia, except:

    A. heart attack.

    B. altered mental status.

    C. chest injury.

    D. hyperventilation syndrome.

___*A*___ **5.** The brain stem normally triggers breathing by increasing respirations when:

    A. carbon dioxide levels increase.

    B. oxygen levels increase.

    C. carbon dioxide levels decrease.

    D. nitrogen levels decrease.

___*A*___ **6.** Which of the following is not a sign of abnormal breathing?

    A. warm, dry skin

    B. speaking in two- or three-word sentences

    C. unequal breath sounds

    D. skin pulling in around the ribs during inspiration

___*C*___ **7.** The proper technique for sizing an oropharyngeal airway before insertion is to:

    A. measure the device from the tip of the nose to the earlobe.

    B. measure the device from the bridge of the nose to the tip of the chin.

    C. measure the device from the corner of the mouth to the earlobe.

    D. measure the device from the center of the jaw to the earlobe.

D **8.** What is the most common problem you may encounter when using a BVM device?

A. volume of the BVM device

B. positioning of the patient's head

C. environmental conditions

D. maintaining an airtight seal

C **9.** When ventilating a patient with a BVM device, you should:

A. look for inflation of the cheeks.

B. look for signs of the patient breathing on his or her own.

C. look for rise and fall of the chest.

D. listen for gurgling.

D **10.** Suctioning the oral cavity of an adult should be accomplished within:

A. 10 seconds.

B. 5 seconds.

C. 20 seconds.

D. 15 seconds.

Labeling

Label the following diagrams with the correct terms.

1. Upper and Lower Airways

A. Nasopharynx

B. _____

C. _____

D. Mouth

E. Oropharynx

F. Epiglottis

G. Apex of Lung

H. Carina

I. Base of Lung

J. Larynx

K. Trachea

L. Alveoli

M. Small Bronci

N. Main

O. Diaphram

P. Upper Airway

Q. Lower Airway

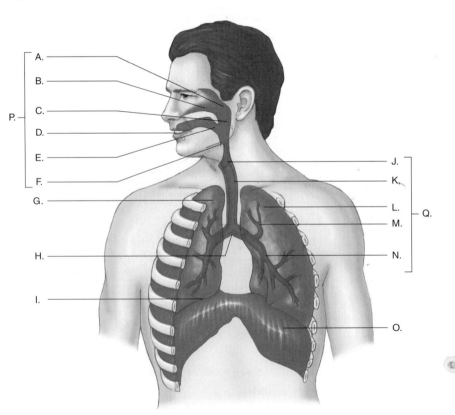

2. The Thoracic Cage

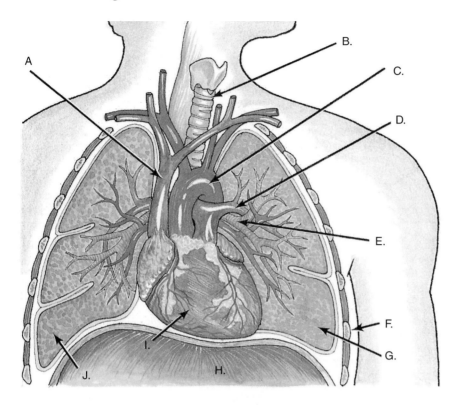

A. _Superior Vena Cava_
B. _Trachea_
C. _Aorta_
D. _Pulmonary Artery_
E. _Bronchus_
F. ~~Arte~~ _Rib_
G. _Left Lower Lobe_
H. _Diaphram_
I. _Right Ventricle_
J. _Right Lower Lobe_

2

3. Cellular Exchange

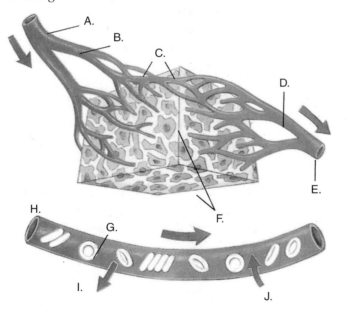

A. _____

B. _____

C. _____

D. _____

E. _____

F. _____

G. _____

H. _____

I. _____

J. _____

4. Pulmonary Exchange
Where arrows are shown, indicate which molecules are moving in that direction.

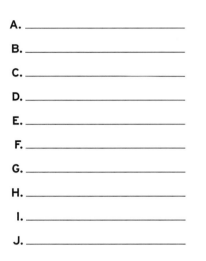

A. _____

B. _____

C. _____

D. _____

E. _____

F. _____

G. _____

H. _____

I. _____

J. _____

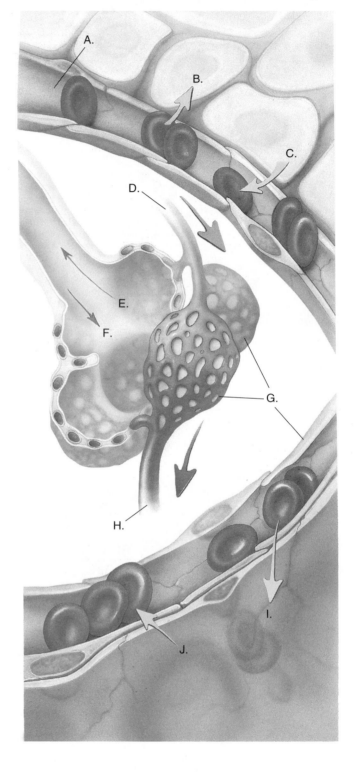

Vocabulary

Define the following terms using the space provided.

1. Gag reflex:

_____

_____

_____

2. Gastric distention:

_____

_____

_____

3. Stoma:

_____

_____

_____

Fill-in

Read each item carefully, then complete the statement by filling in the missing word(s).

1. Air enters the body through the _____.

2. In exhalation, air pressure in the lungs is _____ than the pressure outside.

3. The air we breathe contains _____ percent oxygen and _____ percent nitrogen.

4. The primary mechanism for triggering breathing is the level of

_____ in the blood.

5. During inhalation, the _____lungs_____ and ___Diapha___ contract, causing the thorax to enlarge.

6. The drive to breathe is triggered by _____ or _____ levels in arterial blood.

7. Insufficient oxygen in the cells and tissues is called ___Hypoxia___.

True/False

If you believe the statement to be more true than false, write the letter "T" in the space provided. If you believe the statement to be more false than true, write the letter "F."

1. _____ Nasal airways keep the tongue from blocking the upper airway and facilitate suctioning of the oropharynx.

2. _____ Nasal cannulas can deliver a maximum of 50% oxygen at 6 L/min.

3. _____ Oral airways should be measured from the tip of the nose to the earlobe.

4. _____ Compressed gas cylinders pose no unusual risk.

5. _____ The pin-indexing system is used to ensure compatibility between pressure regulators and oxygen flowmeters.

Short Answer

Complete this section with short written answers using the space provided.

1. List the five early signs of hypoxia.

_____

_____

_____

_____

_____

2. What are the normal respiratory rates for adults, children, and infants?

_____

_____

_____

3. How can you avoid gastric distention while performing artificial ventilation?

_____

_____

_____

4. List the six steps for providing one-rescuer artificial ventilation with a BVM device.

_____

_____

_____

_____

_____

_____

5. List six signs of inadequate breathing.

_____

_____

_____

_____

_____

_____

6. What are accessory muscles? Name three.

_____

_____

_____

7. When should medical control be consulted before inserting a nasal airway?

_____

_____

_____

8. List the three steps in nasal airway insertion.

_____

_____

_____

9. What is the best suction tip for suctioning the pharynx, and why?

_____

_____

_____

10. What is the time limit for each episode of suctioning an adult?

_____

_____

_____

Word Fun

The following crossword puzzle is an activity provided to reinforce correct spelling and understanding of medical terminology associated with emergency care and the EMT-B. Use the clues in the column to complete the puzzle.

CLUES

Across

3. Requires more than normal effort

5. Dying, gasping respirations

7. Larynx, nose, mouth, throat

9. Not enough O_2

11. Face piece attached to a reservoir

12. Exchange of air between lungs and outside

Down

1. Opening in neck connected to trachea

2. Mechanism that causes retching

4. Limits exposure to body fluids

6. Airway inserted in nostril

8. Airway inserted in mouth

10. Diaphragm relaxes

Ambulance Calls

The following real case scenarios provide an opportunity to explore the concerns associated with patient management. Read each scenario, then answer each question in detail.

1. You are dispatched to a crash with multiple patients early one morning near the end of your shift. Your patient was the unrestrained driver of one of the vehicles. She is 38 years old and struck her face against the steering wheel and windshield. There is a large laceration on her nose and several teeth are missing. Though unconscious, she has vomited a large amount of food and blood, which is pooling in her mouth. You note gurgling noises as she attempts to breathe. How would you best manage this patient?

_____

_____

_____

2

_____

_____

_____

_____

_____

_____

_____

2. You are dispatched to a private residence in a quiet neighborhood for "trouble breathing." You arrive to find an 83-year-old male sitting upright on the edge of his bed gasping for air. A family member states that the patient woke this morning feeling "ill" and having a difficult time breathing. The patient acknowledges a history of emphysema, but takes no prescriptions for the problem. He states that he "can't get enough air."

How would you best manage this patient?

_____

_____

_____

_____

_____

_____

_____

_____

_____

3. You have been called to the home of a frantic mother whose 11-month-old daughter is not breathing. When you arrive, the mother meets you at the door with the child in her arms. The mother states that she left the child's nursery to answer the telephone and returned to find the child motionless on the floor next to a spilled jar of marbles. You note that the child remains motionless and there are no signs of respirations. The child is cyanotic.

How would you best manage this patient?

_____

_____

_____

Skill Drills

Test your knowledge of skill drills by placing the photos below in the correct order. Number the first step with a "1," the second step with a "2," etc.

Skill Drill 7-1: Positioning an Unconscious Patient

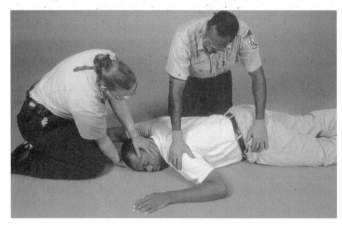

Have your partner place his or her hand on the patient's far shoulder and hip.

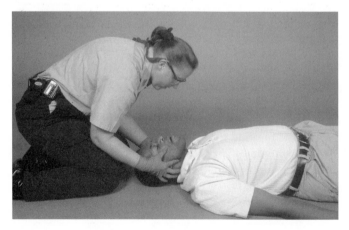

Open and assess the patient's airway and breathing status.

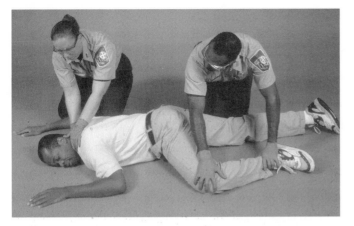

Support the head while your partner straightens the patient's legs.

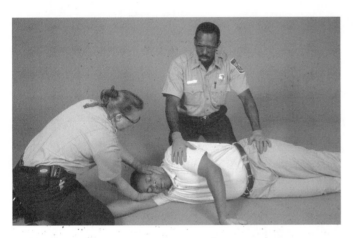

Roll the patient as a unit with the person at the head calling the count to begin the move.

2

Workbook Activities

The following activities have been designed to help you. Your instructor may require you to complete some or all of these activities as a regular part of your EMT-B training program. You are encouraged to complete any activity that your instructor does not assign as a way to enhance your learning in the classroom.

Chapter Review

The following exercises provide an opportunity to refresh your knowledge of this chapter.

Matching

Match each of the terms in the left column to the appropriate definition in the right column.

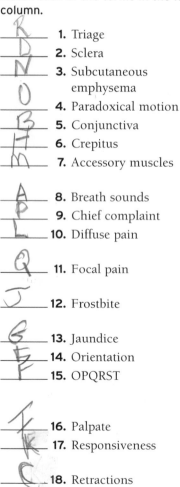

R **1.** Triage	**A.** indication of air movement in the lungs
D **2.** Sclera	**B.** lining of the eyelid
N **3.** Subcutaneous emphysema	**C.** movements in which the skin pulls in around the ribs during inspiration
O **4.** Paradoxical motion	**D.** white of the eyes
B **5.** Conjunctiva	**E.** the mental status of a patient
H **6.** Crepitus	**F.** the six pain questions
M **7.** Accessory muscles	**G.** yellow skin color due to liver disease or dysfunction
A **8.** Breath sounds	**H.** a crackling sound
P **9.** Chief complaint	**I.** examine by touch
L **10.** Diffuse pain	**J.** damage to tissues as the result of exposure to cold
Q **11.** Focal pain	**K.** the way in which a patient responds to external stimuli
J **12.** Frostbite	**L.** pain not identified as being specific to a single location
G **13.** Jaundice	**M.** secondary muscles of respiration
E **14.** Orientation	**N.** air under the skin
F **15.** OPQRST	**O.** motion of a segment of chest wall that is opposite the normal movement during breathing
I **16.** Palpate	**P.** the reason a patient called for help
K **17.** Responsiveness	**Q.** pain easily identified as being specific to a single location
C **18.** Retractions	**R.** a process of identifying the severity of each patient's condition; sorting

Patient Assessment

Multiple Choice

Read each item carefully, then select the best response.

_C_ **1.** The scene size-up consists of all of the following, except:

    A. determining mechanism of injury.

    B. requesting additional assistance.

    C. determining level of responsiveness.

    D. PPE/BSI.

_D_ **2.** Possible dangers you may observe during the scene size-up include:

    A. oncoming traffic.

    B. unstable surfaces.

    C. downed electrical lines.

    D. all of the above

_D_ **3.** With a traumatic injury, the body has been exposed to some force or energy that has resulted in:

    A. a temporary injury.

    B. permanent damage.

    C. death.

    D. all of the above

_B_ **4.** With _____, the force of the injury occurs over a broad area, and the skin is usually not broken.

    A. motor vehicle crashes

    B. blunt trauma

    C. penetrating trauma

    D. gunshot wounds

_C_ **5.** With _____, the force of the injury occurs at a small point of contact between the skin and the object.

    A. motor vehicle crashes

    B. blunt trauma

    C. penetrating trauma

    D. falls

D 6. You can use the mechanism of injury as a kind of guide to predict the potential for a serious injury by evaluating:

   A. the amount of force applied to the body.

   B. the length of time the force was applied.

   C. the areas of the body that are involved.

   D. all of the above

A 7. In penetrating trauma, the severity of injury depends on all of the following, except:

   A. the geographic location. *A*

   B. the characteristics of the penetrating object.

   C. the amount of force or energy.

   D. the part of the body affected.

C 8. In motor vehicle crashes, the amount of force that is applied to the body is directly related to the _____ of the crash.

   A. distance *C*

   B. location

   C. speed

   D. time

D 9. Your evaluation should also be based, to some extent, on:

   A. the patient's position in the car.

   B. the use of seat belts.

   C. how the patient's body shifts during the crash.

   D. all of the above *D*

D 10. Ejection from the vehicle dramatically increases the risk of:

   A. head injury. *D*

   B. spinal cord injury.

   C. death.

   D. all of the above

B 11. In falls, the amount of force that is applied to the body is directly related to the:

   A. surface landed on. *B*

   B. distance fallen.

   C. past medical history.

   D. how the patient landed.

D 12. In order to quickly determine the nature of illness, talk with:

   A. the patient.

   B. family members. *A*

   C. bystanders.

   D. all of the above

B 13. When considering the need for additional resources, questions to ask include all of the following, except:

   A. How many patients are there? *B*

   B. Is it raining?

   C. Who contacted EMS?

   D. Does the scene pose a threat to you or your patient's safety?

_____ **14.** The initial assessment includes evaluation of all of the following, except:

    A. mental status.

    B. pupils.

    C. airway.

    D. circulation.

_____ **15.** The best indicator of brain function is the patient's:

    A. pulse rate.

    B. pupillary response.

    C. mental status.

    D. respiratory rate and depth.

_____ **16.** An altered mental status may be caused by:

    A. head trauma.

    B. hypoxemia.

    C. hypoglycemia.

    D. all of the above

_____ **17.** All of the following are signs of inadequate breathing except:

    A. tightness in the chest.

    B. two- to three-word dyspnea.

    C. use of accessory muscles.

    D. nasal flaring.

_____ **18.** Airway obstruction in an unconscious patient is most commonly due to:

    A. vomitus.

    B. the tongue.

    C. dentures.

    D. food.

_____ **19.** Signs of airway obstruction in an unconscious patient include:

    A. obvious trauma, blood, or other obstruction.

    B. noisy breathing.

    C. extremely shallow or absent breathing.

    D. all of the above

_____ **20.** The _____ of the patient's pulse will give you a general idea of the overall status of the patient's cardiac function.

    A. rate

    B. rhythm

    C. strength

    D. all of the above

_____ **21.** The AED should be used on medical patients who are at least _____ years old and weigh more than 55 lb and who have been assessed to be unresponsive, apneic, and pulseless.

    A. 7

    B. 8

    C. 9

    D. 10

_____ **22.** In almost all instances, controlling external bleeding is accomplished by:

    A. direct pressure.

    B. elevation.

    C. pressure points.

    D. tourniquet.

3

23. Assessing the _____ is one of the most important and readily accessible ways of evaluating circulation.

A. pulse

B. respirations

C. skin

D. capillary refill

24. Skin color depends on:

A. pigmentation.

B. blood oxygen levels.

C. the amount of blood circulating through the vessels of the skin.

D. all of the above

25. In deeply pigmented skin, you should look for changes in color in areas of the skin that have less pigment, including:

A. the sclera.

B. the conjunctiva.

C. the mucous membranes of the mouth.

D. all of the above

26. Other conditions, not related to the body's circulation, such as _____, may slow capillary refill.

A. local circulatory compromise

B. hypothermia

C. age

D. all of the above

27. If a patient has inadequate circulation, you must take immediate action to do all of the following, except:

A. restore or improve circulation.

B. apply an AED.

C. control severe bleeding.

D. improve oxygen delivery to the tissues.

28. Any patient with impaired circulation should receive high-flow oxygen via a nonrebreathing mask or assisted ventilations to improve oxygen delivery at the _____ level.

A. alveoli

B. capillary

C. cellular

D. pulmonary

29. While initial treatment is important, it is essential to remember that immediate _____ is one of the keys to the survival of any high-priority patient.

A. airway control

B. bleeding control

C. transport

D. application of oxygen

30. Goals of the focused history and physical exam include:

A. identifying the patient's chief complaint.

B. understanding the specific circumstances surrounding the chief complaint.

C. directing further physical examination.

D. all of the above

_B_ **31.** Understanding the _____ helps you to understand the severity of the patient's problem and provide invaluable information to hospital staff as well.
   A. chief complaint
   B. mechanism of injury
   C. physical exam
   D. focused history

_C_ **32.** Seat belts that are worn improperly, across the abdomen rather than across the pelvic bones, increase the potential for:
   A. down-and-under pathway injuries.
   B. ejections.
   C. internal injuries.
   D. lumbar spine fractures.

_B_ **33.** An integral part of the rapid trauma assessment is evaluation using the mnemonic:
   A. AVPU.
   B. DCAP-BTLS.
   C. OPQRST.
   D. SAMPLE.

_C_ **34.** It is particularly important to evaluate the neck before:
   A. log rolling the patient.
   B. examining the chest.
   C. covering it with a cervical collar.
   D. checking for the presence of a carotid pulse.

_A_ **35.** To check for motor function, you should ask the patient:
   A. to wiggle his or her fingers or toes.
   B. to identify which extremity you are touching.
   C. if they can feel you touching them.
   D. all of the above

_C_ **36.** The "E" in SAMPLE stands for:
   A. eating habits.
   B. emergency medications.
   C. events leading up to the episode.
   D. episodes experienced previously.

_C_ **37.** When assessing a complaint of dizziness, you should evaluate all of the following, except:
   A. pulse.
   B. blood pressure.
   C. movement.
   D. skin.

_B_ **38.** Patients who require a complete rapid trauma assessment, coupled with short scene time and immediate transport to the hospital, include all of the following, except:
   A. any patient who experienced a significant mechanism of injury.
   B. any patient hit from the rear complaining of neck pain.
   C. any patient who is unresponsive or disoriented.
   D. any patient who is extremely intoxicated from drugs or alcohol.

39. The "S" in the mnemonic OPQRST stands for:
   A. signs.
   B. symptoms.
   C. severity.
   D. syncope.

40. A patient who points to a single place for his or her pain has what is known as:
   A. diffuse pain.
   B. focal pain.
   C. radiating pain.
   D. referred pain.

41. When assessing a chief complaint of chest pain, you should evaluate:
   A. skin color.
   B. pulse.
   C. breath sounds.
   D. all of the above

42. Baseline vital signs provide useful information about the:
   A. overall functions of the patient's heart.
   B. overall functions of the patient's lungs.
   C. patient's stability.
   D. all of the above

43. If you have successfully stabilized the ABCs on any patient who is unconscious, confused, or unable to relate the chief complaint adequately, you should:
   A. try to obtain information from family members.
   B. perform a rapid assessment.
   C. look for clues from medication bottles.
   D. transport immediately.

44. When performing a detailed physical exam, depending on what is learned, you should be prepared to:
   A. return to the initial assessment if a potentially life-threatening condition is identified.
   B. provide treatment for problems that were identified during the exam.
   C. modify any treatment that is underway on the basis of any new information.
   D. all of the above

45. When performing a detailed exam, check the neck for:
   A. subcutaneous emphysema.
   B. jugular vein distention.
   C. crepitus.
   D. all of the above

46. The purpose of the ongoing assessment is to ask and answer the following questions, except:
   A. Is treatment improving the patient's condition?
   B. What is the patient's diagnosis?
   C. Has an already identified problem gotten better? Worse?
   D. What is the nature of any newly identified problems?

_____47. When reevaluating any interventions you started, take a moment to ensure that:
- **A.** oxygen is still flowing.
- **B.** backboard straps are still tight.
- **C.** bleeding has been controlled.
- **D.** all of the above

Vocabulary emtb. vocab explorer

Define the following terms using the space provided.

1. Blunt trauma:

_____

_____

_____

2. Penetrating trauma:

_____

_____

_____

3. Mechanism of injury:

_____

_____

_____

4. Capillary refill:

_____

_____

_____

5. Golden Hour:

_____

_____

_____

Fill-in

Read each item carefully, then complete the statement by filling in the missing word(s).

1. From a practical point of view, prehospital emergency care is simply a series of

   _____ about treatment and transport.

2. The best way to reduce your risk of exposure is to follow _____ (BSI) precautions.

3. You cannot help your patient if you become a _____ yourself.

4. You should park your unit in a place that will offer you and your partner the

   greatest _____ but also rapid access to the patient and your equipment.

5. Risk for serious injury also varies depending on whether seat belts are used and

   whether they are worn _____.

6. Any patient who has fallen more than _____ times his or her own height should be considered at risk for serious injury.

7. In gunshot wounds, the area that is involved may be predicted by creating an

   imaginary line between the _____ and the _____, if one exists, although this is only an approximation at best.

8. _____ is a process of identifying the severity of each patient's condition.

9. The _____ _____ is based on your immediate assessment of the environment, the presenting signs and symptoms, mechanism of injury in a trauma patient, and the patient's chief complaint.

10. The first steps in caring for any patient focus on finding and treating the most

    _____ illnesses and injuries.

11. With an unresponsive patient or one with a decreased level of consciousness, you

    should immediately assess the _____ of the airway.

12. If a patient seems to have difficulty breathing, you should immediately

    _____ the airway.

13. Assess the skin temperature by touching the patient's skin with your

    _____ or the back of your hand.

14. Correct identification of high-priority patients is an essential aspect of the

    _____ and helps to improve patient outcome.

True/False

If you believe the statement to be more true than false, write the letter "T" in the space provided. If you believe the statement to be more false than true, write the letter "F."

1. ___F___ Responsiveness is evaluated with the mnemonic DCAP-BTLS.
2. ___T___ The detailed physical exam is normally performed en route to the hospital.
3. ___F___ The damage associated with a gunshot wound is easily evaluated by drawing an imaginary line between the entrance and exit wounds.
4. ___T___ Capillary refill can be checked in children by squeezing an entire arm or leg at a distal point.
5. ___F___ An ongoing assessment is not necessary for stable patients.
6. ___T___ Distinguishing between medical and trauma patients is less important than identifying and treating their problems appropriately.
7. ___F___ Because capillary refill is checked in different parts of the body for adults and children, add 1 to 2 seconds to the normal time frame when assessing capillary refill in children.
8. ___T___ The apparent absence of a palpable pulse in a responsive patient is not caused by cardiac arrest.
9. ___T___ A patient with a poor general impression is considered a priority patient.
10. ___F___ A rapid trauma assessment is not necessary for a patient without a significant mechanism of injury.
11. ___F___ Airbags prevent steering wheel injuries.

Short Answer

Complete this section with short written answers using the space provided.

1. List the three factors affecting the degree of injury resulting from penetrating trauma.

_____

_____

_____

2. What is the single goal of initial assessment?

_____

_____

_____

3. What is the general impression based on?

_____

_____

_____

NOTES

3

4. What do the letters ABC stand for in the assessment process?

_____

_____

_____

5. Describe the difference between diffuse and focal pain.

_____

_____

_____

6. What four questions are asked when assessing orientation and what purpose do these questions serve?

_____

_____

_____

_____

7. What are the three goals of the focused history and physical exam?

_____

_____

_____

8. List the elements of DCAP-BTLS.

Deformities Contusions Abrasions Penetration
Punctures Burns Tenderness Lacerations Swelling

9. List at least five significant mechanisms of injury in an adult.

_____

_____

_____

_____

_____

Word Fun

The following crossword puzzle is an activity provided to reinforce correct spelling and understanding of medical terminology associated with emergency care and the EMT-B. Use the clues in the column to complete the puzzle.

Across

1. Involuntary muscle contraction, to protect
3. Rattling, moist sounds
6. Motion in opposite direction
9. Distal discomfort or pain
10. Skin pulls in around ribs
12. Below 95°F
14. Formation of clots
15. Pain not associated with a single site

Down

2. Coarse breath sounds
4. Acronym for history
5. Broad area trauma without broken skin
7. Times four
8. Pain assessment acronym
11. Grating or grinding
13. Pain in a single area

3

Ambulance Calls

The following real case scenarios provide an opportunity to explore the concerns associated with patient management. Read each scenario, then answer each question in detail.

1. You are dispatched to a motor vehicle collision where you find a 32-year-old male with extensive trauma to the face and gurgling in his airway. He is responsive only to pain. You also note that the windshield is spider-webbed and there is deformity to the steering wheel. He is not wearing a seat belt.

   How would you best manage this patient? What clues tell you the transport status?

_____

_____

_____

_____

_____

_____

2. You are called to the scene of a 75-year-old female who is "weak." Family members tell you she is not "acting right." She does not seem to notice as you approach. She is breathing adequately and her pulse is strong at a rate of 72. She appears to be in no obvious distress.

How would you best manage this patient? What would you ask family members?

3. You are working a wreck involving a 24-year-old male unrestrained driver who is confused and anxious. You have him immobilized on a long backboard and have him on oxygen via nonrebreathing mask. His ABCs are within normal limits. He has a hematoma just above his left orbit, but no signs of bleeding or shock.

How would you transport this patient? What would you do en route?

Rapid Transport

Skill Drills

Skill Drill 8-1: Performing a Rapid Trauma Assessment
Test your knowledge of this skill drill by placing the photos below in the correct order.
Number the first step with a "1," the second step with a "2," etc.

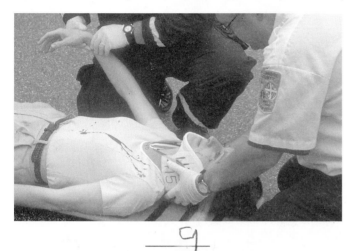

_____9_____

Assess baseline vitals and SAMPLE history.

_____3_____

Apply a cervical collar.

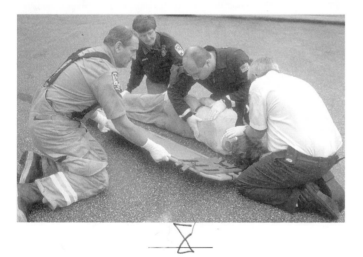

_____8_____

Log roll the patient and assess the back.

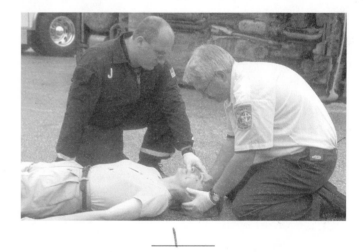

_____1_____

Check ABCs, continue spinal immobilization, and assess mental status.

Assess the head.

(continued)

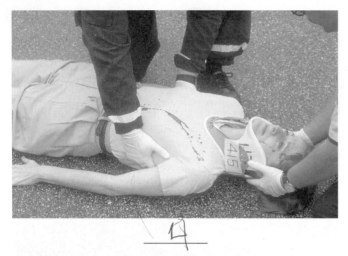

_4_

Assess the chest, including breath sounds.

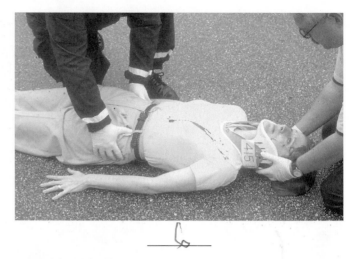

_6_

Assess the pelvis.

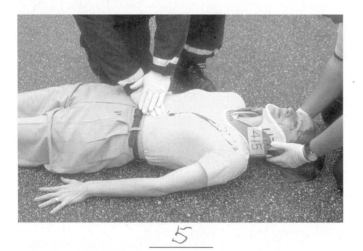

_5_

Assess the abdomen.

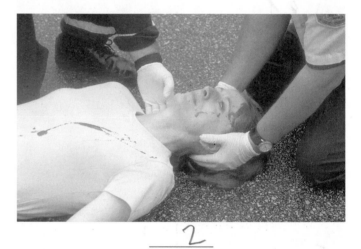

_2_

Assess the neck.

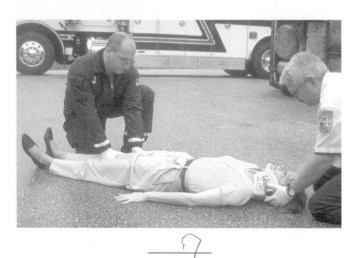

_7_

Assess the extremities.

Skill Drill 8-2: Performing a Rapid Medical Assessment: Unresponsive Patient
Test your knowledge of this skill drill by placing the photos below in the correct order.
Number the first step with a "1," the second step with a "2," etc.

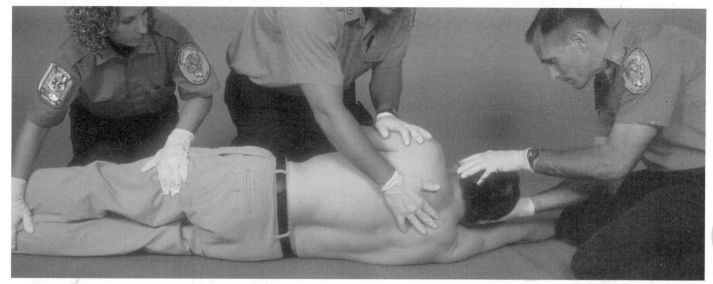

3

_____

Assess the back.

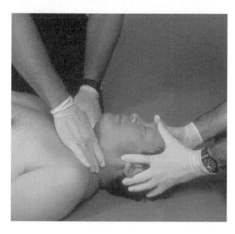

_____

Assess the neck.

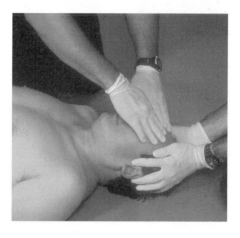

_____

Assess the head.

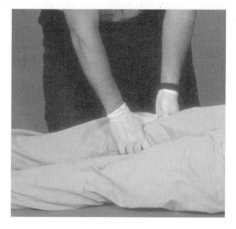

_____

Assess the extremities.

(continued)

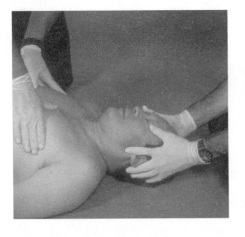

Assess the chest.

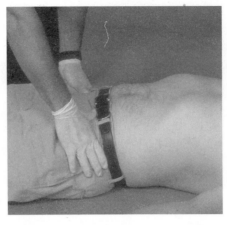

Assess the pelvis.

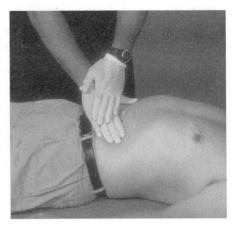

Assess the abdomen.

Skill Drill 8-3: Performing the Detailed Physical Exam

Test your knowledge of this skill drill by placing the photos below in the correct order. Number the first step with a "1," the second step with a "2," etc.

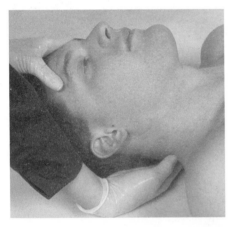

Palpate the neck, front and back.

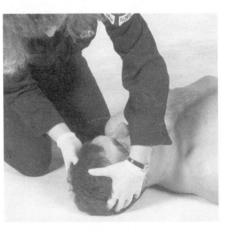

Observe and palpate the head.

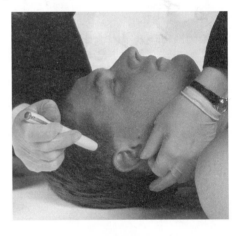

Check the ears for drainage or blood.

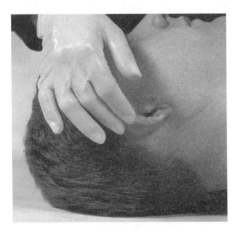

Look behind the ear for Battle's sign.

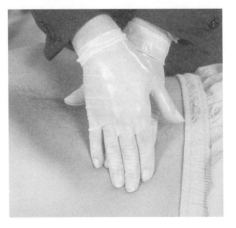

Gently palpate the abdomen.

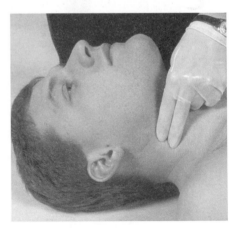

Observe for jugular vein distention.

(continued)

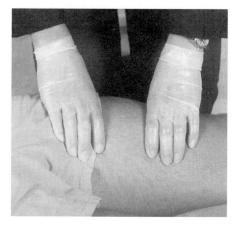

Inspect the extremities; assess distal circulation and motor sensory function.

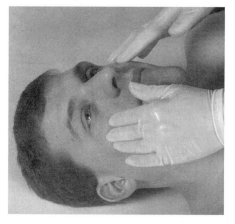

Palpate the maxillae.

Inspect the chest and observe breathing motion.

3

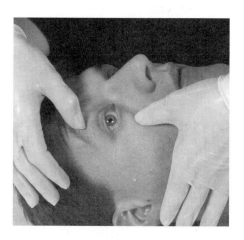

Log roll the patient and inspect the back.

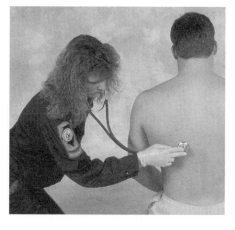

Listen to posterior breath sounds (bases, apices).

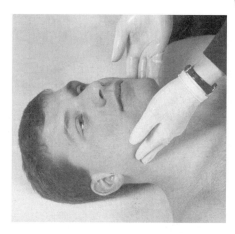

Palpate the mandible.

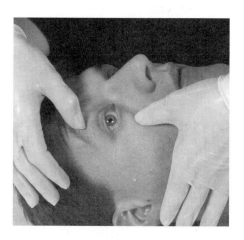

Inspect the eyelids and the area around the eyes.

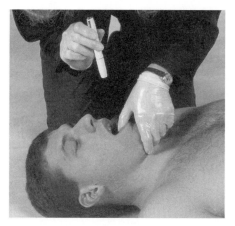

Assess the mouth.

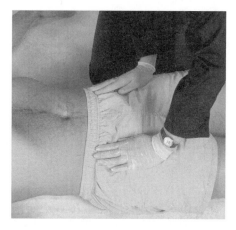

Gently press the iliac crests.

(continued)

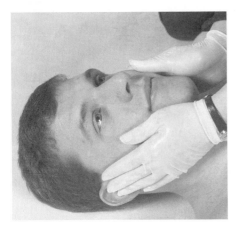

Palpate the zygomas.

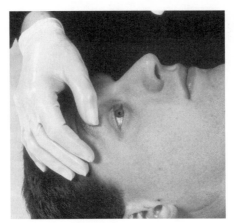

Examine the eyes for redness, contact lenses. Check pupil function.

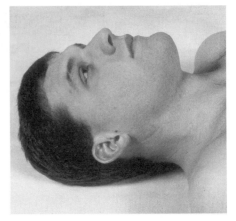

Inspect the neck.

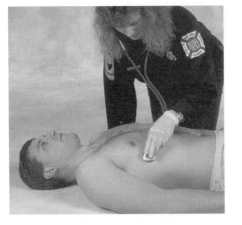

Listen to anterior breath sounds (midaxillary, midclavicular).

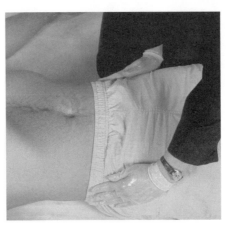

Gently compress the pelvis from the sides.

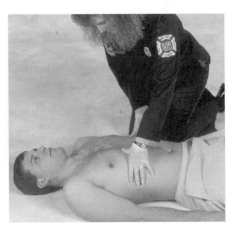

Gently palpate the ribs.

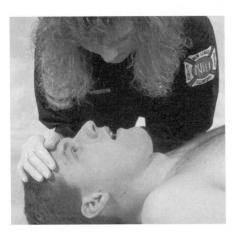

Check for unusual breath odors.

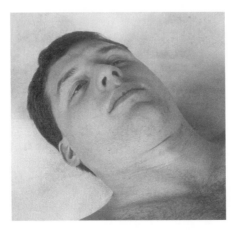

Observe the face.

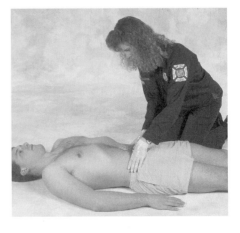

Observe the abdomen and pelvis.

Notes

Workbook Activities

The following activities have been designed to help you. Your instructor may require you to complete some or all of these activities as a regular part of your EMT-B training program. You are encouraged to complete any activity that your instructor does not assign as a way to enhance your learning in the classroom.

Chapter Review

The following exercises provide an opportunity to refresh your knowledge of this chapter.

■ NOTES ■

Matching

Match each of the terms in the left column to the appropriate definition in the right column.

_____ 1. Base station

_____ 2. Mobile radio

_____ 3. Portable radio

_____ 4. Repeater

_____ 5. Telemetry

_____ 6. UHF range

_____ 7. VHF range

_____ 8. Cellular telephone

_____ 9. Dedicated line

_____ 10. MED channels

_____ 11. Scanner

_____ 12. Channel

_____ 13. Rapport

A. "hot line"

B. a trusting relationship built with your patient

C. communication through an interconnected series of repeater stations

D. assigned frequency used to carry voice and/or data communications

E. radio receiver that searches across several frequencies until the message is completed

F. VHF and UHF channels designated exclusively for EMS use

G. vehicle-mounted device that operates at a lower frequency than a base station

H. a process in which electronic signals are converted into coded, audible signals

I. radio frequencies between 30 and 300 MHz

J. hand-carried or hand-held devices that operate at 1 to 5 watts

K. special base station radio that receives messages and signals on one frequency and then automatically retransmits them on a second frequency

L. radio frequencies between 300 and 3,000 MHz

M. radio hardware containing a transmitter and receiver that is located in a fixed location

Communications and Documentation

Multiple Choice

Read each item carefully, then select the best response.

_____ **1.** The base station may be used:

    A. in a single place by an operator speaking into a microphone that is connected directly to the equipment.

    B. remotely through telephone lines.

    C. by radio from a communication center.

    D. all of the above

_____ **2.** The transmission range of a(n) _____ is more limited than that of mobile or base station radios.

    A. portable radio

    B. 800 MHz radio

    C. cellular phone

    D. UHF radio

_____ **3.** Base stations:

    A. usually have more power than mobile or portable radios.

    B. have higher, more efficient antenna systems.

    C. allow for communication with field units at much greater distances.

    D. all of the above

_____ **4.** _____ are helpful when you are away from the ambulance and need to communicate with dispatch, another unit, or medical control.

    A. Base stations

    B. Portable radios

    C. Mobile radios

    D. Cellular phones

_____ **5.** Digital signals are also used in some kinds of paging and tone alerting systems because they transmit _____ and allow more choices and flexibility.

    A. numerically

    B. faster

    C. alphanumerically

    D. encoded messages

_____ 6. Once connected to a cellular network, you can call any other telephone in the world and can send _____ signals.

    A. voice

    B. data

    C. telemetry

    D. all of the above

_____ 7. As with all repeater-based systems, a cellular telephone is useless if the equipment:

    A. fails.

    B. loses power.

    C. is damaged by severe weather or other circumstances.

    D. all of the above

_____ 8. In the simplex mode, all of the following are true, except:

    A. when one party transmits, the other must wait to reply.

    B. you must push a button to talk.

    C. it is called a "pair of frequencies."

    D. radio transmissions can occur in either direction, but not simultaneously in both.

_____ 9. Principle EMS-related responsibilities of the FCC include:

    A. monitoring radio operations.

    B. establishing limitations for transmitter power output.

    C. allocating specific radio frequencies for use by EMS providers.

    D. all of the above

_____ 10. Responsibilities of the dispatcher include all of the following, except:

    A. properly screening and assigning priority to each call.

    B. selecting and alerting the appropriate EMS response units.

    C. dispatching and directing EMS response units to the correct location.

    D. providing emergency medical care to the telephone caller.

_____ 11. In order to dispatch the appropriate unit to the necessary location, the dispatcher must find out:

    A. the exact location of the patient.

    B. the nature of the problem.

    C. the severity of the problem.

    D. all of the above

_____ 12. Determination of the level and type of response necessary is based on:

    A. the dispatcher's perception of the nature and severity of the problem.

    B. the anticipated response time to the scene.

    C. the level of training.

    D. all of the above

_____ 13. Information given to the responding unit(s) should include all of the following, except:

    A. the number of patients.

    B. the time the unit will arrive.

    C. the exact location of the incident.

    D. responses by other public safety agencies.

_____ **14.** You must consult with medical control to:

    A. notify the hospital of an incoming patient.

    B. request advice or orders from medical control.

    C. advise the hospital of special situations.

    D. all of the above

_____ **15.** The patient report commonly includes all of the following, except:

    A. a list of the patient's medications.

    B. the patient's age and gender.

    C. a brief history of the patient's current problem.

    D. your estimated time of arrival.

_____ **16.** In most areas, medical control is provided by the _____ who work at the receiving hospital.

    A. nurses

    B. physicians

    C. interns

    D. staff

_____ **17.** For _____ reasons, the delivery of sophisticated care, such as assisting patients in taking medications, must be done in association with physicians.

    A. logical

    B. ethical

    C. legal

    D. all of the above

_____ **18.** Standard radio operating procedures are designed to:

    A. reduce the number of misunderstood messages.

    B. keep transmissions brief.

    C. develop effective radio discipline.

    D. all of the above

_____ **19.** Be sure that you report all patient information in a(n) _____ manner.

    A. objective

    B. accurate

    C. professional

    D. all of the above

_____ **20.** Medical control guides the treatment of patients in the system through all of the following, except:

    A. hands-on care.

    B. protocols.

    C. direct orders.

    D. post-call review.

_____ **21.** Depending upon how the protocols are written, you may need to call medical control for direct orders to:

    A. administer certain treatments.

    B. transport a patient.

    C. request assistance from other agencies.

    D. immobilize a patient.

_____ **22.** The delivery of EMS involves an impressive array of:

    A. assessments.

    B. stabilization.

    C. treatments.

    D. all of the above

_____23. During transport:

    A. you must periodically reassess the patient's overall condition.

    B. it is not necessary to report changes in the patient's condition.

    C. you are required to check vital signs once.

    D. once treatment is provided, it is safe to finish your paperwork, since the patient's condition will remain stable.

_____24. While en route to and from the scene, you should report all of the following to the dispatcher, except:

    A. any special hazards.

    B. traffic delays.

    C. abandoned vehicles in the median.

    D. road construction.

_____25. Situations that might require special preparation on the part of the hospital include:

    A. HazMat situations.

    B. mass-casualty incidents.

    C. rescues in progress.

    D. all of the above

_____26. The _____ officially occurs during your oral report at the hospital, not as a result of your radio report en route.

    A. patient report

    B. transfer of care

    C. termination of services

    D. all of the above

_____27. Effective communication between the EMT-B and health care professionals in the receiving facility is an essential cornerstone of _____ patient care.

    A. efficient

    B. effective

    C. appropriate

    D. all of the above

_____28. Components that must be included in the oral report during transfer of care include:

    A. the patient's name.

    B. any important history.

    C. vital signs assessed.

    D. all of the above

_____29. Your _____ are critically important in gaining the trust of both the patient and family.

    A. gestures

    B. body movements

    C. attitude toward the patient

    D. all of the above

_____30. If the patient is hearing impaired, you should:

    A. stand on the patient's left side.

    B. shout.

    C. speak clearly and distinctly.

    D. use baby talk.

31. The functional age relates to the person's:
  A. ability to function in daily activities.
  B. mental state.
  C. activity pattern.
  D. all of the above

32. When caring for a visually impaired patient, you should:
  A. use sign language.
  B. touch the patient only when necessary to render care.
  C. try to avoid sudden movements.
  D. never walk them to the ambulance.

33. When attempting to communicate with non-English-speaking patients, you should:
  A. use short, simple questions and simple words whenever possible.
  B. always use medical terms.
  C. shout.
  D. position yourself so the patient can read your lips.

34. The patient information that is included in the minimum data set includes all of the following, except:
  A. chief complaint.
  B. the time that the EMS unit arrived at the scene.
  C. respirations and effort.
  D. skin color and temperature.

35. The administrative information that is included in the minimum data set includes:
  A. the time that patient care was transferred.
  B. chief complaint.
  C. skin color and temperature.
  D. systolic blood pressure for patients older than 3 years.

36. Functions of the prehospital care report include:
  A. continuity of care.
  B. education.
  C. research.
  D. all of the above

37. A good prehospital care report documents:
  A. the care that was provided.
  B. the patient's condition on arrival.
  C. any changes.
  D. all of the above

38. When completing the narrative section, be sure to:
  A. describe what you see and what you do.
  B. only include positive findings.
  C. record your conclusions about the incident.
  D. use appropriate radio codes.

39. Instances in which you may be required to file special reports with appropriate authorities include:
  A. gunshot wounds.
  B. dog bites.
  C. suspected physical, sexual, or substance abuse.
  D. all of the above

3

Vocabulary

Define the following terms using the space provided.

1. Simplex:

_____

_____

_____

2. Standing orders:

_____

_____

_____

3. Federal Communications Commission (FCC):

_____

_____

_____

4. Duplex:

_____

_____

_____

Fill-in

Read each item carefully, then complete the statement by filling in the missing word.

1. Written communications, in the form of a written _____, provide you with an opportunity to communicate the patient's story to others who may participate in the patient's care in the future.

2. A two-way radio consists of two units: a _____ and a _____.

3. A _____, also known as a hot line, is always open or under the control of the individuals at each end.

4. With _____, electronic signals are converted into coded, audible signals.

5. Low-power portable radios that communicate through a series of interconnected repeater stations called "cells" are known as _____.

6. _____ are commonly used in EMS operations to alert on- and off-duty personnel.

7. When the first call to 9-1-1 comes in, the dispatcher must try to judge its relative

 _____ to begin the appropriate EMS response using emergency medical dispatch protocols.

8. The principal reason for radio communication is to facilitate communication

 between you and _____ .

9. You could be successfully sued for _____ if you describe a patient in a way that injures his or her reputation.

10. Regardless of your system's design, your link to _____ is vital to maintain the high quality of care that your patient requires and deserves.

11. To ensure complete understanding, once you receive an order from medical

 control, you must _____ the order back, word for word, and then receive confirmation.

12. By their very nature, _____ do not require direct communication with medical control.

13. Maintaining _____ with your patient builds trust and lets the patient know that he or she is your first priority.

14. Children can easily see through lies or deception, so you must always be

 _____ with them.

15. If the patient does not speak any English, find a family member or friend to act

 as a(n) _____ .

16. The national EMS community has identified a _____ that should enable communication and comparison of EMS runs between agencies, regions, and states.

17. _____ adult patients have the right to refuse treatment.

True/False

If you believe the statement to be more true than false, write the letter "T" in the space provided. If you believe the statement to be more false than true, write the letter "F."

1. _____ The two-way radio is actually at least two units: a transmitter and a receiver.

2. _____ Base stations typically have more power and much higher and more efficient antenna systems than mobile or portable radios.

3. _____ A cellular telephone is just another kind of portable radio that is available for EMS use.

4. _____ The transmission range of a mobile radio is more limited than that of a portable radio.

5. _____ A dedicated line, a special telephone line used for specific point-to-point communications, is always open or under the control of the individuals at each end.

6. _____ The written report is a vital part of providing emergency medical care and ensuring the continuity of patient care.

7. _____ EMS systems that use repeaters are unable to get good signals from portable radios.

8. _____ Small changes in your location will not significantly affect the quality of your transmission.

9. _____ When used improperly or not understood, codes create confusion rather than clear it up.

10. _____ Radio equipment that is operating properly should be serviced at least every 2 years.

11. _____ Your reporting responsibilities end when you arrive at the hospital.

12. _____ Patients deserve to know that you can provide medical care and that you are concerned about their well-being.

Short Answer

Complete this section with short written answers using the space provided.

1. List the five principal FCC responsibilities related to EMS.

_____

_____

_____

_____

_____

2. List five guidelines for effective radio communications.

_____

_____

_____

_____

3. List the six functions of a prehospital care report.

_____

_____

_____

_____

_____

_____

4. Describe the two types of written report forms generally in use in EMS systems.

_____

_____

_____

Word Fun emtb vocab explorer

The following crossword puzzle is an activity provided to reinforce correct spelling and understanding of medical terminology associated with emergency care and the EMT-B. Use the clues in the column to complete the puzzle.

CLUES

Across

2. Frequencies between 300 and 3,000 MHz

4. Electronic signals converted into coded audible signals

7. Point-to-point

8. Agency with jurisdiction over radios

12. Radio receiver that searches

13. Transmit in both ways, but not at the same time

14. Hardware in a fixed location

Down

1. Transmit and receive simultaneously

3. Outline specific directions, protocols

5. Frequencies for exclusive EMS usage

6. Frequencies between 30 and 300 MHz

9. Trusting relationship

10. Receives on one, transmits on another

11. Assigned frequency

Ambulance Calls

The following real case scenarios provide an opportunity to explore the concerns associated with patient management. Read each scenario, then answer each question in detail.

1. You are in the dispatch office filling in for the EMS dispatcher who needed to use the restroom. The phone rings and you answer it to find a hysterical female screaming about a child falling in an old well. All you can get out of her is that he is 5 years old and is not making any noise. The address is on the computer display.

   How would you best manage this situation and what additional help would you call for?

   _____

   _____

   _____

   _____

   _____

   _____

   _____

   _____

2. You are called to a special school for deaf children where a 6-year-old female fell on the playground and twisted her ankle. Her mother has been notified and will meet you at the emergency room. The patient is sitting beside the monkey bars when you arrive.

   How would you best manage this patient?

   _____

   _____

   _____

   _____

   _____

   _____

   _____

   _____

3. You are dispatched to the local park where a collision occurred between a bicycle and a pedestrian. Your patient, a 25-year-old male, was walking his dog when the cyclist swerved to miss a squirrel and collided with him, knocking him to the ground. You notice he has deformity to his right forearm and the dog is a seeing-eye dog.

How would you best manage this patient? The dog?

_____

_____

_____

_____

_____

_____

_____

_____

NOTES

Workbook Activities

The following activities have been designed to help you. Your instructor may require you to complete some or all of these activities as a regular part of your EMT-B training program. You are encouraged to complete any activity that your instructor does not assign as a way to enhance your learning in the classroom.

Chapter Review

The following exercises provide an opportunity to refresh your knowledge of this chapter.

NOTES

Matching

Match each of the terms in the left column to the appropriate definition in the right column.

_I_ **1.** Absorption

_J_ **2.** Contraindication

_F_ **3.** Side effect

_G_ **4.** Adsorption

_D_ **5.** Dose

_H_ **6.** Indication

_B_ **7.** Action

_C_ **8.** Pharmacology

_E_ **9.** Capsules

_A_ **10.** Topical medications

A. lotions, creams, ointments

B. effect that a drug is expected to have

C. the study of the properties and effects of drugs and medications

D. amount of medication given

E. gelatin shells filled with powdered or liquid medication

F. any action of a drug other than the desired one

G. to bind or stick to a surface

H. therapeutic use for a particular medication

I. process by which medications travel through body tissues

J. situation in which a drug should not be given

Multiple Choice

Read each item carefully, then select the best response.

_D_ **1.** Medication(s) that an EMT-B may help patients to self-administer include:

A. metered-dose inhaler medications.

B. nitroglycerin.

C. epinephrine.

D. all of the above

_C_ **2.** The proper dose of a medication depends on all of the following, except:

A. the patient's age.

B. the patient's size.

C. generic substitutions.

D. the desired action.

General Pharmacology

_____A_____ **3.** Nitroglycerin relieves the squeezing or crushing pain associated with angina by:

   A. dilating the arteries to increase the oxygen supply to the heart muscle.

   B. causing the heart to contract harder and increase cardiac output.

   C. causing the heart to beat faster to supply more oxygen to the heart.

   D. all of the above

_____A_____ **4.** The brand name that a manufacturer gives to a medication is called the _____ name.

   A. trade

   B. generic

   C. chemical

   D. prescription

_____A_____ **5.** The fastest way to deliver a chemical substance is by the _____ route.

   A. intravenous

   B. oral

   C. sublingual

   D. intramuscular

_____B_____ **6.** Medications that have the prefix "depo" in their names form a _____ in the muscle after being injected.

   A. deposition

   B. depository

   C. depot

   D. deponent

_____C_____ **7.** Insulin is a medication that is given by the _____ route.

   A. intravenous

   B. oral

   C. subcutaneous

   D. intramuscular

____D____ **8.** The form the manufacturer chooses for a medication ensures:

    A. the proper route of the medication.

    B. the timing of its release into the bloodstream.

    C. its effects on target organs or body systems.

    D. all of the above

____D____ **9.** Solutions may be given:

    A. orally.

    B. intramuscularly.

    C. rectally.

    D. all of the above

____B____ **10.** In the prehospital setting, _____ is the preferred method of giving oxygen to patients who have sufficient tidal volume, and can provide up to 95% inspired oxygen.

    A. a nasal cannula

    B. a nonrebreathing mask

    C. a bag-valve mask

    D. any of the above

____A____ **11.** Characteristics of epinephrine include:

    A. dilating passages in the lungs.

    B. constricting blood vessels.

    C. increasing the heart rate and blood pressure.

    D. all of the above

____C____ **12.** "Reactive airway disease," the other name for _____, can be a life-threatening condition.

    A. COPD

    B. bronchitis

    C. asthma

    D. emphysema

____A____ **13.** The release of histamine causes:

    A. relaxing of the small blood vessels.

    B. bronchial spasms.

    C. edema of the airway tissues.

    D. all of the above

____A____ **14.** Epinephrine acts as a specific antidote to:

    A. adrenalin.

    B. histamine.

    C. asthma.

    D. bronchitis.

____A____ **15.** Nitroglycerin relieves pain because its purpose is to increase blood flow by relieving the spasms or causing the arteries to:

    A. dilate.

    B. constrict.

    C. thicken.

    D. contract.

____C____ **16.** Nitroglycerin affects the body in the following ways (select all that apply):

    A. It decreases blood pressure.

    B. It relaxes veins throughout the body.

    C. It often causes a mild headache after administration.

    D. It increases blood return to the heart.

Vocabulary

Define the following terms using the space provided.

1. Trade name:

_____

_____

_____

2. Generic name:

_____

_____

_____

3. OTC:

_____

_____

_____

4. Solution:

_____

_____

_____

5. Suspension:

_____

_____

_____

6. Sublingual:

_____

_____

_____

7. Metered-dose inhaler (MDI):

_____

_____

_____

Fill-in

Read each item carefully, then complete the statement by filling in the missing word.

1. _____ is a simple sugar that is readily absorbed by the bloodstream.

2. _____ is the main hormone that controls the body's fight-or-flight response.

3. Nitroglycerin is usually taken _____.

4. In all but the _____ route, the medication is absorbed into the bloodstream through various body tissues.

5. When given by mouth, _____ may be absorbed from the stomach fairly quickly because the medication is already dissolved.

6. A _____ is a chemical substance that is used to treat or prevent disease or relieve pain.

True/False

If you believe the statement to be more true than false, write the letter "T" in the space provided. If you believe the statement to be more false than true, write the letter "F."

1. _____ Oxygen is a flammable substance.
2. _____ Glucose may be administered to an unconscious patient in order to save his or her life.
3. _____ Epinephrine is a hormone produced by the body to aid in digestion.
4. _____ Nitroglycerin decreases blood pressure.
5. _____ Sublingual medications are rapidly absorbed into the digestive tract.
6. _____ Vital signs should be taken before and after a medication is given.
7. _____ Hypoglycemia can be caused by an excess of insulin.
8. _____ Even though medications can react with each other, this is not a potentially harmful condition for the patient.
9. _____ Nitroglycerin should only be administered when the patient's systolic blood pressure is below 100 mm Hg.

Short Answer

Complete this section with short written answers using the space provided.

1. List seven routes of medication administration.

_____

_____

_____

_____

_____

_____

_____

2. Describe the general steps of administering medication.

_____

_____

_____

3. Describe the action of activated charcoal and the steps of administration that are specific to this medication.

_____

_____

_____

4. List five characteristics of epinephrine.

_____

_____

_____

_____

_____

5. How is an epinephrine auto-injector activated?

_____

_____

_____

6. List five effects of nitroglycerin.

_____

_____

_____

_____

_____

7. Explain why metered-dose inhalers are often used with a spacer.

_____

_____

_____

■ CLUES ■

Across

3. Through the skin

5. Process for medication to travel

6. Into the vein

7. Raises heart rate and blood pressure

8. Into the bone

9. Dilates arteries in angina patients

Down

1. Therapeutic use for medication

2. Under the tongue

4. Binding to or sticking to a surface

Word Fun

The following crossword puzzle is a good way to reinforce correct spelling and understanding of medical terminology associated with emergency care and the EMT-B. Use the clues in the column to complete the puzzle.

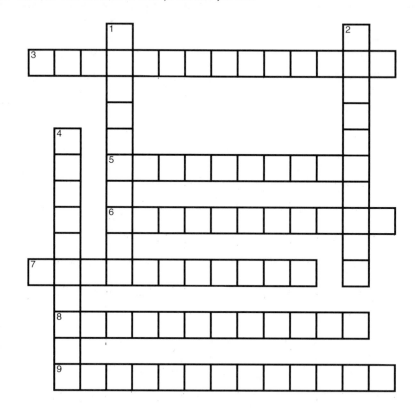

Ambulance Calls

The following real case scenarios provide an opportunity to explore the concerns associated with patient management. Read each scenario, then answer each question in detail.

1. You are dispatched to the home of a 56-year-old man with a history of angina. He is complaining of a "squeezing" chest pain that has lasted for approximately 20 minutes. You find out through your SAMPLE history that the patient takes nitroglycerin and has not taken any today. His blood pressure is 150/90 mm Hg. How would you best manage this patient?

_____

_____

_____

_____

_____

_____

_____

_____

2. You respond to a residence for a 22-year-old female complaining of dyspnea and audible wheezing. The patient's mother tells you that the patient is an asthmatic and is also allergic to shellfish. The girl ate artificial crabmeat for lunch because she thought it was safe. She has an MDI and also an EpiPen for severe allergic reactions. How would you best manage this patient?

_____

_____

_____

_____

_____

_____

_____

_____

_____

3. You are dispatched to the residence of a 68-year-old male who is complaining of a "crushing" chest pain radiating down his left arm for the past hour. He is pale, cool, diaphoretic, and is very nauseated. He tells you he had a heart attack several years ago and takes nitroglycerin as needed. He took two tablets prior to your arrival and reports no relief. How would you best manage this patient?

_____

_____

_____

_____

_____

_____

_____

_____

_____

4

Workbook Activities

The following activities have been designed to help you. Your instructor may require you to complete some or all of these activities as a regular part of your EMT-B training program. You are encouraged to complete any activity that your instructor does not assign as a way to enhance your learning in the classroom.

Chapter Review

The following exercises provide an opportunity to refresh your knowledge of this chapter.

Matching

Match each of the terms in the left column to the appropriate definition in the right column.

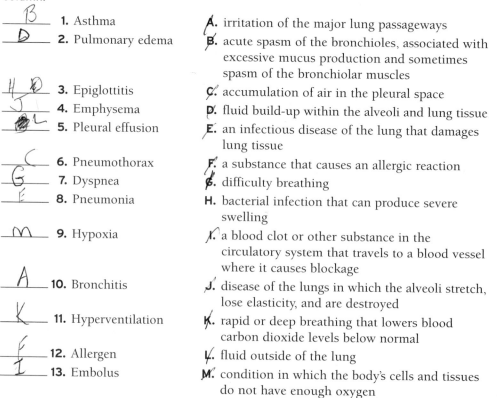

_B_ **1.** Asthma

_D_ **2.** Pulmonary edema

_H D_ **3.** Epiglottitis

_J_ **4.** Emphysema

_L_ **5.** Pleural effusion

_C_ **6.** Pneumothorax

_G_ **7.** Dyspnea

_E_ **8.** Pneumonia

_M_ **9.** Hypoxia

_A_ **10.** Bronchitis

_K_ **11.** Hyperventilation

_F_ **12.** Allergen

_I_ **13.** Embolus

A. irritation of the major lung passageways

B. acute spasm of the bronchioles, associated with excessive mucus production and sometimes spasm of the bronchiolar muscles

C. accumulation of air in the pleural space

D. fluid build-up within the alveoli and lung tissue

E. an infectious disease of the lung that damages lung tissue

F. a substance that causes an allergic reaction

G. difficulty breathing

H. bacterial infection that can produce severe swelling

I. a blood clot or other substance in the circulatory system that travels to a blood vessel where it causes blockage

J. disease of the lungs in which the alveoli stretch, lose elasticity, and are destroyed

K. rapid or deep breathing that lowers blood carbon dioxide levels below normal

L. fluid outside of the lung

M. condition in which the body's cells and tissues do not have enough oxygen

C H A P T E R 11

Respiratory Emergencies

Multiple Choice

Read each item carefully, then select the best response.

_____ 1. When treating a patient with dyspnea, you must be prepared to treat:
- **A.** the symptoms.
- **B.** the underlying problem.
- **C.** the patient's anxiety.
- **D.** all of the above

_____ 2. The oxygen-carbon dioxide exchange takes place in the:
- **A.** trachea.
- **B.** bronchial tree.
- **C.** alveoli.
- **D.** blood.

_____ 3. Oxygen-carbon dioxide exchange may be hampered if:
- **A.** the pleural space is filled with air or excess fluid.
- **B.** the alveoli are damaged.
- **C.** the air passages are obstructed.
- **D.** all of the above

_____ 4. If carbon dioxide levels drop too low, the person automatically breathes:
- **A.** normally.
- **B.** rapidly and deeply.
- **C.** slower, less deeply.
- **D.** fast and shallow.

_____ 5. If the level of carbon dioxide in the arterial blood rises above normal, the patient breathes:
- **A.** normally.
- **B.** rapidly and deeply.
- **C.** slower, less deeply.
- **D.** fast and shallow.

_A_   **6.** _____ is (are) a sign(s) of adequate breathing.
   A. A normal rate and depth
   B. Pale or cyanotic skin
   C. Pursed lips and nasal flaring
   D. Cool, damp skin

_D_   **7.** The level of carbon dioxide in the arterial blood can rise due to:
   A. emphysema.
   B. chronic bronchitis.
   C. cardiovascular disease.
   D. all of the above

_AB_   **8.** The second stimulus that develops in patients with normally high levels of carbon dioxide responds to:
   A. increased oxygen levels.
   B. decreased oxygen levels.
   C. increased carbon dioxide levels.
   D. decreased carbon dioxide levels.

_A_   **9.** _____ is a sign of hypoxia to the brain.
   A. Altered mental status
   B. Decreased heart rate
   C. Decreased respiratory rate
   D. Delayed capillary refill time

_B_   **10.** An obstruction to the exchange of gases between the alveoli and the capillaries may result from:
   A. epiglottitis.
   B. pneumonia.
   C. colds.
   D. all of the above

_A_   **11.** Pulmonary edema can develop quickly after a major:
   A. heart attack.
   B. episode of syncope.
   C. brain injury.
   D. all of the above

_BD_   **12.** The _____ is the narrowest point in a child's airway.
   A. carina
   B. trachea
   C. epiglottis
   D. larynx

_D_   **13.** In addition to a major heart attack, pulmonary edema may also be produced by:
   A. inhaling large amounts of smoke.
   B. traumatic injuries to the chest.
   C. inhaling toxic chemical fumes.
   D. all of the above

_C_   **14.** Chronic oxygenation problems from bronchitis can lead to:
   A. cerebral edema.
   B. upper airway obstruction.
   C. right-sided heart failure.
   D. fluid retention.

_A_ 15. _____ is a loss of the elastic material around the air spaces as a result of chronic stretching of the alveoli when bronchitic airways obstruct easy expulsion of gases.

  A. Emphysema

  B. Bronchitis

  C. Pneumonia

  D. Diphtheria

_D_ 16. Most patients with COPD will:

  A. chronically produce sputum.

  B. have a chronic cough.

  C. have difficulty expelling air from their lungs.

  D. all of the above

_B_ 17. The patient with COPD usually presents with:

  A. an increased blood pressure.

  B. a green or yellow productive cough.

  C. a decreased heart rate.

  D. all of the above

_C_ 18. A pneumothorax caused by a medical condition without any injury is known as:

  A. a tension pneumothorax.

  B. a subcutaneous pneumothorax.

  C. spontaneous.

  D. none of the above

_C_ 19. Asthma produces a characteristic _____ as patients attempt to exhale through partially obstructed air passages.

  A. rhonchi

  B. stridor

  C. wheezing

  D. rattle

_C_ 20. An allergic response to certain foods or some other allergen may produce an acute:

  A. bronchodilation.

  B. asthma attack.

  C. vasoconstriction.

  D. insulin release.

_D_ 21. Treatment for anaphylaxis and acute asthma attacks include:

  A. epinephrine.

  B. high-flow oxygen.

  C. antihistamines.

  D. all of the above

_C_ 22. A collection of fluid outside the lungs on one or both sides of the chest is called a:

  A. spontaneous pneumothorax.

  B. subcutaneous emphysema.

  C. pleural effusion.

  D. tension pneumothorax.

4

23. Always consider _____ in patients who were eating just before becoming short of breath:

   A. upper airway obstruction

   B. anaphylaxis

   C. lower airway obstruction

   D. bronchoconstriction

24. Pulmonary emboli may occur as a result of:

   A. damage to the lining of the vessels.

   B. a tendency for blood to clot unusually fast.

   C. slow blood flow in the lower extremity.

   D. all of the above

25. _____ is defined as overbreathing to the point that the level of arterial carbon dioxide falls below normal.

   A. Reactive airway syndrome

   B. Hyperventilation

   C. Tachypnea

   D. Pleural effusion

26. Slowing of respirations after administration of oxygen to a COPD patient does not necessarily mean that the patient no longer needs the oxygen; he or she may need:

   A. insulin.

   B. even more oxygen.

   C. mouth-to-mouth resuscitation.

   D. none of the above

27. Aspiration of vomit into the lungs may cause:

   A. croup.

   B. alkalosis.

   C. pneumonia.

   D. bronchitis.

28. Questions to ask during the focused history and physical examination include:

   A. What has the patient already done for the breathing problem?

   B. Does the patient use a prescribed inhaler?

   C. Does the patient have any allergies?

   D. all of the above

29. The problem in asthma is getting the air:

   A. to diffuse through mucus.

   B. past the carina.

   C. into the narrowed trachea.

   D. out of the lungs.

30. Generic names for popular inhaled medications include:

   A. ventolin.

   B. metaprel.

   C. terbutaline.

   D. all of the above

31. Contraindications to helping a patient self-administer any MDI medication include:

   A. not obtaining permission from medical control.

   B. noticing that the inhaler is not prescribed for this patient.

   C. noticing that the patient has already met the maximum prescribed dose.

   D. all of the above

32. Possible side effects of over-the-counter cold medications may include:

   A. agitation.

   B. increased heart rate.

   C. increased blood pressure.

   D. all of the above

33. A prolonged asthma attack that is unrelieved by epinephrine may progress into a condition known as:

   A. pleural effusion.

   B. status epilepticus.

   C. status asthmaticus.

   D. reactive airway disease.

Labeling emtb.com anatomy review

Label the following diagrams with the correct terms.

The Upper Airway

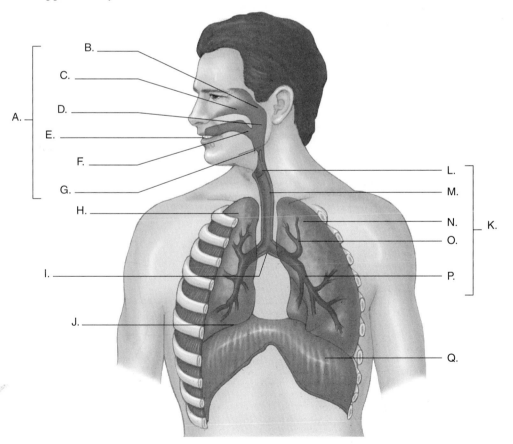

A. _____

B. _____

C. _____

D. _____

E. _____

F. _____

G. _____

H. _____

I. _____

J. _____

K. _____

L. _____

M. _____

N. _____

O. _____

P. _____

Q. _____

Vocabulary

Define the following terms using the space provided.

1. Stridor:

_____

_____

_____

2. Croup:

_____

_____

_____

3. Rales:

_____

_____

_____

4. Rhonchi:

_____

_____

_____

5. Diphtheria:

_____

_____

_____

6. Chronic obstructive pulmonary disease (COPD):

_____

_____

_____

Fill-in

Read each item carefully, then complete the statement by filling in the missing word(s).

1. The level of _____ bathing the brain stem stimulates respiration.

2. The level of _____ in the blood is a secondary stimulus for respiration.

3. _____ passes from the blood through capillaries, which can carry

_____ back to the heart.

4. Carbon dioxide and oxygen are exchanged in the _____.

5. Air enters the body through the _____.

6. Abnormal breathing is indicated by a rate slower than _____ breaths per

minute or faster than _____ breaths per minute.

7. During respiration, oxygen is provided to the blood, and _____ is removed
from it.

True/False

If you believe the statement to be more true than false, write the letter "T" in the space provided. If you believe the statement to be more false than true, write the letter "F."

1. __F__ Chronic bronchitis is characterized by spasm and narrowing of the bronchioles due to exposure to allergens.

2. __T__ With pneumothorax, the lung collapses because the negative vacuum pressure in the pleural space is lost.

3. __F__ Anaphylactic reactions occur only in patients with a previous history of asthma or allergies.

4. __F__ Decreased breath sounds in asthma occur because fluid in the pleural space has moved the lung away from the chest wall.

5. __T__ Pulmonary emboli are difficult to diagnose.

6. __T__ A patient with aspirin poisoning may hyperventilate in response to acidosis.

7. __F__ The distinction between hyperventilation and hyperventilation syndrome is straightforward and should guide the EMT-B's treatment choices.

8. __T__ COPD most often results from cigarette smoking.

9. __F__ Asthma and COPD are characterized by long inspiratory times.

Short Answer

Complete this section with short written answers using the space provided.

1. List five characteristics of normal breathing.

_____

_____

_____

_____

_____

2. List the five most common mechanisms occurring in lung disorders.

_____

4

_____

_____

_____

_____

3. Under what conditions should you not assist a patient with a metered-dose inhaler?

_____

_____

_____

4. Describe chronic bronchitis.

_____

_____

_____

5. List five signs of inadequate breathing.

_____

_____

_____

_____

_____

6. Explain carbon dioxide retention.

_____

_____

_____

Word Fun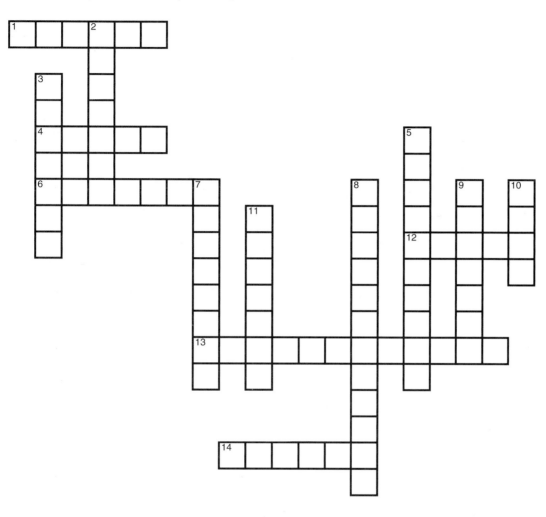

The following crossword puzzle is an activity provided to reinforce correct spelling and understanding of medical terminology associated with emergency care and the EMT-B. Use the clues provided to complete the puzzle.

Across

- **1.** High-pitched, whistling sound
- **4.** Crackling, rattling sounds
- **6.** Difficulty breathing
- **12.** Barking cough
- **13.** Inflammation of leaf-shaped airway cover
- **14.** Muscle spasm in small airways

Down

- **2.** Traveling clot
- **3.** Harsh, high-pitched sound
- **5.** Irritation of major airways
- **7.** Substance causing a reaction
- **8.** Air in pleural space
- **9.** Coarse sounds from mucus in airways
- **10.** Slow process of chronic disruption of airways
- **11.** Low oxygen

Ambulance Calls

The following real case scenarios will give you an opportunity to explore the concerns associated with patient management. Read each scenario, then answer each question in detail.

1. You are dispatched to the home of a 56-year-old male complaining of a sudden onset of severe dyspnea with sharp, localized chest pain. His history reveals surgery for a hip fracture last week. He has been completely immobile until today.

How would you best manage this patient?

_____

_____

_____

_____

_____

_____

2. You are dispatched to a residence where you find a 24-year-old female who is visibly upset and breathing rapidly. She is complaining of numbness and tingling in her hands and feet, as well as dyspnea. Family members tell you that they found her this way and that she has a history of panic attacks. The patient assures you that this is not a panic attack and that it is hard for her to breathe. How would you best manage this patient?

_____

_____

_____

_____

_____

_____

3. You are called to the home of a 73-year-old male complaining of severe dyspnea. The patient has a history of COPD and is on home oxygen at 2 L/min via nasal cannula. His family tells you he has a long history of breathing problems and emphysema. He is cyanotic around his lips and his respirations are 36 and shallow. How would you best manage this patient?

_____

_____

_____

_____

_____

Skill Drills [emt-b video clips]

Test your knowledge of this skill drill by filling in the correct words in the photo captions.

Skill Drill 11-1: Assisting a Patient with a Metered-Dose Inhaler

1. Ensure inhaler is at room temperature or _____.

2. Remove oxygen mask.
Hand inhaler to patient. Instruct about breathing and _____ _____.
Use a _____ if patient has one.

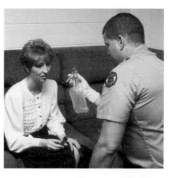

3. Instruct patient to press inhaler and inhale. Instruct about _____ _____.

4. Reapply _____.
After a few _____, have patient repeat _____ if order/protocol allows.

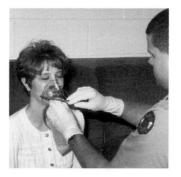

4

Workbook Activities

The following activities have been designed to help you. Your instructor may require you to complete some or all of these activities as a regular part of your EMT-B training program. You are encouraged to complete any activity that your instructor does not assign as a way to enhance your learning in the classroom.

Chapter Review

The following exercises provide an opportunity to refresh your knowledge of this chapter.

Matching

Match each of the terms in the left column to the appropriate definition in the right column.

M	1. Atria	A. absence of all heart electrical activity
D	2. Coronary arteries	B. swelling of feet and ankles
P	3. Atrioventricular (AV) node	C. calcium and cholesterol buildup inside blood vessels
H	4. Myocardium	D. blood vessels that supply blood to the myocardium
O	5. Sinus node	E. abnormal heart rhythm
L	6. Venae cavae	F. unusually slow heart rhythm, less than 60 beats/min
I	7. Ventricles	G. lack of oxygen
	8. Aorta	H. Heart muscle
C	9. Atherosclerosis	I. lower chambers of the heart
E	10. Arrhythmia	J. tissue death
J	11. Ischemia	K. rapid heart rhythm, more than 100 beats/min
B	12. Pedal edema	L. carry oxygen-poor blood back to the heart
G	13. Infarction	M. upper chambers of the heart
K	14. Tachycardia	N. body's main artery
	15. Asystole	O. electrical impulses begin here
F	16. Bradycardia	P. electrical impulses slow here to allow blood to move from the atria to the ventricles

Cardiovascular Emergencies

Multiple Choice

Read each item carefully, then select the best response.

___D___ 1. We can help to reduce the number of deaths attributed to cardiovascular disease with:
- **A.** better public awareness.
- **B.** early access.
- **C.** public access defibrillation.
- **D.** all of the above

___D___ 2. The aorta receives its blood supply from the:
- **A.** right atria.
- **B.** left atria.
- **C.** right ventricle.
- **D.** left ventricle.

___A___ 3. Blood enters into the right atrium from the body through the:
- **A.** vena cava.
- **B.** aorta.
- **C.** pulmonary artery.
- **D.** pulmonary vein.

___B___ 4. The only veins in the body to carry oxygenated blood are the:
- **A.** external jugular veins.
- **B.** pulmonary veins.
- **C.** subclavian veins.
- **D.** inferior vena cava.

___A___ 5. Normal electrical impulses originate in the sinus node, just above the:
- **A.** atria.
- **B.** ventricles.
- **C.** AV junction.
- **D.** Bundle of His.

B 6. Dilation of the coronary arteries will _____ blood flow.
   A. shut off
   B. increase
   C. decrease
   D. regulate

C 7. At the level of the navel, the aorta divides into two main branches called the right and left _____ arteries.
   A. femoral
   B. renal
   C. iliac
   D. sacral

C 8. The _____ are tiny blood vessels about one cell thick.
   A. arterioles
   B. venules
   C. capillaries
   D. ventricles

A 9. _____ carry oxygen to the body's tissues and then remove carbon dioxide.
   A. Red blood cells
   B. White blood cells
   C. Platelets
   D. Veins

C 10. _____ is (are) a mixture of water, salts, nutrients, and proteins.
   A. Platelets
   B. Cerebrospinal fluid
   C. Plasma
   D. All of the above

C 11. _____ is the maximum pressure exerted by the left ventricle as it contracts.
   A. Cardiac output
   B. Diastolic blood pressure
   C. Systolic blood pressure
   D. Stroke volume

A 12. Atherosclerosis can lead to a complete _____ of a coronary artery.
   A. occlusion
   B. disintegration
   C. dilation
   D. contraction

B 13. The lumen of an artery may be partially or completely blocked by the blood-clotting system due to a _____ that exposes the inside of the atherosclerotic wall.
   A. tear
   B. crack
   C. clot
   D. rupture

_B_ **14.** Tissues downstream from a blood clot will suffer from lack of oxygen. If blood flow is resumed in a short time, the _____ tissues will recover.
  A. dead
  B. ischemic
  C. necrosed
  D. dry

_D_ **15.** Risk factors for myocardial infarction include all of the following except:
  A. male gender.
  B. high blood pressure.
  C. stress.
  D. increased activity level.

_B_ **16.** When, for a brief period of time, heart tissues do not get enough oxygen, the pain is called:
  A. AMI.
  B. angina.
  C. ischemia.
  D. CAD.

_D_ **17.** Angina pain may be felt in the:
  A. arms.
  B. midback.
  C. epigastrium.
  D. all of the above

_D_ **18.** Angina may be associated with:
  A. shortness of breath.
  B. nausea.
  C. sweating.
  D. all of the above

_C_ **19.** Because oxygen supply to the heart is diminished with angina, the _____ can be compromised and the person is at risk for significant cardiac rhythm problems.
  A. circulation
  B. cardiac output
  C. electrical system
  D. vasculature

_C_ **20.** About _____ minutes after blood flow is cut off, some heart muscle cells begin to die.
  A. 10
  B. 20
  C. 30
  D. 40

_C_ **21.** An acute myocardial infarction is more likely to occur in the larger, thick-walled left ventricle, which needs more _____ than in the right ventricle.
  A. oxygen and glucose
  B. force to pump
  C. blood and oxygen
  D. electrical activity

C 22. The pain of AMI differs from the pain of angina because:

A. it may be caused by exertion.

B. it is usually relieved by rest.

C. it does not resolve in a few minutes.

D. all of the above

D 23. Consequences of AMI may include:

A. cardiogenic shock.

B. congestive heart failure.

C. sudden death.

D. all of the above

D 24. Sudden death is usually the result of _____, in which the heart fails to generate an effective blood flow.

A. AMI

B. atherosclerosis

C. PVCs

D. cardiac arrest

A 25. Disorganized, ineffective quivering of the ventricles is known as:

A. ventricular fibrillation.

B. asystole.

C. ventricular stand still.

D. ventricular tachycardia.

B 26. In _____, often caused by a heart attack, the problem is that the heart lacks enough power to force the proper volume of blood through the circulatory system.

A. asystole

B. cardiogenic shock

C. ventricular fibrillation

D. angina

A 27. Causes of congestive heart failure include all of the following except:

A. chronic hypotension.

B. heart valve damage.

C. a myocardial infarction.

D. longstanding high blood pressure.

D 28. Signs and symptoms of shock include all of the following except:

A. elevated heart rate.

B. pale, clammy skin.

C. air hunger.

D. elevated blood pressure.

B 29. In patients with CHF, changes in heart function occur, including:

A. a decrease in heart rate.

B. enlargement of the left ventricle.

C. enlargement of the right ventricle.

D. a decrease in blood pressure.

A 30. Fluid that collects in the feet and legs is called:

A. pedal edema.

B. pulmonary edema.

C. cerebral edema.

D. tibial edema.

_____ C _____ **31.** Physical findings of AMI include skin that is _____ because of poor cardiac output and the loss of perfusion.

   A. pink

   B. white

   C. gray

   D. red

_____ C _____ **32.** All patient assessments begin by determining whether or not the patient:

   A. is breathing.

   B. can talk.

   C. is responsive.

   D. has a pulse.

_____ B _____ **33.** In order for you to use an AED, the patient must weigh at least:

   A. 55 kg.

   B. 25 lb.

   C. 110 lb.

   D. 25 kg.

_____ B _____ **34.** To assess chest pain, use the mnemonic:

   A. AVPU.

   B. OPQRST.

   C. SAMPLE.

   D. CHART.

_____ C _____ **35.** When using the mnemonic OPQRST, the "P" stands for:

   A. parasthesia.

   B. pain.

   C. provocation.

   D. predisposing factors.

_____ D _____ **36.** Nitroglycerin may be in the form of a:

   A. skin patch.

   B. spray.

   C. pill.

   D. all of the above

_____ D _____ **37.** When administering nitroglycerin to a patient, you should check the patient's _____ within 5 minutes after each dose.

   A. level of consciousness

   B. breathing

   C. pulse

   D. blood pressure

_____ C _____ **38.** In general, a maximum of _____ dose(s) of nitroglycerin are given for any one episode of chest pain.

   A. one

   B. two

   C. three

   D. four

_____ B _____ **39.** _____ are inserted when the electrical control system of the heart is so damaged that it cannot function properly.

   A. Stents

   B. Pacemakers

   C. Balloon angioplasties

   D. Defibrillation

4

_____D_____ **40.** When the battery wears out in a pacemaker, the patient may experience:

A. syncope.

B. dizziness.

C. weakness.

D. all of the above

_____C_____ **41.** The computer inside the AED is specifically programmed to recognize rhythms that require defibrillation to correct, most commonly:

A. asystole.

B. ventricular tachycardia.

C. ventricular fibrillation.

D. supraventricular tachycardia.

_____C_____ **42.** You should apply the AED only to unresponsive patients with no:

A. significant medical problems.

B. cardiac history.

C. pulse.

D. brain activity.

_____B_____ **43.** _____ usually refers to a state of cardiac arrest despite an organized electrical complex.

A. Asystole

B. Pulseless electrical activity

C. Ventricular fibrillation

D. Ventricular tachycardia

_____C_____ **44.** The links in the chain of survival include all of the following except:

A. early access and CPR.

B. early ACLS.

C. early administration of nitroglycerin.

D. early defibrillation.

_____D_____ **45.** An AED may fail to function properly due to:

A. the batteries not working.

B. improper maintenance.

C. operator error.

D. all of the above

Labeling

Label the following diagrams with the correct terms.

1. The Right and Left Sides of the Heart
Where arrows appear, indicate the substance and its origin and destination.

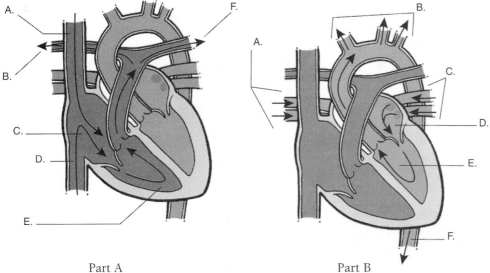

Part A   Part B

Part A.

A. _____

B. _____

C. _____

D. _____

E. _____

F. _____

Part B.

A. _____

B. _____

C. _____

D. _____

E. _____

F. _____

2. Electrical Conduction

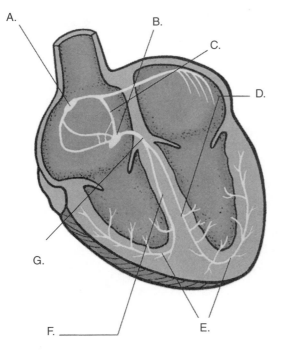

A. _____

B. _____

C. _____

D. _____

E. _____

F. _____

G. _____

4

3. Pulse Points

State the name of the artery that is being assessed at each pulse point below.

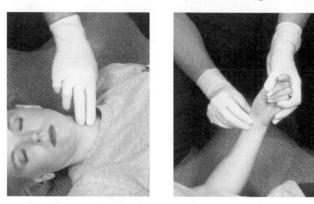

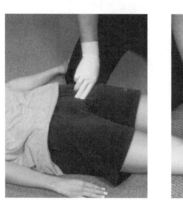
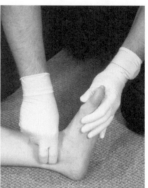

_____

_____

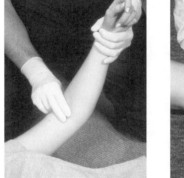
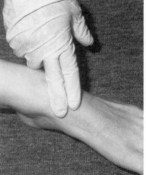

_____

_____

Vocabulary

Define the following terms using the space provided.

1. Angina pectoris:

_____

_____

_____

2. Ventricular fibrillation:

_____

_____

_____

3. Cardiogenic shock:

_____

_____

_____

4. Acute myocardial infarction (AMI):

_____

_____

_____

5. Cardiac arrest:

_____

_____

_____

6. Syncope:

_____

_____

_____

7. Congestive heart failure (CHF):

_____

_____

_____

Fill-in

Read each item carefully, then complete the statement by filling in the missing word.

1. The heart is divided down the middle by a wall called the _____.

2. The _____ is the largest artery.

3. The _____ ventricle pumps blood in through the pulmonary circulation.

4. Electrical impulses spread from the _____ node to the ventricles.

5. Blood supply to the heart is increased by _____ of the coronary arteries.

6. _____ cells remove carbon dioxide from the body's tissues.

7. _____ blood pressure reflects the pressure on the walls of the arteries when the ventricle is at rest.

8. The heart has _____ chambers.

9. The _____ side of the heart is more muscular because it must pump blood into the aorta and all the other arteries of the body.

True/False

If you believe the statement to be more true than false, write the letter "T" in the space provided. If you believe the statement to be more false than true, write the letter "F."

1. __F__ The right side of the heart pumps oxygen-rich blood to the body.

2. _____ In the normal heart, the need for increased blood flow to the myocardium is easily met by an increase in heart rate.

3. __T__ Atherosclerosis results in narrowing of the lumen of coronary arteries.

4. __F__ Infarction is a temporary interruption of the blood supply to the tissues.

5. __T__ Angina can result from a spasm of the artery.

6. _____ The pain of angina and the pain of AMI are easily distinguishable.

7. __F__ Nitroglycerin works in most patients within 5 minutes to relieve the pain of AMI.

8. _____ If an AED malfunctions during use, you must report that problem to the manufacturer and the Department of Human Resources.

9. _____ Angina occurs when the heart's need for oxygen exceeds its supply.

10. __F__ White blood cells are the most numerous and help the blood to clot.

Short Answer

Complete this section with short written answers in the space provided.

1. Name and describe the two basic types of defibrillators.

_____

_____

_____

2. What are the three most common errors of AED use?

_____

_____

_____

3. If ALS is not responding to the scene, what are the three points at which transport should be initiated for a cardiac arrest patient?

_____

_____

_____

4. List six safety considerations for operating an AED.

_____

_____

_____

_____

_____

_____

5. What is the procedure for assisting a patient with nitroglycerin administration?

_____

_____

_____

6. List three ways in which AMI pain differs from angina pain.

_____

_____

_____

7. List three serious consequences of AMI.

_____

_____

_____

8. Name at least five signs and symptoms associated with AMI.

_____

_____

_____

_____

4

■■■ N O T E S ■■■

9. Describe the technique for AED pad placement.

_____

_____

_____

Word Fun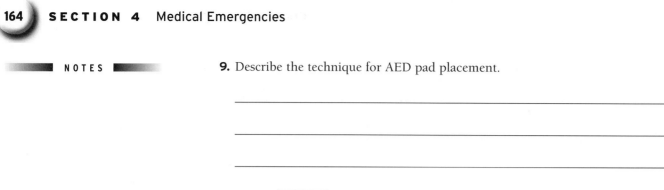

The following crossword puzzle is an activity provided to reinforce correct spelling and understanding of medical terminology associated with emergency care and the EMT-B. Use the clues in the column to complete the puzzle.

■■■ C L U E S ■■■

Across

1. Less than 60 beats/min
5. Greater than 100 beats/min
6. Lower chamber
7. Lack of oxygen to tissues
8. Widening
9. Absence of electrical activity

Down

2. Irregular heart rhythm
3. To shock the heart
4. Blockage
9. Main artery of body

Ambulance Calls

The following real case scenarios provide an opportunity to explore the concerns associated with patient management. Read each scenario, then answer each question in detail.

1. You are dispatched to the residence of a 58-year-old male complaining of chest pain. He states that it feels like "somebody is standing on my chest." He sat down when it started and took a nitroglycerin tablet. He is still a little nauseated and sweaty, but feels better. He is very anxious.

   How would you best manage this patient?

   _____

   _____

   _____

_____

_____

_____

_____

_____

_____

2. You are called to a business office for a 38-year-old male complaining of indigestion. Coworkers tell you that the pain started suddenly and he clutched his chest and vomited. He insists that he will be all right, and that it is probably something he ate. He is pale and diaphoretic.

How would you best manage this patient?

_____

_____

_____

_____

_____

_____

_____

_____

3. You are dispatched to a local personal care home for an elderly woman complaining of shortness of breath. She is gasping for breath and is extremely anxious. Her legs are very swollen and she says she took her "water pill" this morning. The staff tells you that she has a history of diabetes, congestive heart failure, and hypertension.

How would you best manage this patient?

_____

_____

_____

_____

_____

Skill Drills emt-b

Skill Drill 12-1: Caring for a Conscious Patient with Chest Pain
Test your knowledge of this skill drill by filling in the correct words in the photo captions.

1. _____ the patient as you perform the initial assessment.
Apply _____.
Place the patient in a _____ position.

2. Measure and record the _____ _____ _____, obtain a _____ history, and _____ the patient closely.
Obtain a _____ history and _____ _____.
Ask about the patient's discomfort using _____.

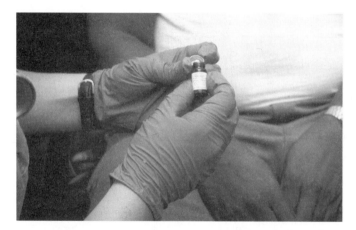

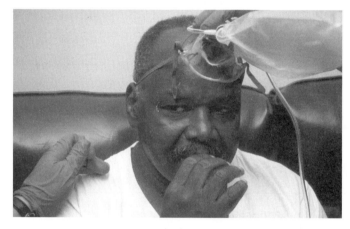

3. Check medication and _____ _____.

4. Help the patient administer _____.

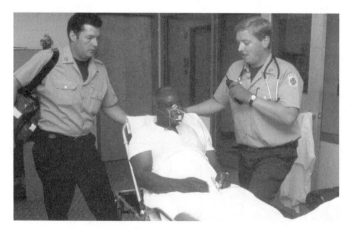

5. Prepare to _____ the patient.
Report to _____ _____.

Skill Drill 12-2: AED and CPR

Test your knowledge of this skill drill by placing the photos below in the correct order. Number the first step with a "1," the second step with a "2," etc.

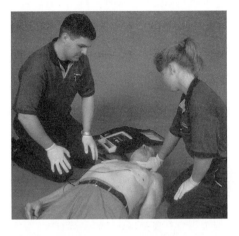

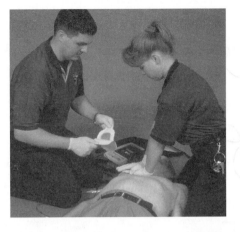

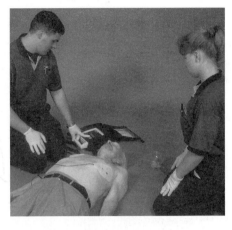

Check pulse.

If pulse is present, check breathing.

If pulseless, begin CPR.

Prepare the AED pads.

Turn on the AED; begin narrative if needed.

Apply AED pads.

Stop CPR.

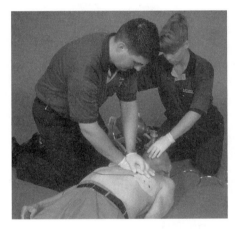

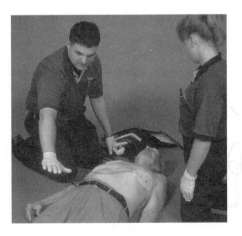

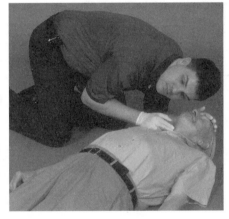

If breathing adequately, give oxygen and transport. If not, open airway, ventilate, and transport.

If no pulse, perform CPR for 1 minute.

Clear the patient and analyze again.

If necessary, repeat one cycle of up to three shocks.

Transport and call medical control.

Continue to support breathing or perform CPR, as needed.

Verbally and visually clear the patient.

Push the Analyze button if there is one.

Wait for the AED to analyze rhythm.

If no shock advised, perform CPR for 1 min.

If shock advised, recheck that all are clear and push the Shock button.

Push the Analyze button, if needed, to analyze rhythm again.

Press Shock if advised (second shock).

Push the Analyze button, if needed, to analyze rhythm again.

Press Shock if advised (third shock).

Stop CPR if in progress.

Assess responsiveness.

Check breathing and pulse.

If unresponsive and not breathing adequately, give two slow ventilations.

Workbook Activities

The following activities have been designed to help you. Your instructor may require you to complete some or all of these activities as a regular part of your EMT-B training program. You are encouraged to complete any activity that your instructor does not assign as a way to enhance your learning in the classroom.

Chapter Review

The following exercises provide an opportunity to refresh your knowledge of this chapter.

■ NOTES ■

Matching

Match each of the terms in the left column to the appropriate definition in the right column.

_____ **1.** Brain stem

_____ **2.** Foramen magnum

_____ **3.** Spinal nerves

_____ **4.** Cerebrum

_____ **5.** Cranial nerves

_____ **6.** Cerebellum

_____ **7.** Embolism

_____ **8.** Aneurysm

_____ **9.** Status epilepticus

_____ **10.** Hypoglycemia

_____ **11.** Aphasia

_____ **12.** Berry aneurysm

_____ **13.** Dysarthria

A. slurred, hard to understand speech

B. the back part of this area of the brain processes sight

C. weakness in an artery wall

D. controls most basic functions of the body

E. exit at each vertebra and carry messages to and from the body

F. this area of the brain controls muscle and body coordination

G. hole in the base of the skull

H. low blood glucose

I. innervate eyes, ears, face

J. clot that forms elsewhere and travels to the site of damage

K. weakness in a blood vessel that resembles a tiny balloon

L. inability to produce or understand speech

M. seizures that recur every few minutes

Multiple Choice

Read each item carefully, then select the best response.

_____ **1.** Seizures may occur as a result of:

A. metabolic problems.

B. brain tumor.

C. a recent or old head injury.

D. all of the above

Neurologic Emergencies

_A_ **2.** The _____ controls the most basic functions of the body, such as breathing, blood pressure, swallowing, and pupil constriction.

   A. brain stem

   B. cerebellum

   C. cerebrum

   D. spinal cord

_A_ **3.** At each vertebra in the neck and back, _____ nerves, called spinal nerves, branch out from the spinal cord and carry signals to and from the body.

   A. two

   B. three

   C. four

   D. five

_C_ **4.** Brain disorders include all of the following except:

   A. coma.

   B. infection.

   C. hypoglycemia.

   D. tumor.

_C_ **5.** When blood flow to a particular part of the brain is cut off by a blockage inside a blood vessel, the result is:

   A. a hemorrhagic stroke.

   B. atherosclerosis.

   C. an ischemic stroke.

   D. a cerebral embolism.

_A_ **6.** Patients who are at the highest risk of hemorrhagic stroke are those who have:

   A. untreated hypertension.

   B. an aneurysm.

   C. a berry aneurysm.

   D. atherosclerosis.

_B_ 7. Patients with a subarachnoid hemorrhage typically complain of a sudden severe:

A. bout of dizziness.

B. headache.

C. altered mental status.

D. thirst.

_B_ 8. The plaque that builds up in atherosclerosis obstructs blood flow, and interferes with the vessel's ability to:

A. constrict.

B. dilate.

C. diffuse.

D. exchange gases.

_D_ 9. A TIA, or mini-stroke, is the name given to a stroke when symptoms go away on their own in less than:

A. half an hour.

B. 1 hour.

C. 12 hours.

D. 24 hours.

_A_ 10. Seizures characterized by unconsciousness and a generalized severe twitching of all the body's muscles that lasts several minutes or longer is called a:

A. grand mal seizure.

B. petit mal seizure.

C. focal motor seizure.

D. febrile seizure.

_D_ 11. Metabolic seizures may be due to:

A. epilepsy.

B. a brain tumor.

C. high fevers.

D. hypoglycemia.

_D_ 12. When assessing a patient with a history of seizure activity, it is important to:

A. determine whether this episode differs from any previous ones.

B. recognize the postictal state.

C. look for other problems associated with the seizure.

D. all of the above

_D_ 13. Signs and symptoms of possible seizure activity include:

A. altered mental status.

B. incontinence.

C. rapid and deep respirations.

D. all of the above

_C_ 14. Common causes of altered mental status include all of the following, except:

A. body temperature abnormalities.

B. hypoxemia.

C. unequal pupils.

D. hypoglycemia.

_____ **15.** The principle difference between a patient who has had a stroke and a patient with hypoglycemia almost always has to do with the:

    A. pupillary response.

    B. mental status.

    C. blood pressure.

    D. capillary refill time.

_____ **16.** Consider the possibility of _____ in a patient who has had a seizure.

    A. brain injury

    B. hyperglycemia

    C. hypoglycemia

    D. hypertension

_____ **17.** Individuals with chronic alcoholism can have abnormalities in liver function and in their blood-clotting and immune systems, which can predispose them to:

    A. intracranial bleeding.

    B. brain and bloodstream infections.

    C. hypoglycemia.

    D. all of the above

_____ **18.** Low oxygen levels in the bloodstream will affect the entire brain, causing:

    A. anxiety.

    B. restlessness.

    C. confusion.

    D. all of the above

_____ **19.** Patients with _____ may have trouble understanding speech but can speak clearly.

    A. aphasia

    B. receptive aphasia

    C. expressive aphasia

    D. dysarthria

_____ **20.** High blood pressure in stroke patients should not be treated in the field because:

    A. the brain is raising the blood pressure in an attempt to force more oxygen into its injured parts.

    B. quite often, blood pressure will return to normal or may drop significantly on its own.

    C. many times it is a response to bleeding in the brain.

    D. all of the above

_____ **21.** The following conditions may simulate a stroke except:

    A. hyperglycemia.

    B. a postictal state.

    C. hypoglycemia.

    D. subdural bleeding.

_____ **22.** When assessing a patient with a possible CVA, you should check the _____ first.

    A. pulse

    B. airway

    C. pupils

    D. blood pressure

4

_____ **23.** Indications that the patient can understand you include:

    A. pressure of the hand.

    B. efforts to speak.

    C. nodding the head.

    D. all of the above

_____ **24.** Key physical tests for patients suspected of having a stroke include tests of:

    A. speech.

    B. neck movement.

    C. leg movement.

    D. all of the above

_____ **25.** A patient with a GCS of 12 has:

    A. no dysfunction.

    B. mild dysfunction.

    C. moderate to severe dysfunction.

    D. severe dysfunction.

_____ **26.** To transport the suspected stroke patient, the patient should be placed in which position?

    A. Trendelenburg's

    B. Fowler's

    C. comfortable

    D. Lithotomy

_____ **27.** Following a major seizure, you should anticipate:

    A. a decreased heart rate.

    B. rapid, deep respirations.

    C. respiratory arrest.

    D. a return to their normal mental status within 5–10 minutes.

_____ **28.** Assess the mental status using the mnemonic:

    A. OPQRST.

    B. SAMPLE.

    C. AVPU.

    D. PEARL.

_____ **29.** Even a patient who has a history of chronic epilepsy that is controlled with medications may have an occasional seizure, commonly referred to as a _____ seizure.

    A. chronic

    B. generalized

    C. absence

    D. breakthrough

Labeling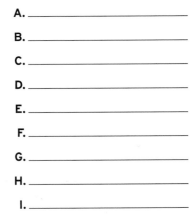

Label the following diagrams with the correct terms.

1. Brain

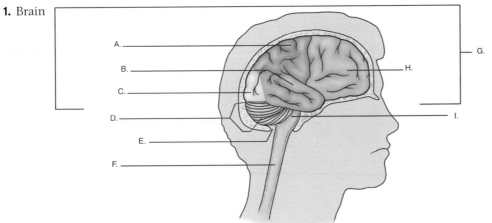

A. _____

B. _____

C. _____

D. _____

E. _____

F. _____

G. _____

H. _____

I. _____

2. Spinal Cord

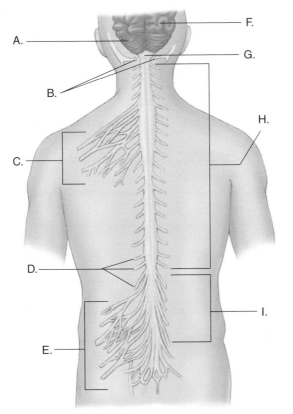

A. _____

B. _____

C. _____

D. _____

E. _____

F. _____

G. _____

H. _____

I. _____

Vocabulary

Define the following terms listed using the space provided.

1. Cerebrovascular accident (CVA):

_____

_____

2. Ischemic stroke:

_____

_____

_____

3. Transient ischemic attack (TIA):

_____

_____

_____

4. Hemorrhagic stroke:

_____

_____

_____

5. Generalized seizure:

_____

_____

_____

6. Absence seizure:

_____

_____

_____

7. Atherosclerosis:

_____

_____

_____

8. Cerebral embolism:

_____

_____

_____

9. Febrile seizure:

_____

_____

_____

10. Thrombosis:

_____

_____

_____

Fill-in

Read each item carefully, then complete the statement by filling in the missing word(s).

1. There are _____ cranial nerves.

2. Playing the piano is coordinated by the _____.

3. The front part of the cerebrum controls _____.

4. The cranial nerves run to the _____.

5. The brain is divided into _____ major parts.

6. All messages traveling to and from the brain travel along _____.

7. Each hemisphere of the cerebrum controls activities on the _____ side of the body and the _____ side of the face.

8. The _____ is the largest part of the brain.

9. _____ is a loss of bowel and bladder control, and can be due to a generalized seizure.

10. The _____ is the body's computer.

11. Onset of _____ bleeding is usually very rapid after injury.

12. Weakness on one side of the body is known as _____.

13. No matter what the cause, you should consider _____ to be an emergency that requires immediate attention, even when it appears that the cause may simply be alcohol intoxication or a minor car crash or fall.

True/False

If you believe the statement to be more true than false, write the letter "T" in the space provided. If you believe the statement to be more false than true, write the letter "F."

1. _____ The postictal state following a seizure commonly lasts only about 3 to 5 minutes.

2. _____ Metabolic seizures result from an area of abnormality in the brain.

3. _____ Febrile seizures result from sudden high fevers and are generally well tolerated by children.

4. _____ Hemiparesis is the inability to speak or understand speech.

5. _____ The dura covers the brain.

6. _____ Unconscious stroke patients are usually unable to speak or hear.

7. _____ Right-sided facial droop is most likely an indication of a problem in the right cerebral hemisphere.

Short Answer

Complete this section with short written answers using the space provided.

1. List and describe the three key tests for assessing stroke.

_____

_____

_____

2. Why is prompt transport of stroke patients critical?

_____

_____

_____

3. What are some techniques for cooling a child with a febrile seizure?

_____

_____

_____

4. Describe the characteristics of a postictal state.

_____

_____

_____

5. What is the difference between infarcted and ischemic cells?

_____

_____

_____

6. List three conditions that may simulate stroke.

_____

_____

_____

Word Fun

The following crossword puzzle is an activity provided to reinforce correct spelling and understanding of medical terminology associated with emergency care and the EMT-B. Use the clues in the column to complete the puzzle.

CLUES

Across

  6. Cholesterol and calcium build-up

  7. One-sided weakness

  8. Loss of brain function

Down

  1. Related to high temperatures, particularly with children

  2. Inability to pronounce clearly

  3. Low blood glucose level

  4. Clotting of vessel

  5. Inability to speak

Ambulance Calls

The following real case scenarios will give you an opportunity to explore the concerns associated with patient management. Read each scenario, then answer each question in detail.

  1. You are called to a residence for an 18-month-old girl who had witnessed seizure-like activity approximately 15 minutes prior to your arrival. The only history is a recent upper respiratory infection, but the mother forgot to give the child her medication for 3 days. The patient is lethargic, flushed, and very hot to touch. The mother does not have a thermometer.

  How would you best manage this patient?

_____

_____

_____

_____

_____

_____

_____

_____

_____

_____

_____

2. You are dispatched to a 36-year-old male who had seizure activity at least an hour ago. The patient is incontinent, cold, clammy, and unresponsive. His friends tell you that the "shaking" stopped and he has not woke up. They thought he might just be tired until they discovered they could not wake him. He has no history of seizure activity. He has diabetes for which he takes medication.

How would you best manage this patient?

_____

_____

_____

_____

_____

_____

_____

_____

_____

3. You are dispatched to the residence of a 77-year-old male complaining of possible CVA. He presents with facial drooping on his left side as well as left-sided weakness. He appears to be in no apparent distress. His blood pressure is 190/110, pulse is 84, and respirations are 16 and nonlabored. He has no history of strokes.

How would you best manage this patient?

_____

_____

_____

_____

_____

_____

_____

_____

_____

_____

4

Workbook Activities

The following activities have been designed to help you. Your instructor may require you to complete some or all of these activities as a regular part of your EMT-B training program. You are encouraged to complete any activity that your instructor does not assign as a way to enhance your learning in the classroom.

Chapter Review

The following exercises provide an opportunity to refresh your knowledge of this chapter.

■ NOTES ■

Matching

Match each of the terms in the left column to the appropriate definition in the right column.

_____ 1. Aneurysm

_____ 2. Colic

_____ 3. Retroperitoneal

_____ 4. Ulcer

_____ 5. Hernia

_____ 6. Ileus

_____ 7. Guarding

_____ 8. Anorexia

_____ 9. Emesis

_____ 10. Referred pain

_____ 11. PID

_____ 12. Cystitis

_____ 13. Strangulation

_____ 14. Peritonitis

_____ 15. Peritoneum

A. paralysis of the bowel

B. pain felt in an area of the body other than the actual source

C. protective, involuntary abdominal muscle contractions

D. acute, intermittent cramping abdominal pain

E. behind the peritoneum

F. vomiting

G. common cause of acute abdomen in women

H. a membrane lining the abdomen

I. swelling or enlargement of a weakened arterial wall

J. loss of hunger or appetite

K. protrusion of a loop of an organ or tissue through an abnormal body opening

L. obstruction of blood circulation resulting from compression or entrapment of organ tissue

M. abrasion of the stomach or small intestine

N. inflammation of the bladder

O. inflammation of the peritoneum

The Acute Abdomen

Multiple Choice

Read each item carefully, then select the best response.

_____ C **1.** Peritonitis, with associated fluid loss, may lead to _____ shock.
   A. hemorrhagic
   B. septic
   C. hypovolemic
   D. metabolic

_____ A **2.** Distention of the abdomen is gauged by:
   A. visualization.
   B. auscultation.
   C. palpation.
   D. the patient's complaint of pain around the umbilicus.

_____ A **3.** A hernia that returns to its proper body cavity is said to be:
   A. reducible.
   B. extractable.
   C. incarcerated.
   D. replaceable.

_____ A **4.** Sensory nerves from the spinal cord to the skin and muscles are part of the:
   A. somatic nervous system.
   B. peripheral nervous system.
   C. autonomic nervous system.
   D. sympathetic nervous system.

_____ D **5.** When an organ of the abdomen is enlarged, rough palpation may cause _____ of the organ.
   A. distention
   B. nausea
   C. swelling
   D. rupture

A **6.** Severe back pain may be associated with which condition?

    A. abdominal aortic aneurysm

    B. PID

    C. appendicitis

    D. mittelschmerz

B **7.** The  are found in the retroperitoneal space.

    A. stomach and gallbladder

    B. kidneys, genitourinary structures, and large vessels

    C. liver and pancreas

    D. adrenal glands and uterus

C **8.** A(n) _____ may occur as a result of a surgical wound that has failed to heal properly.

    A. ectopic pregnancy

    B. strangulation

    C. hernia

    D. ulcer

B **9.** The peritoneal membrane that can perceive the sensations of pain, pressure, and cold is the:

    A. meningeal.

    B. parietal.

    C. retroperitoneal.

    D. visceral.

D **10.** Common disease(s) that produce(s) signs of an acute abdomen include:

    A. diverticulitis.

    B. cholecystitis.

    C. acute appendicitis.

    D. all of the above

C **11.** A patient with peritonitis may present with rapid, shallow breaths due to:

    A. hypovolemia.

    B. ileus.

    C. pain.

    D. inflammation.

C **12.** Common signs and symptoms of irritation or inflammation of the peritoneum may include:

    A. a quiet patient who is resting comfortably.

    B. hypertension and tachycardia.

    C. rebound tenderness and fever.

    D. Kussmaul respirations.

C **13.** In the patient with peritonitis, the degree of pain and tenderness is usually related directly to the severity of:

    A. fever.

    B. distention.

    C. peritoneal inflammation.

    D. bleeding.

D **14.** Pain associated with diverticulitis is usually felt in the:

    A. right upper quadrant.

    B. left upper quadrant.

    C. right lower quadrant.

    D. left lower quadrant.

_____ **15.** When assessing a patient with severe abdominal pain, you should anticipate
the development of _____ and treat the patient when it is evident.
   A. fever
   B. appendicitis
   C. hypovolemic shock
   D. pancreatitis

Labeling

Label the following diagrams with the correct terms.

1. Solid Organs

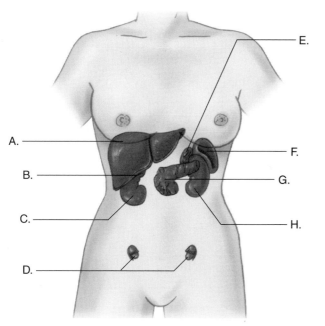

A. _____

B. _____

C. _____

D. _____

E. _____

F. _____

G. _____

H. _____

4

2. Hollow Organs

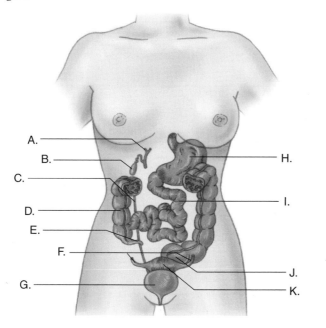

A. _____

B. _____

C. _____

D. _____

E. _____

F. _____

G. _____

H. _____

I. _____

J. _____

K. _____

3. Retroperitoneal Organs

A. _____

B. _____

C. _____

D. _____

E. _____

F. _____

G. _____

H. _____

I. _____

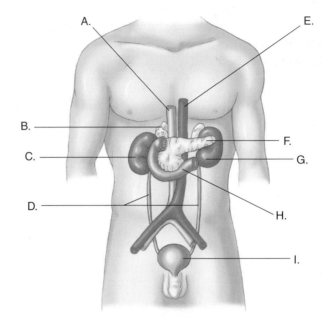

Vocabulary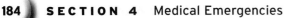

Define the following terms using the space provided.

1. Acute abdomen:

_____

_____

_____

2. Diverticulitis:

_____

_____

_____

3. Ectopic pregnancy:

_____

_____

_____

4. Cholecystitis:

_____

_____

_____

5. Mittelschmerz:

_____

_____

_____

True/False

If you believe the statement to be more true than false, write the letter "T" in the space provided. If you believe the statement to be more false than true, write the letter "F."

1. _____   Referred pain is a result of connection between ligaments in the abdominal and chest cavities.

2. _____   Abdominal pain in women is usually related to the menstrual cycle and is rarely serious.

3. _____   A leading aorta, being retroperitoneal, will not cause peritonitis.

4. _____   It is important to accurately diagnose the cause of acute abdominal pain in order to properly treat the patient.

5. _____   The parietal peritoneum lines the walls of the abdominal cavity.

6. _____   The patient with peritonitis usually reports relief of pain when lying left lateral recumbent with the knees pulled in.

7. _____   When palpating the abdomen, always start with the quadrant where the patient complains of the most severe pain.

8. _____   Massive hemorrhaging is associated with rupture of an abdominal aortic aneurysm.

9. _____   Pneumonia may cause abdominal pain.

Short Answer

Complete this section with short written answers using the space provided.

1. Explain the phenomenon of referred pain.

_____

_____

_____

2. Should an EMT-B attempt to diagnose the cause of abdominal pain? Why or why not?

_____

_____

_____

3. Why does abdominal distention accompany ileus?

_____

_____

_____

4

4. What two conditions may result in hypovolemic shock in the patient with an acute abdomen?

_____

_____

_____

5. List the general EMT-B emergency care for patients with acute abdomen.

_____

_____

_____

_____

Word Fun

The following crossword puzzle is an activity provided to reinforce correct spelling and understanding of medical terminology associated with emergency care and the EMT-B. Use the clues in the column to complete the puzzle.

■ CLUES ■

Across

3. Felt elsewhere in the body

8. Abrasion of stomach or small intestine

10. Complete obstruction

11. Paralysis of the bowels

12. Lining of the abdominal cavity

Down

1. Protrusion of a loop of an organ

2. Involuntary muscle contractions for protection

4. Vomiting

5. Sudden onset of pain below diaphragm

6. Lack of appetite

7. Results in the weakening of an arterial wall

9. Acute cramping abdominal pain

Ambulance Calls

The following real case scenarios provide an opportunity to explore the concerns associated with patient management. Read each scenario, then answer each question in detail.

1. You are called to the local high school nurse's office for a 16-year-old female complaining of severe, abrupt abdominal pain. She is pale, cool, and clammy with absent radial pulses. The school nurse tells you that the patient found out 4 weeks ago that she is pregnant.

   How would you best manage this patient?

   _____

   _____

   _____

   _____

   _____

   _____

   _____

   _____

   _____

   _____

2. You are dispatched to a restaurant where a 28-year-old female is complaining of a sudden onset of nausea and vomiting with severe right lower quadrant pain. The patient is doubled over in pain, pale, and clammy.

   What problem would you suspect and how would you treat this patient?

   _____

   _____

   _____

   _____

   _____

   _____

   _____

   _____

   _____

4

3. You are called to a residence for a 45-year-old male who has a "lump" in his lower abdomen. Upon inspection, you note a mass protruding from his lower abdomen that has a blue discoloration to the area and is tender on palpation. He tells you he first noticed it several days ago after doing some heavy lifting. How would you best manage this patient?

_____

_____

_____

_____

_____

_____

_____

_____

_____

_____

Notes

Workbook Activities

The following activities have been designed to help you. Your instructor may require you to complete some or all of these activities as a regular part of your EMT-B training program. You are encouraged to complete any activity that your instructor does not assign as a way to enhance your learning in the classroom.

Chapter Review

The following exercises provide an opportunity to refresh your knowledge of this chapter.

■ NOTES ■

Matching

Match each of the terms in the left column to the appropriate definition in the right column.

_____ 1. Hormone

_____ 2. Polydipsia

_____ 3. Acetone

_____ 4. Type I diabetes

_____ 5. Acidosis

_____ 6. Insulin

_____ 7. Diabetic coma

_____ 8. Polyuria

_____ 9. Type II diabetes

_____ 10. Polyphagia

_____ 11. Insulin shock

_____ 12. Glucose

_____ 13. Kussmaul respirations

_____ 14. Hyperglycemia

_____ 15. Diabetes

A. altered level of consciousness caused by insufficient glucose in the blood

B. diabetes that usually starts in childhood; requires insulin

C. excessive eating

D. deep, rapid breathing

E. excessive urination

F. excessive thirst persisting for a long period of time

G. diabetes with onset later in life; may be controlled by diet and oral medication

H. chemical produced by a gland that regulates body organs

I. a type of ketone

J. literal meaning: "A passer through; a siphon"

K. extremely high blood glucose level

L. pathologic condition resulting from the accumulation of acids in the body

M. hormone that enables glucose to enter the cells

N. primary fuel, along with oxygen, for cellular metabolism

O. state of unconsciousness resulting from several problems, including ketoacidosis, dehydration, and hyperglycemia

Diabetic Emergencies

Multiple Choice

Read each item carefully, then select the best response.

_____A_____ **1.** Patients with which type of diabetes are more likely to have metabolic problems and organ damage?

A. Type I

B. Type II

C. Sugar diabetes

D. HHNC

_____A_____ **2.** Normal blood glucose levels range from _____ mg/dL.

A. 80 to 120

B. 90 to 140

C. 70 to 110

D. 60 to 100

_____D_____ **3.** Patients with diabetes mellitus and a lack of insulin excrete excess glucose through their:

A. lymphatic system.

B. sweat.

C. respiratory efforts.

D. urine.

_____C_____ **4.** Diabetes is a metabolic disorder in which the hormone _____ is missing or ineffective.

A. estrogen

B. adrenalin

C. insulin

D. epinephrine

_____B_____ **5.** Complications of diabetes include:

A. paralysis.

B. cardiovascular disease.

C. brittle bones.

D. hepatitis.

_A_ 6. The accumulation of ketones and fatty acids in blood tissue can lead to a dangerous condition in diabetic patients known as:

A. diabetic ketoacidosis.

B. insulin shock.

C. HHNC.

D. hypoglycemia.

_C_ 7. The term for excessive eating as a result of cellular "hunger" is:

A. polyuria.

B. polydipsia.

C. polyphagia.

D. polyphony.

_D_ 8. Insulin is produced by the:

A. adrenal glands.

B. hypothalamus.

C. spleen.

D. pancreas.

_D_ 9. Factors that may contribute to diabetic coma include:

A. infection.

B. alcohol consumption.

C. insufficient insulin.

D. all of the above

_B_ 10. The only organ that does not require insulin to allow glucose to enter its cells is the:

A. liver.

B. brain.

C. pancreas.

D. heart.

_B_ 11. The sweet or fruity odor on the breath of a diabetic patient is caused by _____ in the blood.

A. acetone

B. ketones

C. alcohol

D. insulin

_B_ 12. The term for excessive thirst is:

A. polyuria.

B. polydipsia.

C. polyphagia.

D. polyphony.

_D_ 13. Oral diabetic medications include:

A. Micronase.

B. Glucotrol.

C. Diabinase.

D. all of the above

_____ **14.** In order for patients with diabetes to live out a normal life span, they must be willing to adjust their lives to the demands of the disease, especially their eating habits and:

    A. stress levels.

    B. activities.

    C. sleep habits.

    D. antacid intake.

_____ **15.** _____ is one of the basic sugars in the body.

    A. Dextrose

    B. Sucrose

    C. Fructose

    D. Syrup

_____ **16.** _____ is the hormone that is normally produced by the pancreas that enables glucose to enter the cells.

    A. Insulin

    B. Adrenalin

    C. Estrogen

    D. Epinephrine

_____ **17.** The term for excessive urination is:

    A. polyuria.

    B. polydipsia.

    C. polyphagia.

    D. polyphony.

_____ **18.** When fat is used as an immediate energy source, _____ and fatty acids are formed as waste products.

    A. dextrose

    B. sucrose

    C. ketones

    D. bicarbonate

_____ **19.** When using a glucometer, the patient tests his or her glucose level in a drop of:

    A. urine.

    B. blood.

    C. saliva.

    D. cerebrospinal fluid.

_____ **20.** The onset of hypoglycemia can occur within:

    A. seconds.

    B. minutes.

    C. hours.

    D. days.

_____ **21.** Without _____, or with very low levels, brain cells rapidly suffer permanent damage.

    A. epinephrine

    B. ketones

    C. bicarbonate

    D. glucose

_____**22.** _____ is/are a potentially life-threatening complication of insulin shock.

    A. Kussmaul respirations

    B. Hypotension

    C. Seizures

    D. Polydipsia

_____**23.** Blood glucose levels are measured in:

    A. micrograms per deciliter.

    B. milligrams per deciliter.

    C. milliliters per decigram.

    D. microliters per decigram.

_____**24.** Diabetic coma may develop as a result of:

    A. too little insulin.

    B. too much insulin.

    C. overhydration.

    D. metabolic alkalosis.

_____**25.** Always suspect hypoglycemia in any patient with:

    A. Kussmaul respirations.

    B. an altered mental status.

    C. nausea and vomiting.

    D. all of the above

_____**26.** The most important step in caring for the unresponsive diabetic patient is to:

    A. give oral glucose immediately.

    B. perform a focused assessment.

    C. open the airway.

    D. obtain a SAMPLE history.

_____**27.** Determination of diabetic coma or insulin shock should:

    A. be made before transport of the patient.

    B. be made before administration of oral glucose.

    C. be determined by a urine glucose test.

    D. be based upon your knowledge of the signs and symptoms of each condition.

_____**28.** An unresponsive patient involved in a motor vehicle crash with a diabetic identification bracelet should be suspected to be _____ until proven otherwise.

    A. hypoglycemic

    B. hyperglycemic

    C. intoxicated

    D. in shock

_____**29.** Contraindications for the use of oral glucose include:

    A. unconsciousness.

    B. known alcoholic.

    C. insulin shock.

    D. all of the above

_____ **30.** When reassessing the diabetic patient after administration of oral glucose, watch for:

A. airway problems.

B. seizures.

C. sudden loss of consciousness.

D. all of the above

_____ **31.** Signs and symptoms associated with hypoglycemia include:

A. warm, dry skin.

B. rapid, weak pulse.

C. Kussmaul respirations.

D. anxious or combative behavior.

_____ **32.** The patient in insulin shock is experiencing:

A. hyperglycemia.

B. hypoglycemia.

C. diabetic ketoacidosis.

D. a low production of insulin.

_____ **33.** Signs of dehydration include:

A. good skin turgor.

B. elevated blood pressure.

C. sunken eyes.

D. all of the above

_____ **34.** Diabetic patients who complain of "not feeling so well" should:

A. have their glucose level checked.

B. have a rapid trauma assessment completed.

C. be rapidly transported to the closest medical facility.

D. immediately be given oral glucose.

_____ **35.** Causes of insulin shock include:

A. taking too much insulin.

B. vigorous exercise without sufficient glucose intake.

C. nausea, vomiting, anorexia.

D. all of the above

_____ **36.** Insulin shock can develop more often and more severely in children than in adults due to their:

A. high activity level and failure to maintain a strict schedule of eating.

B. genetic makeup.

C. smaller body size.

D. all of the above

_____ **37.** Because diabetic coma is a complex metabolic condition that usually develops over time and involves all the tissues of the body, correcting this condition may:

A. be accomplished quickly through the use of oral glucose.

B. require rapid infusion of IV fluid to prevent permanent brain damage.

C. take many hours in a hospital setting.

D. include a reduction in the amount of insulin normally taken by the patient.

4

_____ 38. A patient in insulin shock or a diabetic coma may appear to be:

    A. having a heart attack.

    B. perfectly normal.

    C. intoxicated.

    D. having a stroke.

_____ 39. If the ill diabetic patient has eaten but has not taken insulin, it is more likely that _____ is developing.

    A. hypoglycemia

    B. diabetic ketoacidosis

    C. insulin shock

    D. metabolic alkalosis

_____ 40. Important questions to ask when questioning the ill diabetic patient include:

    A. Have you eaten normally today?

    B. Do you take medication for diabetes?

    C. Have you taken your normal dose of insulin (or pills) today?

    D. all of the above

_____ 41. When dealing with a diabetic patient who has an altered mental status you should:

    A. place oral glucose gel under the tongue of the unresponsive patient.

    B. only place sugar under the tongue of the unresponsive patient.

    C. give a small dose of insulin to the unresponsive patient.

    D. not attempt to give anything by mouth to an unresponsive patient.

_____ 42. Be careful when administering any type of sugar/glucose to patients in insulin shock. They usually only require:

    A. 1 teaspoon.

    B. a few sips of a sweetened drink.

    C. ¼ of a candy bar.

    D. none of the above

_____ 43. _____ may be substituted for oral glucose for a patient in insulin shock.

    A. Honey

    B. Synthetic sweetening compounds

    C. A diet drink

    D. Sugar-free candy

Vocabulary **emtb** vocab explorer

Define the following terms using the space provided.

1. Diabetes mellitus:

_____

_____

_____

2. Diabetes insipidus:

_____

_____

_____

3. Juvenile diabetes:

_____

_____

_____

4. Ketones:

_____

_____

_____

5. Glucometer:

_____

_____

_____

4

Fill-in

Read each item carefully, then complete the statement by filling in the missing word(s).

1. The full name of diabetes is _____ _____ .

2. Diabetes is considered to be a(n) _____ problem, in which the body becomes allergic to its own tissues and literally destroys them.

3. Diabetes is defined as a lack of or _____ action of insulin.

4. Too much blood glucose by itself does not always cause _____ _____, but on some occasions, it can lead to it.

5. A patient in insulin shock needs _____ immediately and a patient in a

diabetic coma needs _____ and IV fluid therapy.

True/False

If you believe the statement to be more true than false, write the letter "T" in the space provided. If you believe the statement to be more false than true, write the letter "F."

1. _____ When patients use fat for energy, the fat waste products increase the amount of acid in the blood and tissue.

2. _____ The level of consciousness can be affected if a patient has not exercised enough.

3. _____ If blood glucose levels remain low, a patient may lose consciousness or have permanent brain damage.

4. _____ Signs and symptoms can develop quickly in children because their level of activity can exhaust their glucose levels.

5. _____ Diabetic emergencies can occur when a patient's blood glucose level gets too high or when it drops too low.

6. _____ Diabetes is caused by the lack of adequate amounts of insulin.

7. _____ Diabetic patients may require insulin to control their blood glucose.

8. _____ Glucose is a hormone that enables insulin to enter the cells of the body.

9. _____ Insulin is one of the basic sugars essential for cell metabolism in humans.

10. _____ Diabetes can cause kidney failure, blindness, and damage to blood vessels.

11. _____ Most children with diabetes are insulin dependent.

12. _____ Many adults with diabetes can control their blood glucose levels with diet alone.

Short Answer

Complete this section with short written answers using the space provided.

1. What is insulin and what is its role in metabolism?

_____

_____

_____

2. What are two trade names for oral glucose?

_____

_____

_____

3. When should you not give oral glucose to a patient experiencing a suspected diabetic emergency?

_____

_____

_____

4. List the trade names of three oral medications used by diabetics.

_____

_____

_____

5. What are the three problems associated with the development of diabetic coma?

_____

_____

_____

6. List the physical signs of diabetic coma.

_____

_____

_____

7. If a diabetic patient was "fine" 2 hours ago and now is unconscious and unresponsive, which diabetes-related condition would you suspect and why?

_____

_____

_____

8. Why should oral glucose be given to any diabetic patient with an altered level of consciousness?

_____

_____

_____

4

Word Fun

The following crossword puzzle is an activity provided to reinforce correct spelling and understanding of medical terminology associated with emergency care and the EMT-B. Use the clues in the column to complete the puzzle.

Across

1. Unconscious state with high glucose levels

3. Excessive eating

6. Disease that usually starts early in life, generally requires daily injections

8. Excessive thirst

10. Chemical substance produced by a gland

11. May result from hypoglycemia

Down

2. Disease that starts later in life, usually controlled by diet and oral medications

4. Opposite of alkalosis

5. Hormone produced in pancreas

7. Excessive urination

9. One of the basic sugars

Ambulance Calls

The following real case scenarios provide an opportunity to explore the concerns associated with patient management. Read each scenario, then answer each question in detail.

1. You are called to a local residence where you find a 22-year-old female supine in bed, unresponsive to your attempts to rouse her. She is cold and clammy with gurgling respirations. Her mother tells you that her only history is diabetes, which she has had since she was a small child.

   What steps would you take in managing this patient?

   _____

   _____

   _____

   _____

   _____

   _____

   _____

   _____

2. You are dispatched to a minor motor vehicle crash. The police officer on scene tells you it is a one-car wreck involving a tree, and the driver is "not acting right." The driver, a 33-year-old male, appears intoxicated but there is no smell of alcohol. He has no visible injuries.

   How would you best manage this patient?

   _____

   _____

   _____

   _____

   _____

   _____

   _____

   _____

   _____

3. You are called to a personal care home where the staff tells you that one of their residents does not feel well. They take you to the room of a 76-year-old male, a known diabetic patient, who is slumped over in a recliner. He has a fruity smell to his breath and is hot to touch. His mental status is extremely altered.

How would you best manage this patient?

_____

_____

_____

_____

_____

_____

_____

_____

_____

4

Skill Drills emt-b.

Test your knowledge of this skill drill by filling in the correct words in the photo captions.

Skill Drill 15-1: Administering Glucose

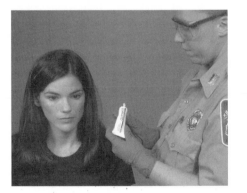

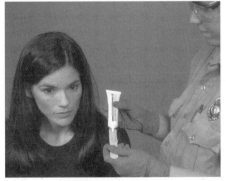

1. Make sure that the tube of glucose is intact and has not _____.

2. Squeeze the entire tube of oral glucose onto the _____ _____ of a _____ _____ or tongue depressor.

3. Open the patient's _____. Place the tongue depressor on the _____ _____ between the cheek and the gum with the _____ _____ next to the cheek.

Comparison Table

Complete the following table on the characteristics of diabetic emergencies.

	Diabetic Coma	Insulin Shock
History Food intake Insulin dosage Onset Skin Infection		
Gastrointestinal Tract Thirst Hunger Vomiting		
Respiratory System Breathing Odor of breath		
Cardiovascular System Blood pressure Pulse		
Nervous System Consciousness		
Urine Sugar Acetone		
Treatment Response		

Notes

4

Workbook Activities

The following activities have been designed to help you. Your instructor may require you to complete some or all of these activities as a regular part of your EMT-B training program. You are encouraged to complete any activity that your instructor does not assign as a way to enhance your learning in the classroom.

Chapter Review

The following exercises provide an opportunity to refresh your knowledge of this chapter.

■ NOTES ■

Matching

Match each of the terms in the left column to the appropriate definition in the right column.

_____ **1.** Allergic reaction

_____ **2.** Leukotrienes

_____ **3.** Wheezing

_____ **4.** Antihistamine

_____ **5.** Urticaria

_____ **6.** Stridor

_____ **7.** Allergen

_____ **8.** Wheal

_____ **9.** Antivenin

_____ **10.** Toxin

A. substance that counteracts the effect of venom from a bite or sting

B. substance made by body; released in anaphylaxis

C. harsh, high-pitched inspiratory sound, usually resulting from upper airway obstruction

D. raised, swollen area on skin resulting from an insect bite or allergic reaction

E. an exaggerated immune response to any substance

F. multiple raised areas on the skin that itch or burn

G. a poison or harmful substance

H. substance that causes an allergic reaction

I. a high-pitched, whistling breath sound usually resulting from blockage of the airway and typically heard on expiration

J. agent that blocks the effects of histamine

Multiple Choice

Read each item carefully, then select the best response.

_____ **1.** It is important to provide prompt transport for any patient who may be having an allergic reaction and to continue to reassess the patient's vital signs en route because:

A. you can only administer epinephrine once.

B. the oxygen level may be too high.

C. signs and symptoms may change rapidly.

D. all of the above

Allergic Reactions and Envenomations

_____ **2.** Side effects of the administration of epinephrine include:

    A. tachycardia.

    B. chest pain.

    C. nausea.

    D. all of the above

_____ **3.** Black widow spiders may be found in:

    A. New Hampshire.

    B. woodpiles.

    C. Georgia.

    D. all of the above

_____ **4.** Steps for assisting a patient with administration of an EpiPen include:

    A. taking body substance isolation precautions.

    B. placing the tip of the auto-injector against the medial part of the patient's thigh.

    C. recapping the injector before placing it in the trash.

    D. all of the above

_____ **5.** Coral snakes may be found in:

    A. Florida.

    B. Kansas.

    C. New Jersey.

    D. all of the above

_____ **6.** The venom of the brown recluse spider is cytotoxic. It causes severe:

    A. nausea and vomiting.

    B. local tissue damage.

    C. headaches.

    D. all of the above

_____ **7.** Since ticks are only a fraction of an inch long and their bite is not painful, they can easily be mistaken for:

    A. dirt.

    B. a freckle.

    C. acne.

    D. all of the above

_____ **8.** Rocky Mountain spotted fever and Lyme disease are both spread through the tick's:

    A. saliva.

    B. blood.

    C. hormones.

    D. all of the above

_____ **9.** Signs of envenomation by a pit viper include:

    A. swelling.

    B. severe burning pain at the site of the injury.

    C. ecchymosis.

    D. all of the above

_____ **10.** Removal of a tick should be accomplished by:

    A. suffocating it with gasoline.

    B. burning it with a lighted match to cause it to release its grip.

    C. using fine tweezers to pull it straight out of the skin.

    D. suffocating it with Vaseline.

_____ **11.** Allergens may include:

    A. food.

    B. animal bites.

    C. semen.

    D. all of the above

_____ **12.** The wasp's stinger is unbarbed, meaning that it can:

    A. be removed easily.

    B. inflict multiple stings.

    C. inject more venom with each sting.

    D. penetrate deeper.

_____ **13.** Anaphylaxis is not always life threatening, but it typically involves:

    A. multiple organ systems.

    B. wheezing.

    C. urticaria.

    D. wheals.

_____ **14.** Signs and symptoms of insect stings or bites include:

    A. swelling.

    B. wheals.

    C. localized heat.

    D. all of the above

_____ **15.** Prolonged respiratory difficulty can cause _____, shock, and even death.

    A. tachypnea

    B. pulmonary edema

    C. tachycardia

    D. airway obstruction

_____ 16. Speed is essential because more than two-thirds of patients who die of anaphylaxis do so within the first:

    A. 10 minutes.

    B. 30 minutes.

    C. hour.

    D. 3 hours.

_____ 17. Questions to ask when obtaining a history from a patient appearing to have an allergic reaction include:

    A. whether the patient has a history of allergies.

    B. what the patient was exposed to.

    C. how the patient was exposed.

    D. all of the above

_____ 18. Systemic symptoms from envenomation by a coelenterate may include:

    A. muscle cramps.

    B. headache.

    C. dizziness.

    D. all of the above

_____ 19. Treatment for a black widow spider bite consists of maintaining:

    A. the airway.

    B. breathing.

    C. circulation.

    D. all of the above

_____ 20. Treatment of a snake bite from a pit viper includes:

    A. calming the patient.

    B. providing BLS as needed if the patient shows no sign of envenomation.

    C. marking the skin with a pen over the swollen area to note whether swelling is spreading.

    D. all of the above

_____ 21. The dosage of epinephrine in an adult EpiPen is:

    A. 0.10 mg.

    B. 0.15 mg.

    C. 0.30 mg.

    D. 0.50 mg.

_____ 22. Epinephrine, whether made by the body or by a drug manufacturer, works rapidly to:

    A. raise the pulse rate and blood pressure.

    B. inhibit an allergic reaction.

    C. dilate the bronchioles.

    D. all of the above

_____ 23. A bite in the abdomen from a black widow spider may cause muscle spasms so severe that the patient may be thought to have:

    A. peritonitis.

    B. gastrointestinal bleeding.

    C. a stomach virus.

    D. all of the above

4

_____24. If a patient suspected of having an allergic reaction has no signs of respiratory distress or shock, you should:

A. place the patient in a supine position.

B. continue with the focused assessment.

C. administer epinephrine via an EpiPen auto-injector before signs and symptoms develop.

D. all of the above

_____25. Systemic signs of envenomation by a pit viper may include:

A. localized swelling.

B. ecchymosis.

C. shock.

D. all of the above

_____26. Often, there are limited or no _____ associated with a coral snake bite.

A. respiratory problems

B. local symptoms

C. bizarre behavior

D. paralysis of the nervous system

_____27. Because the stinger of the honeybee is barbed and remains in the wound, it can continue to inject venom for up to:

A. 1 minute.

B. 15 minutes.

C. 20 minutes.

D. several hours.

_____28. You should not use tweezers or forceps to remove an embedded stinger because:

A. squeezing may cause the stinger to inject more venom into the wound.

B. the stinger may break off in the wound.

C. the tweezers are not sterile and may cause infection.

D. removing the stinger may cause bleeding.

_____29. Your assessment of the patient experiencing an allergic reaction should include evaluations of the:

A. respiratory system.

B. circulatory system.

C. skin.

D. all of the above

_____30. Eating certain foods, such as shellfish or nuts, may result in a relatively _____ reaction that still can be quite severe.

A. mild

B. fast

C. slow

D. rapid

_____31. In dealing with allergy-related emergencies, you must be aware of the possibility of acute _____ and cardiovascular collapse.

A. hypotension

B. tachypnea

C. airway obstruction

D. shock

_____32. Wheezing occurs because excessive _____ and mucus are secreted into the bronchial passages.

A. fluid

B. carbon dioxide

C. blood

D. all of the above

Vocabulary emtb. vocab explorer

Define the following terms using the space provided.

1. Anaphylaxis:

_____

_____

_____

2. Histamine:

_____

_____

_____

3. Epinephrine:

_____

_____

_____

4. Envenomation:

_____

_____

_____

5. Rabies:

_____

_____

_____

6. Nematocysts:

_____

_____

_____

Fill-in

Read each item carefully, then complete the statement by filling in the missing word(s).

1. Wheezing occurs because excessive fluid and mucus are secreted into the

_____ _____.

2. Small areas of generalized itching or burning that appear as multiple, small,

raised areas on the skin are called _____.

3. The stinger of the honeybee is _____, so the bee cannot withdraw it.

4. A reaction involving the entire body is called _____.

5. The presence of _____ or respiratory distress indicates that the patient is having a severe enough allergic reaction to lead to death.

6. The patient in anaphylaxis with dyspnea should be placed in the

_____ position with the head and shoulders elevated.

7. Epinephrine inhibits the allergic reaction and dilates the _____.

8. Your ability to recognize and manage the many signs and symptoms of allergic

reactions may be the only thing standing between a patient and _____

_____.

9. The rabies virus is in the saliva of a _____, or infected animal, and is transmitted through biting or licking an open wound.

True/False

If you believe the statement to be more true than false, write the letter "T" in the space provided. If you believe the statement to be more false than true, write the letter "F."

1. _____ Ice should be promptly applied to any insect sting or snake bite with swelling.

2. _____ The most common type of pit viper is the copperhead.

3. _____ Cottonmouths are known for aggressive behavior.

4. _____ Ticks should be removed by firmly grasping them with tweezers while rotating them counterclockwise.

5. _____ The pain of coelenterate stings may respond to flushing with cold water.

6. _____ Allergic reactions can occur in response to almost any substance.

7. _____ An allergic reaction occurs when the body has an immune response to a substance.

8. _____ Wheezing is a high-pitched breath sound usually resulting from blockage of the airway and heard on expiration.

9. _____ For a patient appearing to have an allergic reaction, give 100% oxygen via nasal cannula.

10. _____ Systemic symptoms of envenomation by coelenterates include headache, dizziness, and hypotension.

Short Answer

Complete this section with short written answers using the space provided.

1. Common side effects of epinephrine include:

_____

_____

_____

2. What are five stimuli that most often cause allergic reactions?

_____

_____

_____

3. What are the steps for administering or assisting with administration of an epinephrine auto-injector?

_____

_____

_____

4. What are the common respiratory and circulatory signs or symptoms of an allergic reaction?

_____

_____

_____

5. What treatments for a snake bite assist with slowing and monitoring the spread of venom?

_____

_____

_____

6. What are the two most common poisonous spiders in the United States and how do their bites differ?

_____

_____

_____

4

7. Dog and human bites, however minor, must be evaluated by a physician if the skin is broken. Why?

_____

_____

_____

8. What are the steps in treating a coelenterate envenomation?

_____

_____

_____

Word Fun

The following crossword puzzle is an activity provided to reinforce correct spelling and understanding of medical terminology associated with emergency care and the EMT-B. Use the clues in the column to complete the puzzle.

CLUES

Across

5. Small spots; itching, raised areas

6. Poison or harmful substance

8. Adrenaline

Down

1. Chemical substance made by the body, leads to anaphylaxis

2. Severe allergic reaction

3. Harsh, high-pitched airway sounds

4. Responsible for allergy symptoms

7. Raised, swollen area from a sting

Ambulance Calls

The following real case scenarios provide an opportunity to explore the concerns associated with patient management. Read each scenario, then answer each question in detail.

1. You are dispatched to the scene of a dog bite where you find a 7-year-old male complaining of several puncture wounds to his left calf. He is visibly upset but bleeding is controlled. The dog, a family pet, is locked inside the house.

How would you best manage this patient?

_____

_____

_____

_____

_____

_____

_____

_____

_____

2. You are called to a possible allergic reaction from the sting of a jellyfish. Your
patient, a 32-year-old female, is having difficulty breathing, has weak radial
pulses, and is covered in hives.

How would you manage this patient?

_____

_____

_____

_____

_____

_____

_____

_____

_____

3. You are dispatched to a local seafood restaurant for a person who is having difficulty
breathing. Upon arrival you find a 22-year-old female with facial edema, cyanosis
around the lips, audible wheezing, and urticaria on her face and upper body. Her
boyfriend tells you she ate shrimp and she is allergic to them. He also tells you she
has some medicine in her purse and hands you an EpiPen prescribed to her.

How would you best manage this patient?

_____

_____

_____

_____

_____

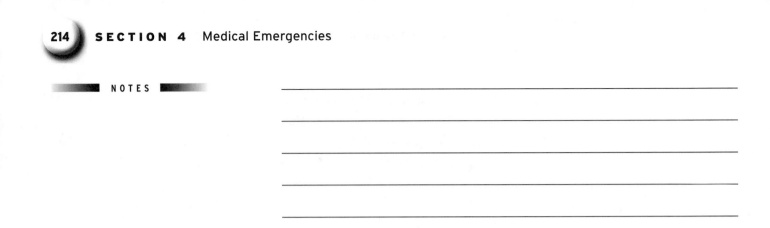

Skill Drills

Skill Drill 16-1: Using an Auto-Injector

Test your knowledge of this skill drill by filling in the correct words in the photo captions.

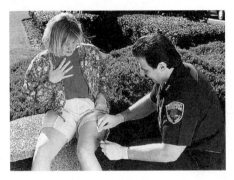

1. Remove the auto-injector's

_____ _____

and quickly wipe the thigh with

_____.

2. Place the tip of the auto-injector
against the _____
thigh.

3. Push the _____
firmly against the

_____ and hold it in
place until all the medication is
injected.

Skill Drill 16-2: Using an AnaKit

Test your knowledge of this skill drill by placing the following photos in the correct order. Number the first step with a "1," the second step with a "2," etc.

If available, apply a cold pack to the sting site.

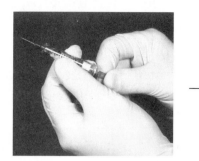

Turn the plunger one-quarter turn.

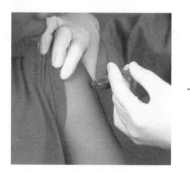

Hold the syringe steady and push the plunger until it stops.

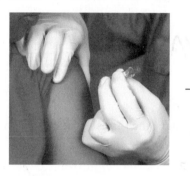

Quickly insert the needle into the muscle.

Hold the syringe upright and carefully use the plunger to remove air.

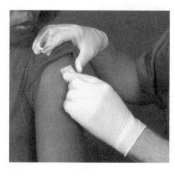

Prepare the injection site with antiseptic and remove the needle cover.

Have the patient take the Chlo-Amine tablets provided in the kit.

Workbook Activities

The following activities have been designed to help you. Your instructor may require you to complete some or all of these activities as a regular part of your EMT-B training program. You are encouraged to complete any activity that your instructor does not assign as a way to enhance your learning in the classroom.

Chapter Review

The following exercises provide an opportunity to refresh your knowledge of this chapter.

NOTES

Matching

Match each of the terms in the left column to the appropriate definition in the right column.

_____ **1.** Poison
_____ **2.** Substance abuse
_____ **3.** Antidote
_____ **4.** Tolerance
_____ **5.** Cholinergic

_____ **6.** Ingestion
_____ **7.** Hematemesis

_____ **8.** Stimulant

_____ **9.** Opioid

_____ **10.** Sedative-hypnotic
_____ **11.** Botulism

_____ **12.** Anticholinergic

A. Xanax, Librium, Valium
B. drug or agent with actions similar to morphine
C. Atropine, Benadryl, some cyclic antidepressants
D. may result from eating improperly canned food
E. need for increasing amounts of a drug to obtain the same effect
F. agent that produces an excited state
G. any substance whose chemical action can damage body structures or impair body functions
H. a substance that will counteract the effects of a particular poison
I. the misuse of any substance to produce a desired effect
J. taking a substance by mouth
K. overstimulates body functions controlled by parasympathetic nerves
L. vomiting blood

Multiple Choice

Read each item carefully, then select the best response.

_____ **1.** Activated charcoal is in the form of a(n):
A. elixir.
B. suspension.
C. syrup.
D. emulsion.

CHAPTER 17

Substance Abuse and Poisoning

___B___ **2.** The presence of burning or blistering of the mucous membranes suggests:

    A. ingestion of depressants.

    B. ingestion of poison.

    C. overdose of heroin.

    D. the patient may be a heavy smoker.

___D___ **3.** Treatment for ingestion of poisonous plants includes:

    A. assessing the ABCs.

    B. taking the plant to the emergency department.

    C. prompt transport.

    D. all of the above

___C___ **4.** The most important consideration in caring for a patient who has been exposed to an organophosphate insecticide or some other cholinergic agent is to:

    A. maintain the airway.

    B. apply high-flow oxygen.

    C. avoid exposure yourself.

    D. initiate CPR.

___D___ **5.** Objects that may provide clues to the nature of the poison include:

    A. a needle or syringe.

    B. scattered pills.

    C. chemicals.

    D. all of the above

___C___ **6.** The most worrisome avenue of poisoning is:

    A. ingestion.

    B. inhalation.

    C. injection.

    D. absorption.

___C___ **7.** The major side effect of ingesting activated charcoal is:

    A. depressed respirations.

    B. overproduction of stomach acid.

    C. black stools.

    D. increased blood pressure.

___B___ **8.** Alcohol is a powerful CNS depressant. It:

  A. sharpens the sense of awareness.

  B. slows reflexes.

  C. increases reaction time.

  D. all of the above

___C___ **9.** Frequently abused synthetic opioids include:

  A. heroin.

  B. morphine.

  C. Demerol.

  D. all of the above

___D___ **10.** Treatment of patients who have overdosed with sedative-hypnotics and have respiratory depression is to:

  A. provide airway clearance.

  B. provide ventilatory assistance.

  C. provide prompt transport.

  D. all of the above

___A___ **11.** Anticholinergic medications have properties that block the _____ nerves.

  A. parasympathetic

  B. sympathetic

  C. adrenergic

  D. parasympatholytic

___C___ **12.** _____ crack produces the most rapid means of absorption and therefore the most potent effect.

  A. Injected

  B. Absorbed

  C. Smoked

  D. Ingested

___D___ **13.** "Nerve gases" overstimulate normal body functions that are controlled by parasympathetic nerves causing:

  A. increased salivation.

  B. increased heart rate.

  C. increased urination.

  D. all of the above

___D___ **14.** Medicine containers can provide critical information such as:

  A. the number of pills originally in the bottle.

  B. the prescribed dose.

  C. the name and concentration of the drug.

  D. all of the above

___B___ **15.** Signs and symptoms of staphylococcal food poisoning include:

  A. difficulty in speaking.

  B. nausea, vomiting, diarrhea.

  C. blurred vision.

  D. all of the above

_A_ **16.** Inhalant effects range from mild drowsiness to coma, but unlike most other sedative-hypnotics, these agents may often cause:

   A. seizures.

   B. vomiting.

   C. swelling of the tongue.

   D. all of the above

_D_ **17.** Cocaine may be taken via:

   A. inhalation.

   B. injection.

   C. absorption.

   D. all of the above

_D_ **18.** Abusable substances include:

   A. vitamins.

   B. nasal decongestants.

   C. food.

   D. all of the above

_D_ **19.** Charcoal is not indicated for patients who:

   A. have ingested an acid, alkali, or petroleum product.

   B. have a decreased level of consciousness.

   C. are unable to swallow.

   D. all of the above

_A_ **20.** A person who has been using marijuana rarely needs transport to the hospital. Exceptions may include someone who is:

   A. hallucinating.

   B. very anxious.

   C. paranoid.

   D. all of the above

_C_ **21.** Halogenated hydrocarbon solvents can make the heart supersensitive to the patient's own _____, putting the patient at high risk for sudden cardiac death from ventricular fibrillation.

   A. blood

   B. electrical activity

   C. adrenalin

   D. antibodies

_B_ **22.** Sympathomimetics are CNS stimulants that frequently cause:

   A. hypotension.

   B. tachycardia.

   C. pinpoint pupils.

   D. all of the above

_D_ **23.** Carbon monoxide:

   A. is odorless.

   B. produces severe hypoxia.

   C. does not damage or irritate the lungs.

   D. all of the above

4

_____C_____ **24.** Chlorine:

  A. is odorless.

  B. does not damage or irritate the lungs.

  C. causes pulmonary edema.

  D. all of the above

_____B_____ **25.** Localized signs and symptoms of absorbed poisoning include:

  A. a history of exposure.

  B. burns, irritation of the skin.

  C. dyspnea.

  D. all of the above

_____C_____ **26.** Poisoning by injection is almost always the result of:

  A. repetitive bee stings.

  B. pit viper envenomation.

  C. deliberate drug overdose.

  D. homicide.

_____D_____ **27.** When dealing with substances such as phosphorous and elemental sodium, you should:

  A. brush the chemical off the patient.

  B. remove contaminated clothing.

  C. apply a dry dressing to the burn area.

  D. all of the above

_____A_____ **28.** Injected poisons are impossible to dilute or remove, as they are usually _____ or cause intense local tissue destruction.

  A. absorbed quickly into the body

  B. bound to hemoglobin

  C. large compounds

  D. combined with the cerebrospinal fluid

_____D_____ **29.** Medical problems that may cause the patient to present as intoxicated include:

  A. head trauma.

  B. toxic reactions.

  C. uncontrolled diabetes.

  D. all of the above

_____D_____ **30.** Signs and symptoms of alcohol withdrawal include:

  A. agitation and restlessness.

  B. fever, sweating.

  C. seizures.

  D. all of the above

_____A_____ **31.** Treatments for inhaled poisons include:

  A. moving the patient into fresh air.

  B. apply an SCBA to the patient.

  C. covering the patient to prevent spread of the poison.

  D. all of the above

_____D_____ **32.** Signs and symptoms of chlorine exposure include:

  A. cough.

  B. chest pain.

  C. wheezing.

  D. all of the above

___D__ **33.** Ingested poisons include:
   A. contaminated food.
   B. household cleaners.
   C. plants.
   D. all of the above

___A__ **34.** The majority of poisoning is through:
   A. ingestion.
   B. inhalation.
   C. injection.
   D. absorption.

___D__ **35.** Ingestion of an opiate, sedative, or barbituate can cause depression of the CNS and:
   A. paralysis of the extremities.
   B. dilation of the pupils.
   C. carpopedal spasms.
   D. slow breathing.

___A__ **36.** Inhaled poisons include:
   A. chlorine.
   B. venom.
   C. dieffenbachia.
   D. all of the above

___A__ **37.** Activated charcoal works by _____ the stomach.
   A. binding with the poison in
   B. creating turbulence in
   C. flushing the acid out of
   D. pushing the toxin down into

___C__ **38.** The most important treatment for poisoning is _____ and/or physically removing the poisonous agent.
   A. administering a specific antidote
   B. high-flow oxygen
   C. diluting
   D. syrup of ipecac

Vocabulary

Define the following terms using the space provided.

1. Sedative-hypnotic:

_____

_____

_____

2. Anticholinergic:

_____

_____

_____

3. Delirium tremens:

_____

_____

_____

4. Hallucinogen:

_____

_____

_____

5. Addiction:

_____

_____

_____

6. Substance abuse:

_____

_____

_____

7. Hypnotic:

_____

_____

_____

Fill-in
Read each item carefully, then complete the statement by filling in the missing word(s).

1. When dealing with exposure to chemicals, treatment focuses on support:

assessing and maintaining the patient's _____.

2. The most commonly abused drug in the United States is _____.

3. Activated charcoal works by _____, or sticking to, many commonly ingested poisons, preventing the toxin from being absorbed into the body.

4. If the patient has a chemical agent in the eyes, you should irrigate them quickly

and thoroughly, at least _____ for acid substances and

_____ for alkalis.

5. Opioid analgesics are CNS depressants and can cause severe _____.

6. Severe acute alcohol ingestion may cause _____.

7. Your primary responsibility to the patient who has been poisoned is to

_____ that a poisoning occurred.

8. The usual dosage for activated charcoal for an adult or child is _____

of activated charcoal per _____ of body weight.

9. As you irrigate the eyes, make sure that the fluid runs from the bridge of the

nose _____.

10. Approximately 80% of all poisoning is by _____, including plants,
contaminated food, and most drugs.

11. Patients experiencing alcohol withdrawal may develop _____ if they
no longer have their daily source of alcohol.

12. Phosphorus and elemental sodium _____ when they come in contact
with water.

13. Increasing tolerance of a substance can lead to _____.

14. _____ may develop from sweating, fluid loss, insufficient fluid intake,
or vomiting associated with DTs.

True/False

If you believe the statement to be more true than false, write the letter "T" in the space
provided. If you believe the statement to be more false than true, write the letter "F."

1. _____ The usual adult dose of activated charcoal is 25 to 50 g.

2. _____ The general treatment of a poisoned patient is to induce vomiting.

3. _____ Activated charcoal is a standard of care in all ingestions.

4. _____ Inhaled chlorine produces profound hypoxia without lung irritation.

5. _____ Shaking activated charcoal decreases its effectiveness.

6. _____ Opioid overdose typically presents with pinpoint pupils.

7. _____ Cholinergics are chemicals such as nerve gases, organophosphate
insecticides, or certain wild mushrooms.

8. _____ Alcohol is a stimulant.

9. _____ Demerol, Dilaudid, and Vicodin are all examples of opioids.

10. _____ Cocaine is one of the most addicting substances known.

Short Answer

Complete this section with short written answers using the space provided.

1. How does activated charcoal work to counteract ingested poison?

_____

_____

2. What are four routes of contact for poisoning?

_____

_____

_____

3. List the typical signs and symptoms of an overdose of sympathomimetics.

_____

_____

_____

4. What are the two main types of food poisoning?

_____

_____

_____

5. What differentiates the presentation of acetaminophen poisoning from that of other substances? What does this mean to the prehospital caregiver?

_____

_____

_____

6. What condition do the mnemonics DUMBELS and SLUDGE pertain to, and what do they mean?

_____

_____

_____

7. In addition to alcohol and marijuana, what are the seven categories of drugs seen in overdoses/poisoning?

_____

_____

_____

_____

_____

_____

8. What five questions should you ask a possible poisoning victim?

_____

_____

_____

_____

_____

9. Why should phosphorous or elemental sodium poisoning victims not be irrigated?

_____

_____

_____

Word Fun

The following crossword puzzle is an activity provided to reinforce correct spelling and understanding of medical terminology associated with emergency care and the EMT-B. Use the clues in the column to complete the puzzle.

CLUES

Across

1. Induces sleep

4. Overwhelming obsession or physical need

5. Substance whose chemical action causes damage

7. Poison or harmful substance; produced by animals, plants

8. Swallowing

9. Need for increasing amounts of a drug

11. Decreases activity and excitement

12. Vomiting blood

Down

2. Similar to morphine

3. Produces false perceptions

4. Used to neutralize or counteract a poison

6. Misuse of any substance to produce a desired effect

10. Material in emesis

Ambulance Calls

The following real case scenarios provide an opportunity to explore the concerns associated with patient management. Read each scenario, then answer each question in detail.

1. You are dispatched to a scene where a 42-year-old male was found face down in an alley. Police tell you he is a known alcoholic and probably just "fell out" again. He is pain responsive and has blood trickling from his nose where he struck the asphalt.

   How would you best manage this patient?

   _____

   _____

   _____

   _____

   _____

   _____

   _____

   _____

   _____

2. You are called to a local child care center where a toddler has been found chewing on the leaves of an unknown plant. The supervisor tells you she thinks the plant may be poisonous and they were not sure what they should do. His mother has been called and will meet you at the hospital.

   How would you best manage this patient?

   _____

   _____

   _____

   _____

   _____

   _____

   _____

   _____

   _____

3. You are called to a possible suicide attempt. You arrive on the scene to find police and a neighbor in the home of a 25-year-old female who is unresponsive, supine on her bed. The neighbor tells you that the patient recently broke up with her boyfriend and has been very distraught. There is an empty pill bottle on the nightstand. When you look at the label you see that the prescription was filled yesterday and there were 30 tablets dispensed. There is also an empty liquor bottle on the floor.

How would you best manage this patient?

_____

_____

_____

_____

_____

_____

_____

_____

_____

4

Workbook Activities

The following activities have been designed to help you. Your instructor may require you to complete some or all of these activities as a regular part of your EMT-B training program. You are encouraged to complete any activity that your instructor does not assign as a way to enhance your learning in the classroom.

Chapter Review

The following exercises provide an opportunity to refresh your knowledge of this chapter.

■ NOTES ■

Matching

Match each of the terms in the left column to the appropriate definition in the right column.

_____ 1. Conduction

___H___ 2. Air embolism

_____ 3. Evaporation

___F___ 4. Hyperthermia

_____ 5. Diving reflex

___N___ 6. Core temperature

_____ 7. Convection

_____ 8. Laryngospasm

_____ 9. Electrolytes

_____ 10. Radiation

_____ 11. Hypothermia

_____ 12. Ambient temperature

_____ 13. Heat cramps

_____ 14. Drowning

A. slowing of heart rate caused by sudden immersion in cold water

B. salts and other chemicals dissolved in body fluids

C. severe constriction of the larynx and vocal cords

D. death from suffocation after submersion in water

E. heat loss resulting from standing in a cold room

F. core temperature greater than 101°F

G. condition when the entire body temperature falls

H. condition caused by air bubbles in the blood vessels

I. heat loss that occurs from helicopter rotor blade downwash

J. heat loss resulting from sitting on snow

K. heat loss resulting from sweating

L. painful muscle spasms that occur after vigorous exercise

M. temperature of the surrounding environment

N. temperature of the central part of the body

Environmental Emergencies

Multiple Choice

Read each item carefully, then select the best response.

_____ 1. _____ causes body heat to be lost, as warm air in the lungs is exhaled into the atmosphere and cooler air is inhaled.

A. Convection

B. Conduction

C. Radiation

D. Respiration

_____ 2. Evaporation, the conversion of any liquid to a gas, is a process that requires:

A. energy.

B. circulating air.

C. a warmer ambient temperature.

D. all of the above

_____ 3. The rate and amount of heat loss by the body can be modified by:

A. increasing heat production.

B. moving to an area where heat loss is decreased.

C. wearing insulated clothing.

D. all of the above

_____ 4. The characteristic appearance of blue lips and/or fingertips seen in hypothermia is the result of:

A. lack of oxygen in venous blood.

B. frostbite.

C. blood vessels constricting.

D. bruising.

_____ 5. Signs and symptoms of severe systemic hypothermia include all of the following, except:

A. weak pulse.

B. coma.

C. confusion.

D. very slow respirations.

_____ **6.** Hypothermia is more common among:

    A. elderly individuals.

    B. infants and children.

    C. those who are already ill.

    D. all of the above

_____ **7.** To assess a patient's general temperature, pull back your glove and place the back of your hand on the patient's:

    A. abdomen, underneath the clothing.

    B. forehead.

    C. forearm, on the inside of the wrist.

    D. neck, at the area where you check the carotid pulse.

_____ **8.** Never assume that a(n) _____, pulseless patient is dead.

    A. apneic

    B. cyanotic

    C. cold

    D. hyperthermic

_____ **9.** Management of hypothermia in the field consists of all of the following except:

    A. stabilizing vital functions.

    B. removing wet clothing.

    C. preventing further heat loss.

    D. massaging the cold extremities.

_____ **10.** It is necessary to assess the pulse of a hypothermic patient for at least _____, especially before considering CPR.

    A. 10 to 20 seconds

    B. 30 to 45 seconds

    C. 60 to 75 minutes

    D. 2 full minutes

_____ **11.** When exposed parts of the body become very cold but not frozen, the condition is called:

    A. frostnip.

    B. chilblains.

    C. immersion foot.

    D. all of the above

_____ **12.** Important factors in determining the severity of a local cold injury include all of the following except:

    A. the temperature to which the body part was exposed.

    B. a previous history of frostbite.

    C. the wind velocity during exposure.

    D. the duration of the exposure.

_____ **13.** Signs and symptoms of systemic hypothermia include:

    A. blisters and swelling.

    B. hard and waxy skin.

    C. altered mental status.

    D. local tissue damage.

_____ **14.** When the body is exposed to more heat energy than it loses, _____ results.

   A. hyperthermia

   B. heat cramps

   C. heat exhaustion

   D. heatstroke

_____ **15.** Contributing factors to the development of heat illnesses include:

   A. high air temperature.

   B. vigorous exercise.

   C. high humidity.

   D. all of the above

_____ **16.** Keeping yourself hydrated while on duty is very important. Drink at least _____ of water per day, and more when exertion or heat is involved.

   A. 8 glasses

   B. 1 liter

   C. 2 liters

   D. 3 liters

_____ **17.** The following statements concerning heat cramps are true except:

   A. they only occur when it is hot outdoors.

   B. they may be seen in well-conditioned athletes.

   C. the exact cause of heat cramps is not well understood.

   D. dehydration may play a role in the development of heat cramps.

_____ **18.** Signs and symptoms of heat exhaustion and associated hypovolemia include:

   A. cold, clammy skin with ashen pallor.

   B. dizziness, weakness, or faintness.

   C. normal vital signs.

   D. all of the above

_____ **19.** Be prepared to transport the patient to the hospital for aggressive treatment of hyperthermia if:

   A. the symptoms do not clear up promptly.

   B. the level of consciousness improves.

   C. the temperature drops.

   D. all of the above

_____ **20.** Often, the first sign of heatstroke is:

   A. a change in behavior.

   B. an increase in pulse rate.

   C. an increase in respirations.

   D. hot, dry, flushed skin.

_____ **21.** The least common but most serious illness caused by heat exposure, occurring when the body is subjected to more heat than it can handle and normal mechanisms for getting rid of the excess heat are overwhelmed, is:

   A. hyperthermia.

   B. heat cramps.

   C. heat exhaustion.

   D. heatstroke.

4

_____22. _____ is the body's attempt at self-preservation by preventing water from entering the lungs.

A. Bronchoconstriction

B. Laryngospasm

C. Esophageal spasms

D. Swelling in the oropharynx

_____23. Treatment of drowning/near drowning begins with:

A. opening the airway.

B. ventilation with 100% oxygen via BVM device.

C. suctioning the lungs to remove the water.

D. rescue and removal from the water.

_____24. If you are unsure whether or not a spinal injury has occurred, you should:

A. stabilize and protect the patient's spine.

B. provide mouth-to-mouth ventilation as you would in any other situation.

C. ascertain whether or not there is a spinal injury.

D. all of the above

_____25. After removing a near drowning patient from the water, it may be difficult to find a pulse because of:

A. dilation of peripheral blood vessels.

B. body temperature at the core.

C. low cardiac output.

D. all of the above

_____26. If the near drowning victim has evidence of upper airway obstruction by foreign matter, attempt to clear it by:

A. removing the obstruction manually.

B. suction.

C. using abdominal thrusts.

D. all of the above

_____27. You must assume that spinal injury exists if the patient is conscious but complains of:

A. weakness.

B. numbness in the arms or legs.

C. paralysis.

D. all of the above

_____28. You should never give up on resuscitating a cold-water drowning victim because:

A. when the patient is submerged in water colder than body temperature, heat is maintained in the body.

B. the resulting hypothermia can protect vital organs from the lack of oxygen.

C. the resulting hypothermia raises the metabolic rate.

D. all of the above

_____29. The three phases of a dive, in the order they occur, are:

A. ascent, descent, bottom.

B. descent, bottom, ascent.

C. orientation, bottom, ascent.

D. descent, orientation, ascent.

_____**30.** Areas usually affected by descent problems include:
   A. the lungs.
   B. the skin.
   C. the joints.
   D. vision.

_____**31.** Potential problems associated with rupture of the lungs include:
   A. air emboli.
   B. pneumomediastinum.
   C. pneumothorax.
   D. all of the above

_____**32.** The organs most severely affected by air embolism are the:
   A. brain and spinal cord.
   B. brain and heart.
   C. heart and lungs.
   D. brain and lungs.

_____**33.** Treatment of hypothermia caused by cold-water immersion is the same as that of hypothermia caused by cold exposure, with the exception of:
   A. the use of humidified oxygen.
   B. there are no exceptions.
   C. clearing the patient's airway of foreign material.
   D. massaging the extremities.

4

Vocabulary

Define the following terms using the space provided.

1. Hyperbaric chamber:

_____

_____

_____

2. Decompression sickness:

_____

_____

_____

3. Heat exhaustion:

_____

_____

_____

4. Frostbite:

_____

_____

_____

5. Near drowning:

_____

_____

_____

6. Pneumomediastinum:

_____

_____

_____

Fill-in

Read each item carefully, then complete the statement by filling in the missing word.

1. Do not attempt to rewarm patients who have _____ hypothermia, because they are prone to developing arrhythmias unless handled very carefully.

2. Most significant diving injuries occur during _____.

3. When treating a patient with frostbite, never attempt _____ if there is any chance that the part may freeze again before the patient reaches the hospital.

4. As with so many hazards, you cannot help others if you do not

practice _____.

5. _____, a common effect of hypothermia, is the body's attempt to maintain heat.

6. Whenever a person dives or jumps into very cold water, the _____ may cause immediate bradycardia.

7. If the patient is alert and responds appropriately, the hypothermia is _____.

True/False

If you believe the statement to be more true than false, write the letter "T" in the space provided. If you believe the statement to be more false than true, write the letter "F."

1. _____ Normal body temperature is 98.6°F (37.0°C).

2. _____ To assess the skin temperature in a patient experiencing a generalized cold emergency, you should feel the patient's skin.

3. _____ Mild hypothermia occurs when the core temperature drops to 85°F.

4. _____ The body's most efficient heat regulating mechanisms are sweating and dilation of skin blood vessels.

5. _____ People who are at greatest risk for heat illnesses are the elderly and children.

6. _____ The signs and symptoms of exposure to heat can include moist, pale skin.

7. _____ The strongest stimulus for breathing is an elevation of oxygen in the blood.

8. _____ Immediate bradycardia after jumping in cold water is called the diving reflex.

9. _____ The signs and symptoms of exposure to heat can include hot, dry skin.

Short Answer

Complete this section with short written answers using the space provided.

1. What are three ways to modify heat loss? Give an example of each.

   _____

   _____

   _____

2. What are the steps in treating heatstroke?

   _____

   _____

   _____

3. What is an air embolism and how does it occur?

   _____

   _____

   _____

4. For what diving emergencies are hyperbaric chambers used?

   _____

   _____

   _____

5. How should a frostbitten foot be treated?

   _____

   _____

   _____

6. What are four "Do Nots" in relation to local cold injuries?

   _____

   _____

   _____

4

NOTES

7. What are the potential signs and symptoms of an air embolism?

_____

_____

_____

Word Fun

The following crossword puzzle is an activity provided to reinforce correct spelling and understanding of medical terminology associated with emergency care and the EMT-B. Use the clues in the column to complete the puzzle.

CLUES

Across

1. Slowing heart from sudden submersion in cold water

3. Loss of heat by direct contact

5. Less than 60 beats per minute

7. Salts and chemicals in body fluids

8. Used for breathing underwater

9. Loss of heat by air movement

10. Temperature greater than 101°F

Down

2. Loss of heat to colder environment

4. Excessive heat problem, may be fatal

6. Air bubbles in blood vessels

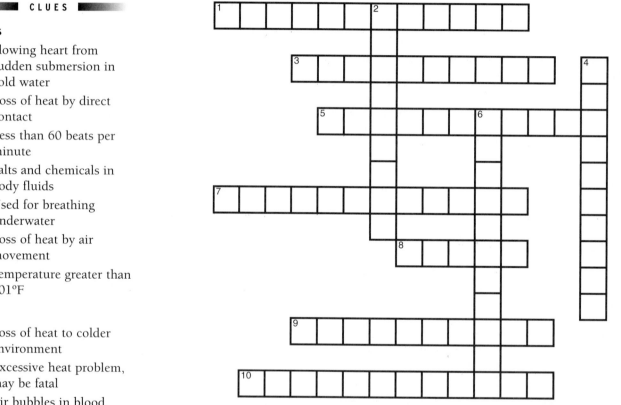

Ambulance Calls

The following real case scenarios provide an opportunity to explore the concerns associated with patient management. Read each scenario, then answer each question in detail.

1. You are called to the local airport for a 52-year-old male who is the pilot of his own aircraft. He tells you he is having severe abdominal pain and joint pain. History reveals that the patient is returning from a dive trip off the coast. He says he has had "the bends" before and this feels similar.

How would you best manage this patient?

_____

_____

_____

_____

_____

_____

_____

_____

_____

_____

2. You are called to the residence of a 74-year-old female. Neighbors tell you that she was found this morning sitting on her porch and the temperature was below freezing. She went outside because she has no heat in her house anyway. She is cold to touch, but responds appropriately to your questions.

How would you best manage this patient?

_____

_____

_____

_____

_____

_____

_____

_____

_____

3. You are dispatched to the local high school for a 15-year-old female complaining of a sudden onset of abdominal cramps. It is very hot inside the gym and she was exercising vigorously. She is sitting on the bleachers, doubled over and crying.

How would you best manage this patient?

_____

_____

_____

_____

_____

_____

Skill Drills

Skill Drill 18-1: Treating for Heat Exhaustion
Test your knowledge of this skill drill by filling in the correct words in the photo captions.

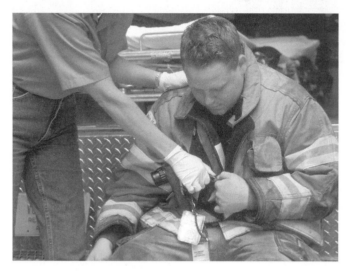

1. Remove _____ _____.

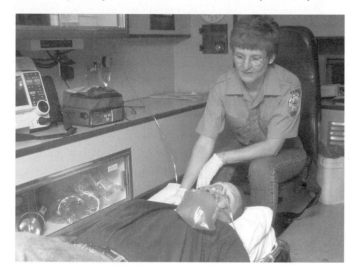

2. Move the patient to a _____ _____.

Give _____.

Place the patient in a _____ position, elevate the legs, and _____ the patient.

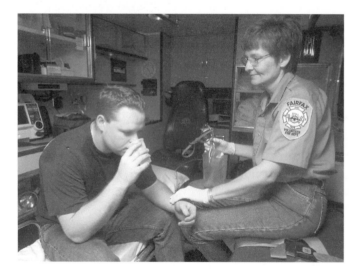

3. If the patient is _____ _____, give water by mouth.

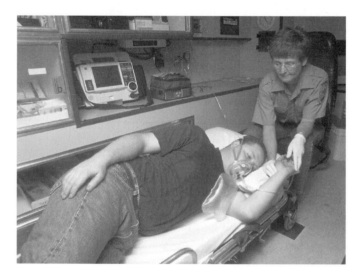

4. If nausea develops, _____ on the side.

Skill Drill 18-2: Stabilizing a Suspected Spinal Injury in the Water

Test your knowledge of this skill drill by placing the photos below in the correct order.
Number the first step with a "1," the second step with a "2," etc.

_____

Secure the patient to the backboard.

_____

Turn the patient to a supine position by rotating the entire upper half of the body as a single unit.

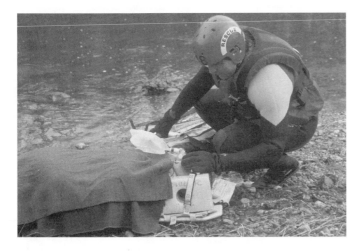

_____

Cover the patient with a blanket and apply oxygen if breathing. Begin CPR if breathing and pulse are absent.

_____

Float a buoyant backboard under the patient.

(continued)

As soon as the patient is turned, begin artificial ventilation using the mouth-to-mouth method or a pocket mask.

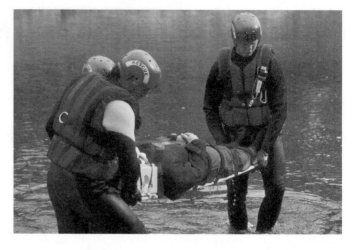

Remove the patient from the water.

Notes

4

Workbook Activities

The following activities have been designed to help you. Your instructor may require you to complete some or all of these activities as a regular part of your EMT-B training program. You are encouraged to complete any activity that your instructor does not assign as a way to enhance your learning in the classroom.

Chapter Review

The following exercises provide an opportunity to refresh your knowledge of this chapter.

■ NOTES ■

Matching
Match each of the terms in the left column to the appropriate definition in the right column.

_____ **1.** Behavioral crisis   **A.** what you can see of a person's response to the environment; his or her actions

_____ **2.** Psychogenic   **B.** temporary or permanent dysfunction of the brain caused by a disturbance in brain tissue function

_____ **3.** Organic brain syndrome   **C.** any reaction to events that interferes with activities of daily living or is unacceptable to the patient or others

_____ **4.** Depression   **D.** a persistent mood of sadness, despair, or discouragement

_____ **5.** Functional disorder   **E.** abnormal operation of an organ that cannot be traced to an obvious change in structure or physiology of the organ

_____ **6.** Behavior   **F.** a symptom or illness caused by mental factors as opposed to physical ones

Multiple Choice
Read each item carefully, then select the best response.

_____ **1.** A psychological or behavioral crisis may be due to:

   A. mind-altering substances.

   B. the emergency situation.

   C. stress.

   D. all of the above

C H A P T E R 19

Behavioral Emergencies

_____ **2.** A normal reaction to a crisis situation would be:

    A. Monday morning blues that last until Friday.

    B. feeling "blue" after the break up of a long-term relationship.

    C. feeling depressed week after week with no discernible cause.

    D. all of the above

_____ **3.** The cause of a behavioral crisis experienced by an unmanageable patient may be:

    A. drug use.

    B. a history of mental illness.

    C. alcohol abuse.

    D. all of the above

_____ **4.** Learning to adapt to a variety of situations in daily life, including stresses and strains, is called:

    A. disruption.

    B. adjustment.

    C. behavior.

    D. functional.

_____ **5.** If the interruption of daily routine tends to recur on a regular basis, the behavior is also considered a _____ problem.

    A. mental health

    B. functional disorder

    C. behavioral

    D. psychogenic

_____ **6.** If an abnormal or disturbing pattern of behavior lasts for at least _____, it is regarded as a matter of concern from a mental health standpoint.

    A. 6 weeks

    B. a month

    C. 6 months

    D. a year

_____ 7. A person who is no longer able to respond appropriately to the environment may be having what is called a psychological or _____ emergency.
   A. psychiatric
   B. behavioral
   C. functional
   D. adjustment

_____ 8. Mental disorders may be caused by a:
   A. social disturbance.
   B. chemical disturbance.
   C. biological disturbance.
   D. all of the above

_____ 9. An altered mental status may arise from:
   A. an oxygen saturation of 98%.
   B. moderate temperatures.
   C. an inadequate blood flow to the brain.
   D. adequate glucose levels in the blood.

_____ 10. Organic brain syndrome may be caused by:
   A. hypoglycemia.
   B. excessive heat or cold.
   C. lack of oxygen.
   D. all of the above

_____ 11. An example of a functional disorder would be:
   A. schizophrenia.
   B. organic brain syndrome.
   C. Alzheimer's.
   D. all of the above

_____ 12. When documenting abnormal behavior, it is important to:
   A. record detailed, subjective findings.
   B. avoid judgmental statements.
   C. avoid quoting the patient's own words.
   D. all of the above

_____ 13. Safety guidelines for behavioral emergencies include:
   A. assessing the scene.
   B. being prepared to spend extra time.
   C. encouraging purposeful movement.
   D. all of the above

_____ 14. In evaluating a situation that is considered a behavioral emergency, the first things to consider are:
   A. airway and breathing.
   B. scene safety and patient response.
   C. history of medications.
   D. respiratory and circulatory status.

_____ 15. Psychogenic circumstances may include:
   A. severe depression.
   B. death of a loved one.
   C. a history of mental illness.
   D. all of the above

_____ **16.** Risk factors for suicide may include:

    A. denial of alcohol use.

    B. recent marriage.

    C. holidays.

    D. all of the above

_____ **17.** Suicidal patients may also be:

    A. homicidal.

    B. hypoxic.

    C. joking.

    D. seeking attention.

_____ **18.** Causes of altered behavior in geriatric patients may include:

    A. constipation.

    B. diabetes.

    C. stroke.

    D. all of the above

_____ **19.** Restraint of a person must be ordered by:

    A. a physician.

    B. a court order.

    C. a law enforcement officer.

    D. all of the above

_____ **20.** When restraining a patient without an appropriate order, legal actions may involve charges of:

    A. abandonment.

    B. negligence.

    C. battery.

    D. breach of duty.

_____ **21.** When restraining a patient face down on a stretcher, it is necessary to constantly reassess the patient's:

    A. level of consciousness.

    B. airway.

    C. emotional status.

    D. pulse rate.

Vocabulary

Complete this section by defining the terms listed using the space provided.

1. Mental disorder:

_____

_____

_____

2. Activities of daily living (ADL):

_____

_____

_____

3. Altered mental status:

_____

_____

_____

4. Implied consent:

_____

_____

_____

Fill-in

Read each item carefully, then complete the statement by filling in the missing word.

1. _____ is what you can see of a person's response to the environment; his or her actions.

2. A _____ or emergency is any reaction to events that interferes with the activities of daily living or has become unacceptable to the patient, family, or community.

3. Chronic _____, or a persistent feeling of sadness or despair, may be a symptom of a mental or physical disorder.

4. _____ is a temporary or permanent dysfunction of the brain caused by a disturbance in the physical or physiologic functioning of the brain.

5. A behavioral crisis puts tremendous stress on a person's _____, including natural abilities and training.

6. Any time you encounter an emotionally depressed patient, you must consider the

possibility of _____.

True/False

If you believe the statement to be more true than false, write the letter "T" in the space provided. If you believe the statement to be more false than true, write the letter "F."

1. _____ Depression lasting 2 to 3 weeks after being fired from a job is a normal mental health response.

2. _____ Low blood glucose or lack of oxygen to the brain may cause behavioral changes, but not to the degree that a psychiatric emergency could exist.

3. _____ From a mental health standpoint, a pattern of abnormal behavior must last at least 3 months to be a matter of concern.

4. _____ A disturbed patient should always be transported with restraints.

5. _____ It is sometimes helpful to allow a patient with a behavioral emergency some time alone to calm down and collect their thoughts.

6. _____ It is important to avoid looking directly at the patient when dealing with a behavioral crisis.

7. _____ A patient should never be asked if he or she is considering suicide.

8. _____ Urinary tract infections can cause behavioral changes in elderly patients.

9. _____ All individuals with mental health disorders are dangerous, violent, or otherwise unmanageable.

10. _____ When completing the documentation, it is important to record detailed, subjective findings that support the conclusion of abnormal behavior.

11. _____ When restraining a patient, at least four people should be present to carry out the restraint.

Short Answer

Complete this section with short written answers using the space provided.

1. What is the distinction between a behavioral crisis and a mental health problem?

_____

_____

_____

2. What three major areas should be considered in evaluating the possible source of a behavioral crisis?

_____

_____

_____

3. What are three factors to consider in determining the level of force required to restrain a patient?

_____

_____

_____

4. List ten safety guidelines for dealing with behavioral emergencies.

_____

_____

_____

5. List ten risk factors for suicide.

_____

_____

_____

4

Word Fun

The following crossword puzzle is a good way to review correct spelling and meaning of medical terminology associated with emergency care and the EMT-B. Use the clues in the column to complete the puzzle.

■ CLUES ■

Across

3. Reactions interfere with normal activities

5. Illness with psychological symptoms

6. Caused by mental factors; not physical

7. Persistent sadness; despair

Down

1. A change in behavior

2. No known physiologic reason for abnormality

4. Basic doings of a normal person

■ NOTES ■

Ambulance Calls

The following real case scenarios provide an opportunity to explore the concerns associated with patient management. Read each scenario, then answer each question in detail.

1. You are dispatched to the residence of a 27-year-old man. Family members called because he has become increasingly more confused and agitated due to excessive street drug use. They tell you he has had a large quantity of alcohol and crack cocaine over the past 24 hours. The patient has violent tattoos covering both forearms and is pacing back and forth across the room clenching and unclenching his fists. He is yelling obscenities at anyone who comes near.

How would you best manage this patient?

_____

_____

_____

_____

_____

_____

_____

_____

_____

_____

2. You are called to the scene of a personal care home where you find a 73-year-old female with erratic behavior. Staff reports that the patient has a history of mild dementia and hypertension. Her mental status is quite altered from normal. Her ABCs are normal.

How would you best manage this patient?

_____

_____

_____

_____

_____

_____

_____

_____

_____

_____

3. You are dispatched to the residence of a 40-year-old female who is upset over the loss of her mother 5 weeks ago. She tells you that she has no family and has cared for her elderly mother for the past 7 years. She has not eaten in several days and is severely depressed.

How would you best manage this patient?

_____

_____

_____

_____

4

 NOTES

Notes

4

Workbook Activities

The following activities have been designed to help you. Your instructor may require you to complete some or all of these activities as a regular part of your EMT-B training program. You are encouraged to complete any activity that your instructor does not assign as a way to enhance your learning in the classroom.

Chapter Review

The following exercises provide an opportunity to refresh your knowledge of this chapter.

Matching

Match each of the terms in the left column to the appropriate definition in the right column.

_____ **1.** Cervix

_____ **2.** Perineum

_____ **3.** Placenta

_____ **4.** Amniotic sac

_____ **5.** Fetus

_____ **6.** Birth canal

_____ **7.** Uterus

_____ **8.** Umbilical cord

_____ **9.** Vagina

_____ **10.** Breech presentation

_____ **11.** Limb presentation

_____ **12.** Multipara

_____ **13.** Nuchal cord

_____ **14.** Presentation

_____ **15.** Miscarriage

A. an umbilical cord that is wrapped around the infant's neck

B. a fluid-filled, bag-like membrane inside the uterus that grows around the developing fetus

C. the area of skin between the vagina and the anus

D. the neck of the uterus

E. Connects mother and infant

F. the outermost part of a woman's reproductive system

G. the part of the infant that appears first

H. the vagina and lower part of the uterus

I. a woman who has had more than one live birth

J. spontaneous abortion

K. delivery in which the presenting part is a single arm, leg, or foot

L. tissue that develops on the wall of the uterus and is connected to the fetus

M. the hollow organ inside the female pelvis where the fetus grows

N. the developing baby in the uterus

O. delivery in which the buttocks come out first

CHAPTER 20

Obstetrics and Gynecological Emergencies

Multiple Choice

Read each item carefully, then select the best response.

_____ **1.** In the event of a nuchal cord, proper procedure is to:

    A. gently slip the cord over the infant's head or shoulder.

    B. clamp the cord and cut it before delivering the infant.

    C. clamp the cord and cut it, then gently unwind it from around the neck if wrapped around more than once.

    D. all of the above

_____ **2.** If the amniotic fluid is greenish instead of clear or has a foul odor, this is called:

    A. nuchal rigidity.

    B. meconium staining.

    C. placenta previa.

    D. bloody show.

_____ **3.** Meconium can cause all of the following, except:

    A. a depressed newborn.

    B. rapid pulse rate.

    C. airway obstruction.

    D. aspiration.

_____ **4.** Once the head is delivered, the baby must be suctioned. This is accomplished by:

    A. suctioning the nose only.

    B. suctioning the nose, then the mouth.

    C. suctioning the mouth, then the nose.

    D. suctioning the mouth only.

_____ **5.** Once the entire infant is delivered, you should immediately:

    A. wrap it in a towel and place it on one side with head lowered.

    B. be sure the head is covered and keep the neck in a neutral position.

    C. use a sterile gauze pad to wipe the infant's mouth, then suction again.

    D. all of the above

_____ **6.** You may help control bleeding by massaging the _____ after delivery of the placenta.

   A. perineum

   B. fundus

   C. lower back

   D. inner thighs

_____ **7.** The APGAR score should be calculated at _____ minutes after birth.

   A. 1 and 5

   B. 3 and 7

   C. 2 and 10

   D. 4 and 8

_____ **8.** Once the infant is delivered, feel for a brachial pulse or the pulsations in the umbilical cord. The pulse rate should be at least _____ beats/min and if not, begin artificial ventilations.

   A. 60

   B. 80

   C. 100

   D. 120

_____ **9.** When assisting ventilations in a newborn with a BVM device, the rate is _____ breaths/min.

   A. 20 to 30

   B. 30 to 50

   C. 35 to 45

   D. 40 to 60

_____ **10.** When performing CPR on a newborn, you should perform a combined total of _____ ventilations and compressions per minute.

   A. 90

   B. 100

   C. 110

   D. 120

_____ **11.** You cannot successfully deliver a _____ presentation in the field.

   A. limb

   B. breech

   C. vertex

   D. all of the above

_____ **12.** Care for a mother with a prolapsed cord includes:

   A. positioning the mother to keep the weight of the infant off the cord.

   B. high-flow oxygen and rapid transport.

   C. use your hand to physically hold the infant's head off the cord.

   D. all of the above

_____ **13.** When handling a delivery of a drug- or alcohol-addicted mother, your first concern should be for:

   A. the airway of the mother.

   B. your personal safety.

   C. the airway of the infant.

   D. the need for CPR for the infant.

_____ **14.** Your first priority when dealing with a sexual assault victim is to:

    A. manage the airway.

    B. preserve all evidence.

    C. not allow the patient to bathe or brush their teeth.

    D. control any bleeding.

_____ **15.** The stages of labor include:

    A. dilation of the cervix.

    B. expulsion of the baby.

    C. delivery of the placenta.

    D. all of the above

_____ **16.** The first stage of labor begins with the onset of contractions and ends when:

    A. the infant is born.

    B. the cervix is fully dilated.

    C. the water breaks.

    D. the placenta is delivered.

_____ **17.** Signs of the beginning of labor include:

    A. bloody show.

    B. contractions of the uterus.

    C. rupture of the amniotic sac.

    D. all of the above

_____ **18.** The second stage of labor begins when the cervix is fully dilated and ends when:

    A. the infant is born.

    B. the water breaks.

    C. the placenta delivers.

    D. the uterus stops contracting.

_____ **19.** You may safely use _____ to treat any heart or lung disease in the mother without harm to the fetus.

    A. epinephrine

    B. antihistamines

    C. oxygen

    D. all of the above

_____ **20.** The third stage of labor begins with the birth of the infant and ends with the:

    A. release of milk from the breasts.

    B. cessation of uterine contractions.

    C. delivery of the placenta.

    D. cutting of the umbilical cord.

_____ **21.** The difference between pre-eclampsia and eclampsia is the onset of:

    A. seeing spots.

    B. seizures.

    C. swelling in the hands and feet.

    D. headaches.

4

_____22. You should consider the possibility of a(n) _____ in women who have missed a menstrual cycle and complain of a sudden stabbing and usually unilateral pain in the lower abdomen.

A. PID

B. ectopic pregnancy

C. miscarriage

D. placenta abruptio

_____23. Once labor has begun, it is important to:

A. hold the mother's legs together.

B. not let the mother go to the bathroom.

C. restrict all movement of the mother.

D. all of the above

_____24. Consider delivery of the fetus at the scene when:

A. delivery can be expected within a few minutes.

B. when a natural disaster, or other problem, makes it impossible to reach the hospital.

C. no transportation is available.

D. all of the above

_____25. When in doubt about the possibility of an imminent delivery:

A. go for help if you are alone.

B. insert two fingers into the vagina to feel for the head.

C. contact medical control for further guidance.

D. all of the above

_____26. _____ is a condition of late pregnancy that also involves headache, visual changes, and swelling of the hands and feet.

A. Pregnancy-induced hypertension

B. Placenta previa

C. Placenta abruptio

D. Supine hypotension syndrome

_____27. Low blood pressure resulting from compression of the inferior vena cava by the weight of the fetus when the mother is supine is called:

A. pregnancy-induced hypertension.

B. placenta previa.

C. placenta abruptio.

D. supine hypotensive syndrome.

_____28. _____ is a situation in which the umbilical cord comes out of the vagina before the infant.

A. Eclampsia

B. Placenta previa

C. Placenta abruptio

D. Prolapsed cord

_____29. Premature separation of the placenta from the wall of the uterus is known as:

A. eclampsia.

B. placenta previa.

C. placenta abruptio.

D. prolapsed cord.

_____30. _____ is a condition in which the placenta develops over and covers the cervix.

A. Eclampsia

B. Placenta previa

C. Placenta abruptio

D. Prolapsed cord

_____ 31. _____ is heralded by the onset of convulsions, or seizures, resulting from severe hypertension in the pregnant woman.

A. Eclampsia

B. Placenta previa

C. Placenta abruptio

D. Supine hypotensive syndrome

_____32. _____ is a condition of infants who are born to alcoholic mothers; it is characterized by physical and mental retardation and a variety of congenital abnormalities.

A. Pregnancy-induced hypertension

B. Ectopic pregnancy

C. Fetal alcohol syndrome

D. Supine hypotensive syndrome

Tables

Complete the following table for the APGAR scoring system by listing the characteristics for each score.

Area of Activity	Score		
	2	1	0
Appearance			
Pulse			
Grimace or Irritability			
Activity or Muscle Tone			
Respiration			

4

Labeling

Label the following diagram with the correct terms.

1. Anatomic structures of the pregnant woman

A. _____

B. _____

C. _____

D. _____

E. _____

F. _____

G. _____

H. _____

I. _____

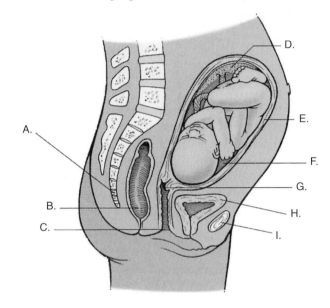

■ NOTES ■

Vocabulary

Define the following terms using the space provided.

1. Primigravida:

_____

_____

_____

2. Multigravida:

_____

_____

_____

3. Ectopic pregnancy:

_____

_____

_____

4. Crowning:

_____

_____

_____

5. APGAR score:

_____

_____

_____

6. Bloody show:

_____

_____

_____

Fill-in

Read each item carefully, then complete the statement by filling in the missing word(s).

1. After delivery, the _____, or afterbirth, separates from the uterus and is delivered.

2. The umbilical cord contains two _____ and one _____.

3. The amniotic sac contains about _____ of amniotic fluid, which helps to insulate and protect the floating fetus as it develops.

4. A full-term pregnancy is from _____ to _____ weeks, counting from the first day of the last menstrual cycle.

5. The pregnancy is divided into three _____ of about 3 months each.

6. There is a high potential of exposure due to _____ released during childbirth.

7. The leading cause of maternal death in the first trimester is internal hemorrhage into the abdomen following rupture of an _____.

8. In serious trauma, the only chance to save the infant is to adequately _____ the mother.

9. During the delivery, be careful that you do not poke your fingers into the infant's eyes or into the two soft spots, called _____, on the head.

True/False

If you believe the statement to be more true than false, write the letter "T" in the space provided. If you believe the statement to be more false than true, write the letter "F."

1. _____ The small mucous plug from the cervix that is discharged from the vagina, often at the beginning of labor, is called a bloody show.

2. _____ Crowning occurs when the baby's head obstructs the birth canal, preventing normal delivery.

3. _____ Labor begins with the rupture of the amniotic sac and ends with the delivery of the baby's head.

4. _____ A woman who is having her first baby is called a multigravida.

5. _____ Once labor has begun, it can be slowed by holding the patient's legs together.

6. _____ Delivery of the buttocks before the baby's head is called a breech delivery.

7. _____ After delivery, the baby should be kept at the same level as the mother's vagina until after the cord is cut.

8. _____ The placenta and cord should be properly disposed of in a biohazard container after delivery.

9. _____ The umbilical cord may be gently pulled to aid in delivery of the placenta.

10. _____ A limb presentation occurs when the baby's arm, leg, or foot is emerging from the vagina first.

11. _____ Multiple births may have more than one placenta.

Short Answer

Complete this section with short written answers using the space provided.

1. What are some possible causes of vaginal hemorrhage in early and late pregnancy?

_____

_____

_____

2. In what position should pregnant patients who are not delivering be transported and why?

_____

_____

_____

3. List three signs that indicate the beginning of labor.

_____

_____

_____

4. Under what three circumstances should you consider delivering the patient at the scene?

_____

_____

_____

5. Once the baby's head emerges, what actions should be taken to prevent too rapid a delivery?

_____

_____

_____

6. Why is it important to avoid pushing on the fontanels?

_____

_____

_____

7. How can you help decrease perineal tearing?

_____

_____

_____

8. What are the two situations in which an EMT-B may insert his or her fingers into a patient's vagina?

_____

_____

_____

9. What are three fetal effects of maternal drug or alcohol addiction?

_____

_____

_____

4

Word Fun

The following crossword puzzle is an activity provided to reinforce correct spelling and understanding of medical terminology associated with emergency care and the EMT-B. Use the clues in the column to complete the puzzle.

■ CLUES ■

Across

3. Fluid-filled bag for developing fetus

6. Umbilicus wrapped around baby's neck

8. Develops over and covers the cervix

9. Rating of newborn on 5 factors

Down

1. Buttocks first

2. Previously given birth

4. Head showing during labor

5. Convulsions from hypertension in pregnant woman

7. Area of skin between anus and vagina, subject to tearing during birth process

■ NOTES ■

Ambulance Calls

The following real case scenarios provide an opportunity to explore the concerns associated with patient management. Read each scenario, then answer each question in detail.

1. You are called to a 27-year-old female, primigravida, in her 28th week of gestation. Her husband tells you she has a history of pre-eclampsia, and suddenly started having convulsions. She is voice responsive and very lethargic. Her blood pressure is 230/180 mm Hg.

How would you best manage this patient?

_____

_____

_____

_____

_____

_____

_____

_____

_____

_____

2. You are dispatched to the residence of a 24-year-old female who is 12 weeks pregnant and complaining of "spotting." The patient tells you that her last pregnancy ended in a miscarriage at 8 weeks. This is her second pregnancy. She denies any pain and vital signs are within normal limits.

How would you best manage this patient?

_____

_____

_____

_____

_____

_____

_____

_____

3. You are on the scene with a 32-year-old female who is 38 weeks pregnant and delivery is imminent. As the infant starts to crown, you notice that the amniotic sac is still intact.

How would you best manage this patient?

_____

_____

_____

_____

_____

_____

_____

Skill Drills emt-b video clips

Skill Drill 20-1: Delivering the Baby

Test your knowledge of this skill drill by placing the photos below in the correct order. Number the first step with a "1," the second step with a "2," etc. Also, fill in the correct words in the photo captions.

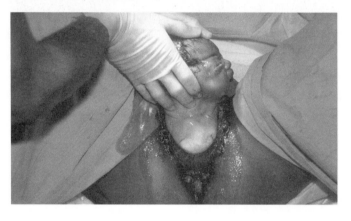

_____

Support the head and upper body as the _____ _____ delivers, guiding the head _____ if needed.

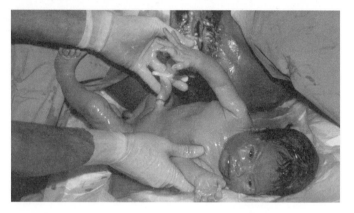

_____

Place clamps _____ inch(es) to _____ inch(es) apart and _____ between them.

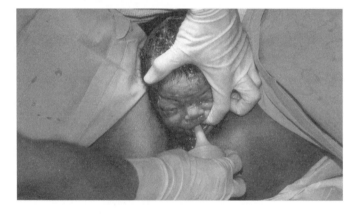

_____

Support the _____ parts of the head with your hands as it emerges.
Suction fluid from the _____, then _____ .

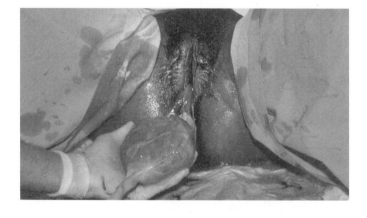

_____

Allow the _____ to deliver itself. Do not pull on the _____ to speed delivery.

(continued)

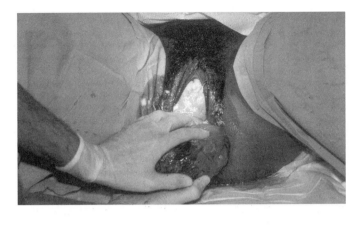

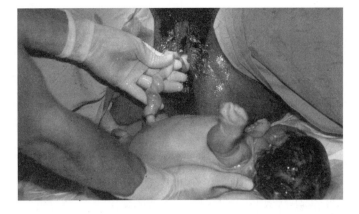

As the _____ _____ appears, guide the head _____ slightly, if needed to deliver the _____.

Handle the slippery delivered infant firmly but gently, keeping the neck in _____ position to _____ the airway.

Skill Drill 20-2: Giving Chest Compressions to an Infant
Test your knowledge of this skill drill by filling in the correct words in the photo captions.

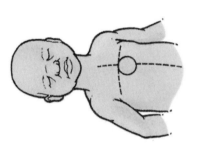

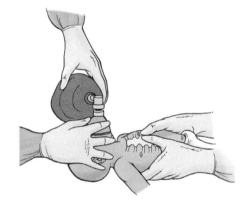

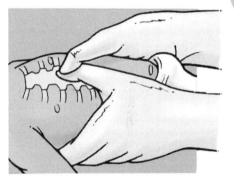

1. Find the proper position: just below the _____ _____, middle of _____ _____ of the sternum.

2. Wrap your hands around the body, with your _____ resting at that position.

3. Press your thumb gently against the sternum, compressing _____ inch(es) to _____ inch(es) deep.

Workbook Activities

The following activities have been designed to help you. Your instructor may require you to complete some or all of these activities as a regular part of your EMT-B training program. You are encouraged to complete any activity that your instructor does not assign as a way to enhance your learning in the classroom.

Chapter Review

The following exercises provide an opportunity to refresh your knowledge of this chapter.

■ NOTES ■

Matching

Match each of the terms in the left column to the appropriate definition in the right column.

_____ **1.** Cavitation

_____ **2.** Deceleration

_____ **3.** Kinetic energy

_____ **4.** Mechanism of injury

_____ **5.** Potential energy

_____ **6.** Blunt trauma

_____ **7.** Penetrating trauma

_____ **8.** Work

A. impact on the body by objects that cause injury without penetrating soft tissue or internal organs and cavities

B. force acting over a distance

C. product of mass, gravity, and height

D. injury caused by objects that pierce the surface of the body

E. how trauma occurs

F. energy of moving object

G. slowing

H. emanation of pressure waves that can damage nearby structures

Multiple Choice

Read each item carefully, then select the best response.

_____ **1.** The following are concepts of energy typically associated with injury, except:
   A. potential energy.
   B. thermal energy.
   C. kinetic energy.
   D. work.

_____ **2.** The energy of a moving object is called:
   A. potential energy.
   B. thermal energy.
   C. kinetic energy.
   D. work.

Kinematics of Trauma

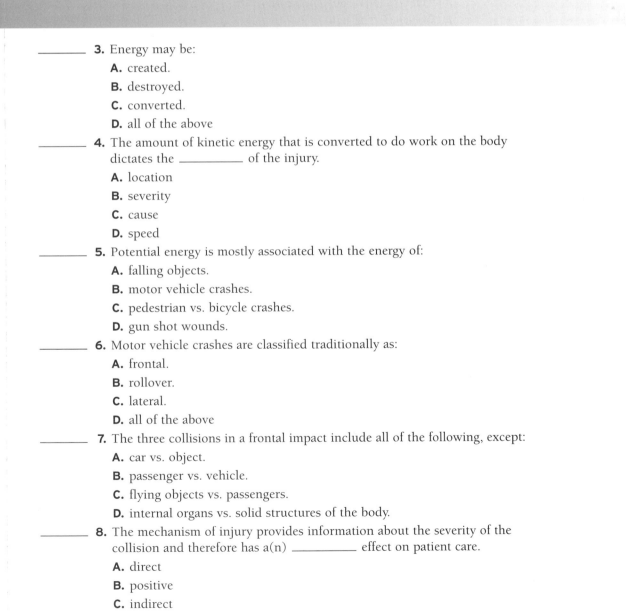

_____ **3.** Energy may be:
   A. created.
   B. destroyed.
   C. converted.
   D. all of the above

_____ **4.** The amount of kinetic energy that is converted to do work on the body
   dictates the _____ of the injury.
   A. location
   B. severity
   C. cause
   D. speed

_____ **5.** Potential energy is mostly associated with the energy of:
   A. falling objects.
   B. motor vehicle crashes.
   C. pedestrian vs. bicycle crashes.
   D. gun shot wounds.

_____ **6.** Motor vehicle crashes are classified traditionally as:
   A. frontal.
   B. rollover.
   C. lateral.
   D. all of the above

_____ **7.** The three collisions in a frontal impact include all of the following, except:
   A. car vs. object.
   B. passenger vs. vehicle.
   C. flying objects vs. passengers.
   D. internal organs vs. solid structures of the body.

_____ **8.** The mechanism of injury provides information about the severity of the
   collision and therefore has a(n) _____ effect on patient care.
   A. direct
   B. positive
   C. indirect
   D. negative

_____ **9.** Your index of suspicion for the presence of life-threatening injuries should automatically increase if you see:

A. seats torn from their mountings.

B. collapsed steering wheels.

C. intrusion into the passenger compartment.

D. all of the above

_____ **10.** In a motor vehicle crash, as the passenger's head hits the windshield, the brain continues to move forward until it strikes the inside of the skull resulting in a _____ injury.

A. compression

B. laceration

C. lateral

D. motion

_____ **11.** Your quick initial assessment of the patient and the evaluation of the _____ can help to direct lifesaving care and provide critical information to the hospital staff.

A. scene

B. index of suspicion

C. mechanism of injury

D. abdominal area

_____ **12.** A contusion to a patient's forehead along with a spiderwebbed windshield suggests possible injury to the:

A. nose.

B. brain.

C. face.

D. heart.

_____ **13.** Significant mechanisms of injury include:

A. moderate intrusions from a lateral impact.

B. severe damage from the rear.

C. collisions in which rotation is involved.

D. all of the above

_____ **14.** Significant clues to the possibility of severe injuries include:

A. death of a passenger.

B. a blown out tire.

C. broken glass.

D. a deployed airbag.

_____ **15.** When properly applied, seat belts are successful in:

A. restraining the passengers in a vehicle.

B. preventing a second collision inside the motor vehicle.

C. decreasing the severity of the third collision.

D. all of the above

_____ **16.** Air bags decrease injury to all of the following, except:

A. chest.

B. heart.

C. face.

D. head.

_____ 17. Signs of most injuries sustained in a motor vehicle crash can be found by simply inspecting the _____ during extrication of the patient.
   A. head and neck
   B. chest
   C. interior of the vehicle
   D. torso

_____ 18. Passengers in the back seat wearing only a lap belt might have a higher incidence of injuries to the thoracic and lumbar spine in a _____ impact.
   A. frontal
   B. lateral
   C. rear-end
   D. rollover

_____ 19. _____ impacts are probably the number one cause of death associated with motor vehicle crashes.
   A. Frontal
   B. Lateral
   C. Rear-end
   D. Rollover

_____ 20. The most common life-threatening event in a rollover is _____ or partial ejection of the passenger from the vehicle.
   A. "sandwiching"
   B. centrifugal force
   C. ejection
   D. spinal cord injury

_____ 21. A fall from more than _____ times the patient's height is considered to be significant.
   A. two
   B. three
   C. four
   D. five

_____ 22. Factors that should be taken into account when evaluating the patient of a fall include:
   A. the height of the fall.
   B. the surface struck.
   C. the part of the body that hit first.
   D. all of the above

_____ 23. Low-energy penetrating trauma may be caused accidentally by impalement or intentionally by a:
   A. knife.
   B. pair of scissors.
   C. ice pick.
   D. all of the above

_____ 24. The area that is damaged by medium- and high-velocity projectiles can be _____ than the diameter of the projectile itself.
   A. slightly larger
   B. many times larger
   C. slightly smaller
   D. many times smaller

5

■ NOTES ■

_____**25.** When you notice a collapsed steering wheel during scene size-up, you should suspect serious _____ injuries even if the driver shows no visible signs of injury.

A. head

B. chest

C. abdominal

D. pelvic

Vocabulary

Define the following terms using the space provided.

1. Newton's First Law:

_____

_____

_____

2. Newton's Second Law:

_____

_____

_____

3. Newton's Third Law:

_____

_____

_____

Fill-in

Read each item carefully, then complete the statement by filling in the missing word(s).

1. _____ are the leading cause of death and disability in the United States among children and young adults.

2. Certain injury _____ occur with certain types of injury _____.

3. _____ occurs to the body when the body's tissues are exposed to energy levels beyond their tolerance.

4. The formula for calculating kinetic energy is _____.

5. _____ of the crash scene may provide valuable information to the staff and treating physicians of the trauma center.

6. Air bags provide the final capture point of the passengers and decrease the

severity of _____ injuries.

7. Seat belts that buckle automatically at the shoulder but require the passengers to

buckle the lap portion can result in the body _____ forward underneath the shoulder restraint when the lap portion is not attached.

True/False

If you believe the statement to be more true than false, write the letter "T" in the space provided. If you believe the statement to be more false than true, write the letter "F."

1. _____ Work is defined as force acting over distance.
2. _____ Energy can be both created and destroyed.
3. _____ The energy of a moving object is called potential energy.
4. _____ Rear-end collisions often cause whiplash injuries.
5. _____ The cervical spine has little tolerance for lateral bending.
6. _____ The injury potential of a fall is related to the height from which the patient fell.
7. _____ Injuries are the leading cause of death and injuries among 1- to 34-year-olds in the United States.

Short Answer

Complete this section with short written answers using the space provided.

1. Describe potential energy.

_____

_____

_____

2. List the three series of collisions typical with motor vehicles.

_____

_____

_____

3. List the three factors to consider when evaluating a fall.

_____

_____

_____

4. Describe the phenomenon of cavitation as it relates to an injury from a bullet.

_____

_____

_____

5

■ N O T E S ■

5. Why is it important to try to determine the type of gun and ammunition used when you are caring for a gunshot victim?

_____

_____

_____

Word Fun

The following crossword puzzle is an activity provided to reinforce correct spelling and understanding of medical terminology associated with emergency care and the EMT-B. Use the clues in the column to complete the puzzle.

■ C L U E S ■

Across

3. Energy from a moving object

6. Impact on the body without penetrating soft tissues

7. Slowing down

8. Force acting over a distance

Down

1. Cause or reason for injury

2. Product of mass + gravity + height

4. The result of body tissues being exposed to energy levels beyond their tolerance

5. Pressure waves from speed

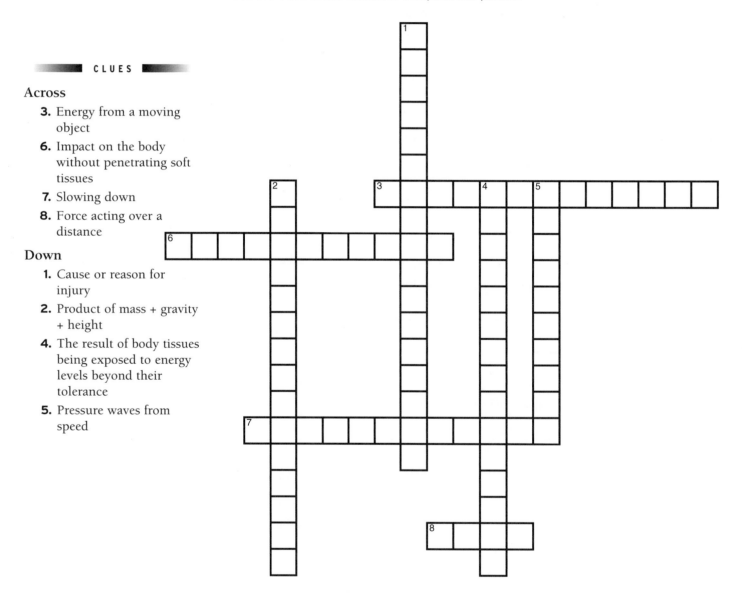

Ambulance Calls

The following real case scenarios provide an opportunity to explore the concerns associated with patient management. Read each scenario, then answer each question in detail.

1. You are called to the scene of a rollover motor vehicle crash. Your patient is a 25-year-old male, restrained driver, complaining of chest pain. He is alert and oriented and his vital signs are within normal limits.

How would you best manage this patient?

_____

_____

_____

_____

_____

_____

_____

_____

_____

2. You are dispatched to the scene of a fall from the roof of a two-story house. Your patient, a 32-year-old male, was found lying on the ground with obvious deformity to his right lower leg. He is alert and oriented, and tells you he slipped on a hammer. Bleeding is controlled and his airway is intact.

How would you best manage this patient?

_____

_____

_____

_____

_____

_____

_____

_____

_____

5

3. You are called to the residence of a 19-year-old male who was stabbed in the abdomen with an ice pick. The scene is safe and the patient is lying on the floor with the ice pick impaled in his left lower quadrant. Bystanders tell you he did not fall. He is alert and complaining of severe pain.

How would you best manage this patient?

_____

_____

_____

_____

_____

_____

_____

_____

_____

Notes

5

Workbook Activities

The following activities have been designed to help you. Your instructor may require you to complete some or all of these activities as a regular part of your EMT-B training program. You are encouraged to complete any activity that your instructor does not assign as a way to enhance your learning in the classroom.

Chapter Review

The following exercises provide an opportunity to refresh your knowledge of this chapter.

Matching

Match each of the terms in the left column to the appropriate definition in the right column.

_____ 1. Pulmonary artery

_____ 2. Heart

_____ 3. Ventricle

_____ 4. Aorta

_____ 5. Atrium

_____ 6. Pulmonary vein

_____ 7. Coagulation

_____ 8. Ecchymosis

_____ 9. Epistaxis

_____ 10. Hematoma

_____ 11. Hemophilia

_____ 12. Hemorrhage

_____ 13. Hypovolemic shock

A. mass of blood in the soft tissues beneath the skin

B. formation of clot to plug opening in injured blood vessel and stop blood flow

C. upper chamber

D. a congenital condition in which a patient lacks one or more of the blood's normal clotting factors

E. hollow muscular organ

F. largest artery in body

G. a condition in which low blood volume results in inadequate perfusion

H. oxygenated blood travels through this

I. bruising

J. deoxygenated blood travels through this

K. lower chamber

L. bleeding

M. nosebleed

Multiple Choice

Read each item carefully, then select the best response.

_____ 1. The function of the blood is to _____ all of the body's cells and tissues.

A. deliver oxygen to

B. deliver nutrients to

C. carry waste products away from

D. all of the above

C H A P T E R 22

Bleeding

_____ **2.** The cardiovascular system consists of:

    A. a pump.

    B. a container.

    C. fluid.

    D. all of the above

_____ **3.** Blood leaves each chamber of a normal heart through a:

    A. vein.

    B. artery.

    C. one-way valve.

    D. capillary.

_____ **4.** Blood enters into the right atrium from the:

    A. coronary arteries.

    B. lungs.

    C. vena cava.

    D. coronary veins.

_____ **5.** Blood enters into the left atrium from the:

    A. coronary arteries.

    B. lungs.

    C. vena cava.

    D. coronary veins.

_____ **6.** The only arteries in the body to carry deoxygenated blood are the:

    A. pulmonary arteries.

    B. coronary arteries.

    C. femoral arteries.

    D. subclavian arteries.

_____ **7.** The _____ is the thickest chamber of the heart.

    A. right atrium

    B. right ventricle

    C. left atrium

    D. left ventricle

_____ **8.** The _____ link(s) the arterioles and the venules.

    A. aorta

    B. capillaries

    C. vena cava

    D. valves

_____ **9.** At the arterial end of the capillaries, the muscles dilate and constrict in response to conditions such as:

    A. fright.

    B. a specific need for oxygen.

    C. a need to dispose of metabolic wastes.

    D. all of the above

_____ **10.** Blood contains all of the following, except:

    A. white cells.

    B. plasma.

    C. cerebrospinal fluid.

    D. platelets.

_____ **11.** _____ is the circulation of blood within an organ or tissue in adequate amounts to meet the cells' current needs for oxygen, nutrients, and waste removal.

    A. Anatomy

    B. Perfusion

    C. Physiology

    D. Conduction

_____ **12.** The _____ only require a minimal blood supply when at rest.

    A. lungs

    B. kidneys

    C. muscles

    D. heart

_____ **13.** The term _____ means constantly adapting to changing conditions.

    A. perfusion

    B. conduction

    C. dynamic

    D. autonomic

_____ **14.** _____ is inadequate tissue perfusion.

    A. Shock

    B. Hyperperfusion

    C. Hypertension

    D. Contraction

_____ **15.** The brain and spinal cord cannot go for more than _____ minutes without perfusion, or the nerve cells will be permanently damaged.

    A. 30 to 45

    B. 12 to 20

    C. 8 to 10

    D. 4 to 6

_____ **16.** An organ or tissue that is considerably _____ is much better able to resist damage from hypoperfusion.

   A. warmer

   B. colder

   C. younger

   D. older

_____ **17.** The body will not tolerate an acute blood loss of greater than _____ of blood volume.

   A. 10%

   B. 20%

   C. 30%

   D. 40%

_____ **18.** If the typical adult loses more than 1 L of blood, significant changes in vital signs such as _____ will occur.

   A. increased heart rate

   B. increased respiratory rate

   C. decreased blood pressure

   D. all of the above

_____ **19.** _____ is a condition in which low blood volume results in inadequate perfusion and even death.

   A. Hypovolemic shock

   B. Metabolic shock

   C. Septic shock

   D. Psychogenic shock

_____ **20.** You should consider bleeding to be serious if all of the following conditions are present, except:

   A. blood loss is rapid.

   B. there is no mechanism of injury.

   C. the patient has a poor general appearance.

   D. assessment reveals signs and symptoms of shock.

_____ **21.** Significant blood loss demands your immediate attention as soon as the _____ has been managed.

   A. fractures

   B. extrication

   C. airway

   D. none of the above

_____ **22.** The process of blood clotting and plugging the hole is called:

   A. conglomeration.

   B. configuration.

   C. coagulation.

   D. coalition.

_____ **23.** Even though the body is very efficient at controlling bleeding on its own, it may fail in situations such as:

   A. when medications interfere with normal clotting.

   B. when damage to the vessel may be so large that a clot cannot completely block the hole.

   C. when sometimes only part of the vessel wall is cut, preventing it from constricting.

   D. all of the above

5

24. A lack of one or more of the blood's clotting factors is called:

A. a deficiency.

B. hemophilia.

C. platelet anomaly.

D. anemia.

25. The first step in controlling bleeding is:

A. direct pressure.

B. maintaining the airway.

C. BSI precautions.

D. elevation.

26. When applying a bandage to hold a dressing in place, stretch the bandage tight enough to control bleeding, but not so tight as to decrease _____ to the extremity.

A. blood flow

B. pulses

C. oxygen

D. CRTs

27. If bleeding continues after applying a pressure dressing, you should do all of the following, except:

A. remove the dressing and apply another sterile dressing.

B. apply manual pressure through the dressing.

C. add more gauze pads over the first dressing.

D. secure both dressings tighter with a roller bandage.

28. When using an air splint to control bleeding in a fractured extremity, you should reassess the _____ frequently.

A. airway

B. breathing

C. circulation in the injured extremity

D. fracture site

29. Contraindications to the use of the PASG include:

A. pulmonary edema.

B. pregnancy.

C. penetrating chest injuries.

D. all of the above

30. A tourniquet is rarely needed to control bleeding and often _____ problems.

A. resolves

B. decreases

C. creates

D. all of the above

31. When applying a tourniquet, make sure you:

A. use the narrowest bandage possible to minimize the area restricted.

B. cover the tourniquet with a bandage.

C. never pad underneath the tourniquet.

D. do not loosen the tourniquet after you have applied it.

_____32. Bleeding from the nose, ears, and/or mouth may be the result of:
  A. a skull fracture.
  B. sinusitis.
  C. coagulation disorders.
  D. all of the above

_____33. You should not attempt to stop the blood flow from the nose or ears following a head injury because:
  A. it should be collected to be reinfused at the hospital.
  B. it could collect within the head and increase the pressure in the brain.
  C. it is contaminated.
  D. you could fracture the skull with the pressure needed to staunch the flow of blood.

_____34. When treating a patient with signs and symptoms of hypovolemic shock and no outward signs of bleeding, always consider the possibility of bleeding into the:
  A. thoracic cavity.
  B. abdomen.
  C. skull.
  D. chest.

_____35. Nontraumatic internal bleeding may be caused by:
  A. an ulcer.
  B. a ruptured ectopic pregnancy.
  C. an aneurysm.
  D. all of the above

_____36. The most common symptom of internal abdominal bleeding is:
  A. bruising around the abdomen.
  B. distention of the abdomen.
  C. rigidity of the abdomen.
  D. acute abdominal pain.

_____37. Signs and symptoms of internal bleeding in both trauma and medical patients include:
  A. hematochezia.
  B. melena.
  C. hemoptysis.
  D. all of the above

_____38. The first sign of hypovolemic shock is a change in:
  A. respirations.
  B. heart rate.
  C. mental status.
  D. blood pressure.

5

Labeling

Label the following diagrams with the correct terms. Where arrows appear, indicate the substance and its origin and destination.

1. The Left and Right Sides of the Heart

A. _____

B. _____

C. _____

D. _____

E. _____

F. _____

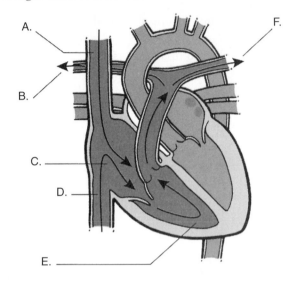

A. _____

B. _____

C. _____

D. _____

E. _____

F. _____

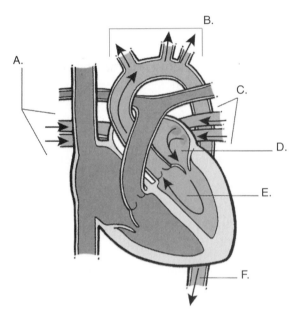

2. Perfusion

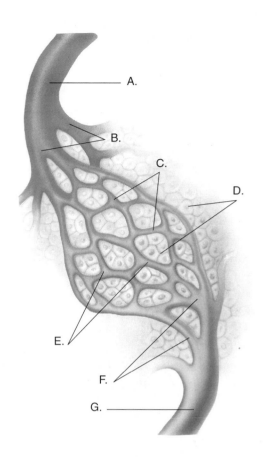

A. _____

B. _____

C. _____

D. _____

E. _____

F. _____

G. _____

5

3. Arterial Pressure Points

A. _____

B. _____

C. _____

D. _____

E. _____

F. _____

G. _____

H. _____

I. _____

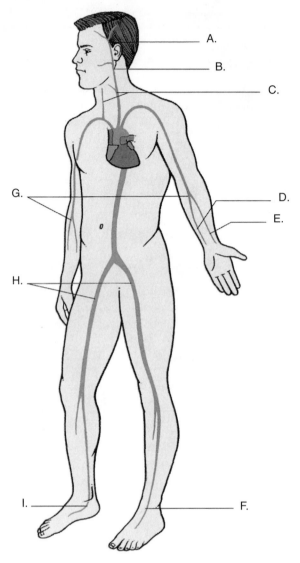

Vocabulary

Define the following terms using the space provided.

1. Perfusion:

_____

_____

_____

2. Shock:

_____

_____

_____

3. Melena:

_____

_____

_____

4. Hematemesis:

_____

_____

_____

5. Hemoptysis:

_____

_____

_____

Fill-in

Read each item carefully, then complete the statement by filling in the missing word.

1. The circulation of blood in adequate amounts to meet cellular needs is called

_____ .

2. The heart is a(n) _____ muscle, which is controlled by the autonomic nervous system.

3. Exchange of oxygen and carbon dioxide occurs in the _____ .

4. Shock is a condition related to _____ .

5. Blood returning to the heart from the lower body is first collected in the

_____ vena cava.

6. Blood contains four major components, which are _____ .

7. The left ventricle receives _____ blood while the right ventricle receives deoxygenated blood.

8. During a state of shock, the autonomic nervous system directs blood away from

some organs and distributes it to the _____ .

9. The brain and spinal cord generally cannot go longer than _____ minutes without adequate perfusion, or permanent nerve cell damage may occur.

5

True/False

If you believe the statement to be more true than false, write the letter "T" in the space provided. If you believe the statement to be more false than true, write the letter "F."

1. _____ Venous blood tends to spurt and is difficult to control.
2. _____ The human body is tolerant of blood losses greater than 20% of blood volume.
3. _____ The first step in controlling external bleeding is application of the PASG.
4. _____ The first step in preparing to treat a bleeding patient is BSI.
5. _____ A properly applied tourniquet should be loosened by the EMT-B every ten minutes.
6. _____ A patient who has swallowed a lot of blood may become nauseated and vomit.
7. _____ You should check with medical control every time a PASG may be indicated.
8. _____ If a wound continues to bleed after it is bandaged, you should remove the bandage and start over again.

Short Answer

Complete this section with short written answers in the space provided.

1. Describe how the autonomic nervous system responds to severe bleeding.

_____

_____

_____

2. Describe the characteristics of bleeding from each type of vessel (artery, vein, capillary).

_____

_____

_____

3. List, in the proper sequence, the methods in which an EMT-B should attempt to control external bleeding.

_____

_____

_____

4. List ten signs and symptoms of hypovolemic shock.

_____

_____

_____

_____

_____

_____

_____

_____

_____

_____

5. List, in the proper sequence, the general EMT-B emergency care for patients with internal bleeding.

_____

_____

_____

Word Fun

The following crossword puzzle is an activity provided to reinforce correct spelling and understanding of medical terminology associated with emergency care and the EMT-B. Use the clues in the column to complete the puzzle.

Across

3. Low blood volume, inadequate perfusion

7. Circulatory system failure

8. Circulation of blood within the body

9. Mass of blood in soft tissues

Down

1. Nosebleed

2. Lacks normal clotting factors

4. Discoloration of skin from closed wound

5. Bleeding

6. Formation of clots to stop bleeding

5

Ambulance Calls

The following real case scenarios provide an opportunity to explore the concerns associated with patient management. Read each scenario, then answer each question in detail.

1. You are dispatched to a school playground for an 8-year-old male complaining of a minor laceration to his left wrist with uncontrollable hemorrhaging. The teacher tells you that he has a history of hemophilia. The blood is steady, but not spurting, and dark in color.

   How would you best manage this patient?

   _____

   _____

   _____

   _____

   _____

   _____

   _____

   _____

   _____

2. You are called to the scene of a motor vehicle crash with major damage to the front of the vehicle where your patient was the restrained driver. She is a 23-year-old female with no pertinent medical history according to a passenger in the vehicle. She is responsive to pain and showing signs of hypovolemic shock. Her only visible injury is a bruise to her right upper quadrant.

   How would you best manage this patient?

   _____

   _____

   _____

   _____

   _____

   _____

   _____

   _____

   _____

3. You are dispatched to aid a 22-year-old male with a history of depression. Upon arrival, police on scene tell you he tried to commit suicide by cutting his wrist. He has a laceration to the right radial artery with bright red spurting blood. He has lost what looks to be possibly 2 pints of blood.

How would you best manage this patient?

_____

_____

_____

_____

_____

_____

_____

_____

_____

_____

Skill Drills emt-b.com video clips

Skill Drill 22-1: Controlling External Bleeding
Test your knowledge of this skill drill by filling in the correct words in the photo captions.

5

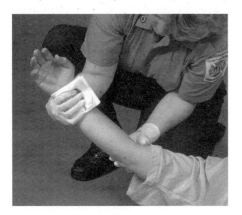
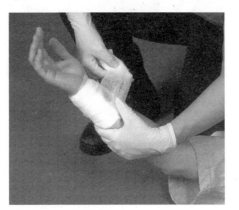
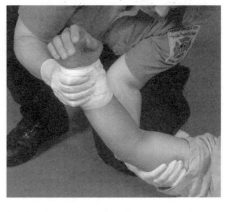

1. Apply _____ _____ over the wound. Elevate the injury above the _____ of the _____ if no _____ is suspected.

2. Apply a _____ _____.

3. Apply pressure at the appropriate _____ _____ while continuing to hold _____ _____ and _____.

Skill Drill 22-2: Applying a Pneumatic Antishock Garment (PASG)

Test your knowledge of this skill drill by placing the photos below in the correct order. Number the first step with a "1," the second step with a "2," etc.

_____

Inflate with the foot pump, and close the stopcocks either when the patient's systolic blood pressure reaches 100 mm Hg or the Velcro crackles.

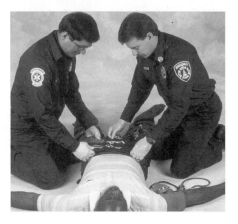

_____

Enclose both legs and the abdomen.

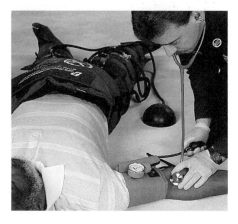

_____

Check the patient's blood pressure again. Monitor vital signs.

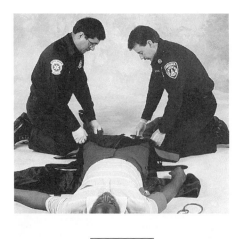

_____

Apply the garment so that the top is below the lowest rib.

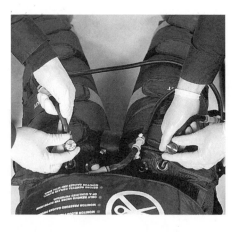

_____

Open the stopcocks.

Notes

5

Workbook Activities

The following activities have been designed to help you. Your instructor may require you to complete some or all of these activities as a regular part of your EMT-B training program. You are encouraged to complete any activity that your instructor does not assign as a way to enhance your learning in the classroom.

Chapter Review

The following exercises provide an opportunity to refresh your knowledge of this chapter.

■ NOTES ■

Matching

Match each of the terms in the left column to the appropriate definition in the right column.

_____ **1.** Shock

_____ **2.** Perfusion

_____ **3.** Sphincters

_____ **4.** Autonomic nervous system

_____ **5.** Blood pressure

_____ **6.** Anaphylaxis

_____ **7.** Septic shock

_____ **8.** Syncope

_____ **9.** Compensated shock

A. severe allergy

B. hypoperfusion

C. regulates involuntary body functions

D. early stage of shock

E. provides a rough measure of perfusion

F. severe bacterial infection

G. sufficient circulation to meet cell needs

H. regulate blood flow in capillaries

I. fainting

Multiple Choice

Read each item carefully, then select the best response.

_____ **1.** Shock:

   A. refers to a state of collapse and failure of the cardiovascular system.

   B. results in the inadequate flow of blood to the body's cells.

   C. results in failure to rid cells of metabolic wastes.

   D. all of the above

_____ **2.** Blood flow through the capillary beds is regulated by:

   A. systolic pressure.

   B. the capillary sphincters.

   C. perfusion.

   D. diastolic pressure.

Shock

_____ **3.** The autonomic nervous system regulates involuntary functions such as:

   A. sweating.

   B. digestion.

   C. constriction and dilation of capillary sphincters.

   D. all of the above

_____ **4.** Regulation of blood flow is determined by:

   A. oxygen intake.

   B. systolic pressure.

   C. cellular need.

   D. diastolic pressure.

_____ **5.** Perfusion requires having a working cardiovascular system as well as:

   A. adequate oxygen exchange in the lungs.

   B. adequate nutrients in the form of glucose in the blood.

   C. adequate waste removal.

   D. all of the above

_____ **6.** The action of the hormones stimulates _____ to maintain pressure in the system and, as a result, perfusion of all vital organs.

   A. an increase in heart rate

   B. an increase in the strength of cardiac contractions

   C. vasoconstriction in nonessential areas

   D. all of the above

_____ **7.** Basic causes of shock include:

   A. poor pump function.

   B. blood or fluid loss.

   C. blood vessel dilation.

   D. all of the above

_____ **8.** Noncardiovascular causes of shock include respiratory insufficiency and:

   A. sepsis.

   B. metabolic.

   C. anaphylaxis.

   D. hypovolemia.

_____ **9.** When the heart no longer functions well, a major effect is the backup of blood into the lungs called pulmonary:

    A. edema.

    B. overload.

    C. cessation.

    D. failure.

_____ **10.** _____ develops when the heart muscle can no longer generate enough pressure to circulate the blood to all organs.

    A. Pump failure

    B. Cardiogenic shock

    C. A myocardial infarction

    D. Congestive heart failure

_____ **11.** Damage to the _____ may cause significant injury to the part of the nervous system that controls the size and muscular tone of the blood vessels.

    A. cervical vertebrae

    B. skull

    C. spinal cord

    D. peripheral nerves

_____ **12.** Neurogenic shock usually results from damage to the spinal cord at the:

    A. cervical level.

    B. thoracic level.

    C. lumbar level.

    D. sacral level.

_____ **13.** Septic shock results from _____, in combination with the loss of plasma through the injured vessel walls.

    A. pump failure

    B. massive vasoconstriction

    C. widespread dilation

    D. increased volume

_____ **14.** In septic shock:

    A. there is an insufficient volume of fluid in the container.

    B. the fluid that has leaked out often collects in the respiratory system.

    C. there is a larger-than-normal vascular bed to contain the smaller-than-normal volume of intravascular fluid.

    D. all of the above

_____ **15.** Neurogenic shock is caused by:

    A. a radical change in the size of the vascular system.

    B. massive vasoconstriction.

    C. low volume.

    D. fluid collecting around the spinal cord causing compression of the cord.

_____ **16.** Hypovolemic shock is a result of:

    A. widespread vasodilation.

    B. low volume.

    C. massive vasoconstriction.

    D. pump failure.

_____ **17.** An insufficient concentration of _____ in the blood can produce shock as rapidly as vascular causes.

A. oxygen

B. hormones

C. epinephrine

D. histamine

_____ **18.** In anaphylactic shock, the combination of poor oxygenation and poor perfusion is a result of:

A. widespread vasodilation.

B. low volume.

C. massive vasoconstriction.

D. pump failure.

_____ **19.** Causes of syncope include:

A. generalized vascular dilation.

B. the sight of blood.

C. cardiac arrhythmias.

D. all of the above

_____ **20.** The last measurable factor to change is normally:

A. mental status.

B. blood pressure.

C. pulse rate.

D. respirations.

_____ **21.** You should suspect shock in all of the following except:

A. an allergic reaction.

B. multiple severe fractures.

C. a severe infection.

D. abdominal or chest injury.

_____ **22.** When treating a suspected shock patient, vital signs should be recorded approximately every _____ minutes.

A. 2

B. 5

C. 10

D. 15

_____ **23.** The Golden Hour refers to the first 60 minutes after:

A. medical help arrives on scene.

B. transport begins.

C. the injury occurs.

D. 9-1-1 is called.

_____ **24.** Signs of cardiogenic shock include all of the following except:

A. cyanosis.

B. strong, bounding pulse.

C. nausea.

D. anxiety.

_____ **25.** Primary treatment of shock includes:

A. securing and maintaining an airway.

B. providing respiratory support.

C. assisting ventilations.

D. all of the above

5

_____26. When assessing the patient who has psychogenic shock, be sure to consider possible _____ if the patient fell.

A. extremity fractures

B. cervical spine injury

C. paralysis

D. pelvic fractures

Vocabulary

Define the following terms using the space provided.

1. Edema:

_____

_____

_____

2. Hypothermia:

_____

_____

_____

3. Shock:

_____

_____

_____

4. Autonomic nervous system:

_____

_____

_____

5. Cyanosis:

_____

_____

_____

6. Dehydration:

_____

_____

_____

7. Sensitization:

_____

_____

_____

Fill-in

Read each item carefully, then complete the statement by filling in the missing word(s).

1. _____ refers to the failure of the cardiovascular system.

2. Pressure in the arteries during cardiac _____ is known as systolic pressure.

3. In shock conditions, the body redirects blood from _____ organs to _____ organs.

4. Blood pressure is a rough measurement of _____.

5. The cardiovascular system consists of the _____, _____, and _____.

6. Inadequate circulation that does not meet the body's needs is known as _____.

7. _____ are circular muscle walls in capillaries, causing the walls to _____ and _____.

8. _____ pressure occurs during cardiac relaxation, while _____ pressure occurs during cardiac contractions.

9. _____ pressure is the pressure in the blood vessels at all times.

10. The autonomic nervous system controls the _____ actions of the body.

True/False

If you believe the statement to be more true than false, write the letter "T" in the space provided. If you believe the statement to be more false than true, write the letter "F."

1. _____ Life-threatening allergic reactions can occur in response to almost any substance that a patient may encounter.

2. _____ Bleeding is the most common cause of shock following an injury.

3. _____ Shock occurs when oxygen and nutrients cannot get to the body's cells.

4. _____ A person in shock, left untreated, will most likely survive.

5. _____ Compensated shock is related to the last stages of shock.

6. _____ An injection of epinephrine is the only really effective treatment for anaphylactic shock.

7. _____ Septic shock is a combination of vessel and content failure.

Short Answer

Complete this section with short written answers using the space provided.

1. List the causes, signs and symptoms, and treatment of anaphylactic shock.

_____

_____

_____

2. List the causes, signs and symptoms, and treatment of cardiogenic shock.

_____

_____

_____

3. List the causes, signs and symptoms, and treatment of hypovolemic shock.

_____

_____

_____

4. List the causes, signs and symptoms, and treatment of metabolic shock.

_____

_____

_____

5. List the causes, signs and symptoms, and treatment of neurogenic shock.

_____

_____

_____

6. List the causes, signs and symptoms, and treatment of psychogenic shock.

_____

_____

_____

7. List the causes, signs and symptoms, and treatment of septic shock.

_____

_____

_____

8. List the three basic physiologic causes of shock.

_____

_____

_____

Word Fun

The following crossword puzzle is an activity provided to reinforce correct spelling and understanding of medical terminology associated with emergency care and the EMT-B. Use the clues in the column to complete the puzzle.

■ CLUES ■

Across

7. Fainting

8. Circular muscles

9. Caused by severe infection

10. Fluid in the extracellular spaces

11. Caused by inadequate heart function

Down

1. Severe allergic reaction

2. Caused by paralysis of nerves

3. Caused by loss of blood and fluids

4. Body temperature below 95°F

5. Circulation of blood in adequate amounts

6. Difficulty breathing

12. Lack of oxygen

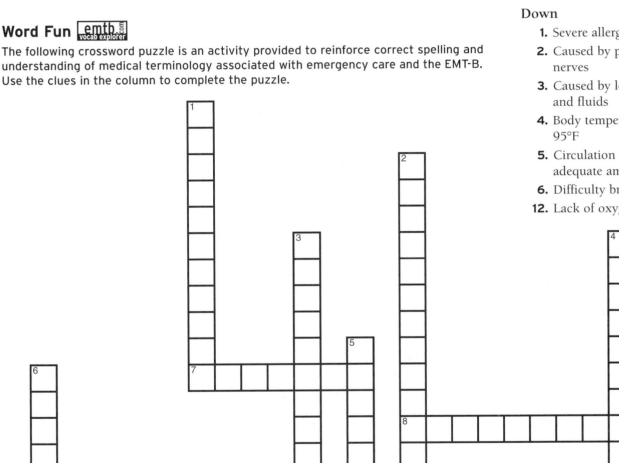

 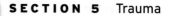
Ambulance Calls

The following real case scenarios provide an opportunity to explore the concerns associated with patient management. Read each scenario, then answer each question in detail.

1. You are dispatched to a rollover motor vehicle crash. Your patient, a 17-year-old male, was ejected and is found supine beside the vehicle. He is pain responsive, respirations are 14 and nonlabored, and he has no radial pulses. His carotid pulse is 72 and weak. His skin is warm and dry. He has no motor function or sensation in his extremities.

How would you best manage this patient?

_____

_____

_____

_____

_____

_____

_____

_____

_____

2. You are called to the scene of a structure fire where your patient is a 72-year-old female who was severely burned in the fire. Her respirations are 28 and shallow, she is unresponsive, and has no radial pulses. She is pale and diaphoretic where she is not burned.

How would you best manage this patient?

_____

_____

_____

_____

_____

_____

_____

_____

3. You are dispatched to a residence where a 16-year-old female was stung by a bee. Her mother tells you she is severely allergic to bees. She is voice responsive, covered in hives, and is wheezing audibly. She has a very weak radial pulse and is blue around the lips.

How would you best manage this patient?

_____

_____

_____

_____

_____

_____

_____

_____

_____

5

Skill Drills

Test your knowledge of this skill drill by placing the photos below in the correct order.
Number the first step with a "1," the second step with a "2," etc.

Skill Drill 23-1: Treating Shock

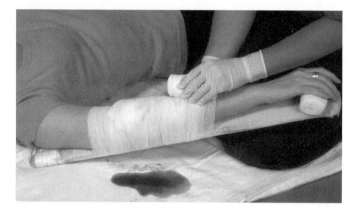

_____

Splint any broken bones or joint injuries.

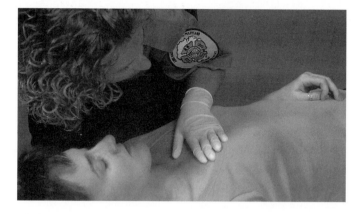

_____

Keep the patient supine, open the airway, and check breathing and pulse.

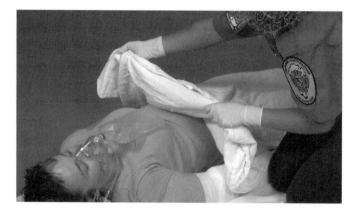

_____

Give high-flow oxygen if you have not already done so, and place blankets under and over the patient.

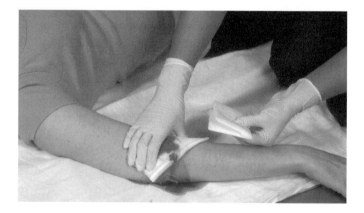

_____

Control obvious external bleeding.

(continued)

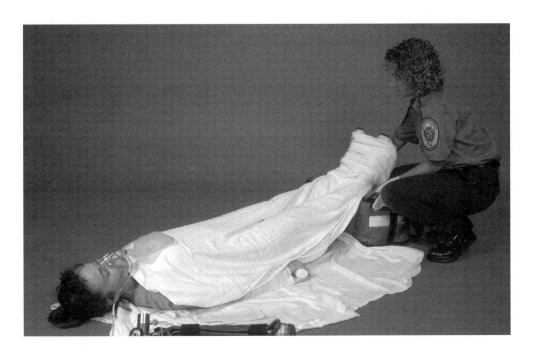

If no fractures are suspected, elevate the legs 6" to 12".

5

Workbook Activities

The following activities have been designed to help you. Your instructor may require you to complete some or all of these activities as a regular part of your EMT-B training program. You are encouraged to complete any activity that your instructor does not assign as a way to enhance your learning in the classroom.

Chapter Review

The following exercises provide an opportunity to refresh your knowledge of this chapter.

NOTES

Matching

Match each of the terms in the left column to the appropriate definition in the right column.

_____ **1.** Dermis

_____ **2.** Sweat glands

_____ **3.** Epidermis

_____ **4.** Mucous membranes

_____ **5.** Sebaceous glands

_____ **6.** Abrasion

_____ **7.** Laceration

_____ **8.** Hemothorax

_____ **9.** Penetrating wound

_____ **10.** Avulsion

_____ **11.** Evisceration

_____ **12.** Pneumothorax

A. gunshot wound

B. cool the body by discharging a substance through the pores

C. tissue hanging as a flap from wound

D. tough external layer forming a watertight covering for the body

E. razor cut

F. secrete a watery substance that lubricates the openings of the mouth and nose

G. inner layer of skin that contains the structures that give skin its characteristic appearance

H. air around lungs

I. produce oil, which waterproofs the skin and keeps it supple

J. blood collecting inside chest

K. exposed intestines

L. skinned knee

Multiple Choice

Read each item carefully, then select the best response.

_____ **1.** The _____ is (are) our first line of defense against external fractures.

    A. extremities

    B. hair

    C. skin

    D. lips

CHAPTER 24

Soft-Tissue Injuries

_____ **2.** The skin covering the _____ is quite thick.
   A. lips
   B. scalp
   C. ears
   D. eyelids

_____ **3.** As the cells on the surface of the skin are worn away, new cells form in the _____ layer.
   A. dermal
   B. germinal
   C. epidermal
   D. subcutaneous

_____ **4.** The hair follicles, sweat glands, and sebaceous glands are found in the:
   A. dermis.
   B. germinal layer.
   C. epidermis.
   D. subcutaneous layer.

_____ **5.** The skin regulates temperature in a cold environment by:
   A. secreting sweat through sweat glands.
   B. constricting the blood vessels.
   C. dilating the blood vessels.
   D. increasing the amount of heat that is radiated from the body's surface.

_____ **6.** Closed soft-tissue injuries are characterized by all of the following except:
   A. pain at the site of injury.
   B. swelling beneath the skin.
   C. damage of the protective layer of skin.
   D. a history of blunt trauma.

_____ **7.** A(n) _____ occurs whenever a large blood vessel is damaged and bleeds.
   A. contusion
   B. hematoma
   C. crushing injury
   D. avulsion

8. A(n) _____ is usually associated with extensive tissue damage.

   A. contusion

   B. hematoma

   C. crushing injury

   D. avulsion

9. A hematoma can result from:

   A. a soft-tissue injury.

   B. a fracture.

   C. any injury to a large blood vessel.

   D. all of the above

10. A(n) _____ occurs when a great amount of force is applied to the body for a long period of time.

   A. contusion

   B. hematoma

   C. crushing injury

   D. avulsion

11. More extensive closed injuries may involve significant swelling and bleeding beneath the skin, which could lead to:

   A. compartment syndrome.

   B. contamination.

   C. hypovolemic shock.

   D. hemothorax.

12. The "S" in ICES stands for:

   A. swelling.

   B. soft-tissue.

   C. splinting.

   D. shock.

13. Open soft-tissue wounds include all of the following, except:

   A. abrasions.

   B. contusions.

   C. lacerations.

   D. avulsions.

14. An abrasion is:

   A. superficial.

   B. deep.

   C. full-thickness.

   D. none of the above

15. A laceration may be:

   A. superficial.

   B. deep.

   C. jagged.

   D. all of the above

16. Bleeding from avulsions can usually be controlled by:

   A. elevation.

   B. pressure dressings.

   C. tourniquets.

   D. pressure points.

_____ **17.** The amount of damage from a gunshot wound is directly related to the:

A. size of the entrance wound.

B. size of the bullet.

C. size of the exit wound.

D. speed of the bullet.

_____ **18.** Because shootings usually end up in court, it is important to factually and completely document:

A. the circumstances surrounding any gunshot injury.

B. the patient's condition.

C. the treatment given.

D. all of the above

_____ **19.** All open wounds are assumed to be _____ and present a risk of infection.

A. contaminated

B. life-threatening

C. minimal

D. extensive

_____ **20.** Before you begin caring for a patient with an open wound, you should:

A. survey the scene.

B. follow BSI precautions.

C. be sure the patient has an open airway.

D. all of the above

_____ **21.** Splinting an extremity even when there is no fracture can help by:

A. reducing pain.

B. minimizing damage to an already-injured extremity.

C. making it easier to move the patient.

D. all of the above

_____ **22.** On inhalation, pressure inside the chest cavity _____, allowing air to enter.

A. increases

B. decreases

C. equalizes

D. stabilizes

_____ **23.** Treatment for an abdominal evisceration includes:

A. pushing the exposed organs back into the abdominal cavity.

B. covering the organs with dry dressings.

C. flexing the knees and legs to relieve pressure on the abdomen.

D. applying moist, adherent dressings.

_____ **24.** An open neck injury may result in _____ if enough air is sucked into a blood vessel.

A. hypovolemic shock

B. tracheal deviation

C. air embolism

D. subcutaneous emphysema

_____ **25.** Burns may result from:

A. heat.

B. toxic chemicals.

C. electricity.

D. all of the above

5

_____ **26.** Factors in helping to determine the severity of a burn include:

A. the depth of the burn.

B. the extent of the burn.

C. whether or not there are critical areas involved.

D. all of the above

_____ **27.** _____ burns involve only the epidermis.

A. Full-thickness

B. Second-degree

C. Superficial

D. Third-degree

_____ **28.** _____ burns cause intense pain.

A. First-degree

B. Second-degree

C. Superficial

D. Third-degree

_____ **29.** _____ burns may involve subcutaneous layers, muscle, bone, or internal organs.

A. Superficial

B. Partial-thickness

C. Full-thickness

D. Second-degree

_____ **30.** With _____ burns, the area is dry and leathery and may appear white, dark brown, or even charred.

A. first-degree

B. second-degree

C. partial-thickness

D. third-degree

_____ **31.** Significant airway burns may be associated with:

A. singeing of the hair within the nostrils.

B. hoarseness.

C. hypoxia.

D. all of the above

_____ **32.** The most important consideration when dealing with electrical burns is:

A. BSI precautions.

B. scene safety.

C. level of responsiveness.

D. airway.

_____ **33.** Treatment of electrical burns includes:

A. maintaining the airway.

B. monitoring the patient closely for respiratory or cardiac arrest.

C. splinting any suspected injuries.

D. all of the above

_____ **34.** All of the following, except _____, may be used as an occlusive dressing:

A. gauze pads

B. Vaseline gauze

C. aluminum foil

D. plastic

_____**35.** Using elastic bandages to secure dressings may result in _____ if the injury swells or if improperly applied.

A. additional tissue damage

B. loss of a limb

C. impaired circulation

D. all of the above

Labeling emtb. anatomy review

Label the following diagrams with the correct terms.

1. The Skin

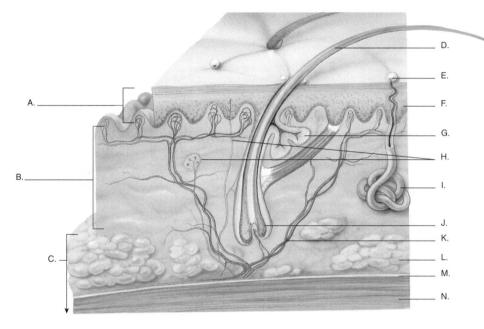

A. _____

B. _____

C. _____

D. _____

E. _____

F. _____

G. _____

H. _____

I. _____

J. _____

K. _____

L. _____

M. _____

N. _____

5

2. The Rule of Nines

A. _____

B. _____

C. _____

D. _____

E. _____

F. _____

G. _____

H. _____

A. _____

B. _____

C. _____

D. _____

E. _____

F. _____

G. _____

H. _____

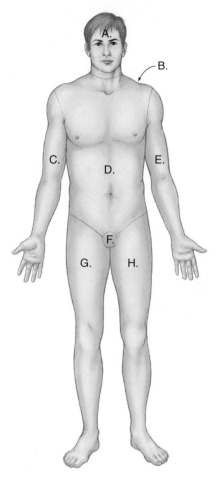

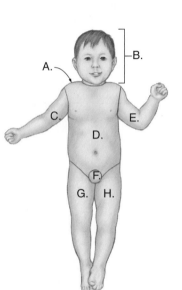

Vocabulary emtb. vocab explorer

Define the following terms using the space provided.

1. Partial-thickness burn:

_____

_____

_____

2. Closed injury:

_____

_____

_____

3. Evisceration:

_____

_____

_____

4. Compartment syndrome:

_____

_____

_____

5. Contamination:

_____

_____

_____

Fill-in

Read each item carefully, then complete the statement by filling in the missing word(s).

1. Mucous membranes are _____.

2. A person will sweat in an effort to _____ the body.

3. Nerve endings are located in the _____.

4. Below the dermis lies the _____ tissue.

5. In cold weather, skin blood vessels will _____.

6. The skin protects the body by keeping _____ out and

   _____ in.

7. Because nerve endings are present, injury to the _____ may be painful.

8. A major function of the skin is regulating body _____ .

9. The external layer of skin is the _____ and the inner layer is the

   _____ .

10. When the vessels of the skin dilate, heat is _____ from the body.

True/False

If you believe the statement to be more true than false, write the letter "T" in the space provided. If you believe the statement to be more false than true, write the letter "F."

1. _____ Partial-thickness burns involve the epidermis and some portion of the dermis.
2. _____ Blisters are commonly seen with superficial burns.
3. _____ Severe burns are usually a combination of superficial, partial-thickness, and full-thickness burns.
4. _____ The Rule of Nines allows you to estimate the percentage of body surface area that has been burned.
5. _____ Two factors, depth and extent, are critical in assessing the severity of a burn.
6. _____ Your first responsibility with a burn patient is to stop the burning process.
7. _____ Burned areas should be immersed in cool water for up to 30 minutes.
8. _____ Electrical burns are always more severe than the external signs indicate.
9. _____ The universal dressing is ideal for covering large open wounds.
10. _____ Occlusive dressings are usually made of Vaseline gauze, aluminum foil, or plastic.
11. _____ Gauze pads prevent air and liquids from entering or exiting the wound.
12. _____ Elastic bandages can be used to secure dressings.
13. _____ Soft roller bandages are slightly elastic and the layers adhere somewhat to one another.
14. _____ Ecchymosis is associated with open wounds.
15. _____ A laceration is considered a closed wound.

Short Answer

Complete this section with short written answers using the space provided.

1. List the three major classifications of depth of burns.

   _____

   _____

   _____

2. List the three general classifications of soft-tissue injuries.

_____

_____

_____

3. Define the acronym ICES.

I: _____

C: _____

E: _____

S: _____

4. Describe the classifications of a critical burn for an infant or child.

_____

_____

_____

5. What treatment should be used with a patient burned by a dry chemical?

_____

_____

_____

5

6. Why are electrical burns particularly dangerous to a patient?

_____

_____

_____

7. Describe a sucking chest wound.

_____

_____

_____

8. List the three primary functions of dressings and bandages.

_____

_____

_____

■ C L U E S ■

Across

4. Displacement of organs outside body

6. Lining of body cavities and passages with contact to outside

7. Torn completely loose or hanging as a flap

8. Presence of infective organisms

9. Inner layer of skin

10. Bruise

11. Blood collected in soft tissues

Down

1. Discoloration associated with closed injury

2. Injury from sharp, pointed object

3. Assigns percentages to burns

5. Scraping wound

9. List the four types of open soft-tissue injuries.

_____

_____

_____

_____

10. List the five factors used to determine the severity of a burn.

_____

_____

_____

_____

_____

Word Fun

The following crossword puzzle is an activity provided to reinforce correct spelling and understanding of medical terminology associated with emergency care and the EMT-B. Use the clues in the column to complete the puzzle.

Ambulance Calls

The following real case scenarios provide an opportunity to explore the concerns associated with patient management. Read each scenario, then answer each question in detail.

1. You are dispatched to a residence where a 10-year-old female fell onto a jagged piece of metal and has a gaping laceration to the right upper arm that is spurting bright red blood. The mother tried to control bleeding with a towel, but it kept soaking through.

   How would you best manage this patient?

   _____

   _____

   _____

   _____

   _____

   _____

   _____

   _____

   _____

2. You are called to an industrial plant for a 26-year-old male complaining of a crush injury of his left foot. Bleeding is moderate and a portion of the dorsal aspect is partially avulsed. He is alert and oriented.

   How would you best manage this patient?

   _____

   _____

   _____

   _____

   _____

   _____

   _____

   _____

   _____

   _____

5

3. You are dispatched to stand by at a structure fire. Suddenly, firefighters emerge from the residence with a 40-year-old female who is coughing violently. She has second-degree burns to both hands and forearms and on her abdomen. She is alert and having mild difficulty breathing.

How would you best manage this patient?

_____

_____

_____

_____

_____

_____

_____

_____

_____

Skill Drills

Skill Drill 24-1: Controlling Bleeding from a Soft-Tissue Injury
Test your knowledge of this skill drill by filling in the correct words in the photo captions.

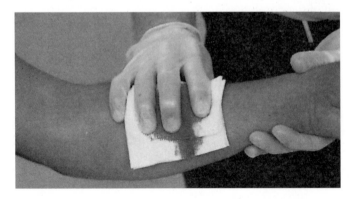

1. Apply _____ _____ with a _____ bandage.

2. Maintain pressure with a _____ bandage.

(continued)

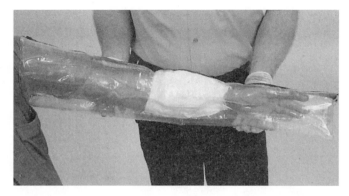

3. If bleeding continues, apply a second _____ and _____ _____bandage over the first.

4. _____ the extremity.

Skill Drill 24-2: Sealing a Sucking Chest Wound
Test your knowledge of this skill drill by filling in the correct words in the photo captions.

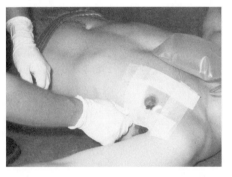
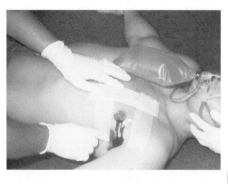

1. Keep the patient _____ and give _____.

2. Seal the wound with a(n) _____ dressing.

3. Follow _____ _____ regarding sealing or _____ _____ the dressing's fourth side.

Skill Drill 24-3: Stabilizing an Impaled Object
Test your knowledge of this skill drill by filling in the correct words in the photo captions.

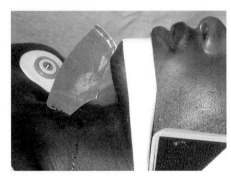
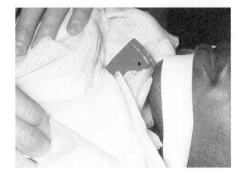
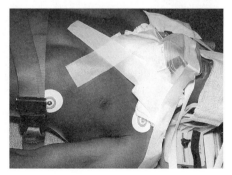

1. Do not attempt to _____ or _____ the object.

2. Control _____ and _____ the object in place using _____ _____, _____, and/or _____.

3. Tape a _____ item over the stabilized object to protect it from _____ during transport.

5

Skill Drill 24-4: Caring for Burns

Test your knowledge of this skill drill by placing the photos below in the correct order. Number the first step with a "1," the second step with a "2," etc.

Prepare for transport.
Treat for shock if needed.

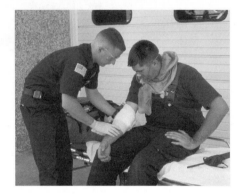

Estimate the severity of the burn, then cover the area with a dry, sterile dressing or clean sheet.
Assess and treat the patient for any other injuries.

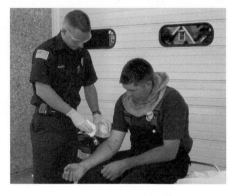

Follow BSI precautions to help prevent infection.
Remove the patient from the burning area; extinguish or remove hot clothing and jewelry as needed.
If the wound(s) is still burning or hot, immerse the hot area in cool, sterile water, or cover with a wet, cool dressing.

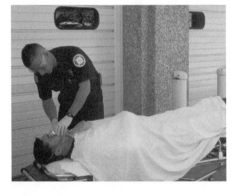

Cover the patient with blankets to prevent loss of body heat.
Transport promptly.

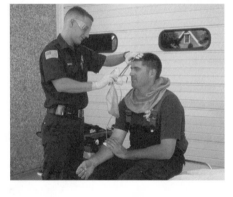

Give supplemental oxygen and continue to assess the airway.

Notes

Workbook Activities

The following activities have been designed to help you. Your instructor may require you to complete some or all of these activities as a regular part of your EMT-B training program. You are encouraged to complete any activity that your instructor does not assign as a way to enhance your learning in the classroom.

Chapter Review

The following exercises provide an opportunity to refresh your knowledge of this chapter.

NOTES

Matching

Match each of the terms in the left column to the appropriate definition in the right column.

_____ 1. Cornea

_____ 2. Iris

_____ 3. Lens

_____ 4. Pupil

_____ 5. Sclera

_____ 6. Orbit

_____ 7. Conjunctiva

_____ 8. Globe

_____ 9. Lacrimal glands

_____ 10. Retina

A. membrane that covers the exposed surface of the eye

B. focuses light

C. muscle behind cornea

D. transparent tissue in front of pupil and iris

E. circular opening in iris

F. the eyeball

G. eye socket

H. the light-sensitive area of the eye where images are projected

I. tear glands

J. white portion of eye

Multiple Choice

Read each item carefully, then select the best response.

_____ 1. The conjunctiva covers the:

  A. outer surface of the eyelids.

  B. exposed surface of the eye.

  C. lens.

  D. iris.

_____ 2. The purpose of the _____, an extremely tough, fibrous tissue, is to help maintain the eye's globular shape.

  A. cornea

  B. lens

  C. retina

  D. sclera

Eye Injuries

_____ **3.** The _____ allows light to enter the eye.
   A. cornea
   B. lens
   C. retina
   D. sclera

_____ **4.** The circular muscle that adjusts the size of the opening behind the cornea to regulate the amount of light that enters the eye is called the:
   A. iris.
   B. lens.
   C. retina.
   D. sclera.

_____ **5.** In a normal, uninjured eye, the pupils:
   A. are round.
   B. are equal in size.
   C. react equally when exposed to light.
   D. all of the above

_____ **6.** Important signs and symptoms to record include all of the following, except:
   A. how the injury occurred.
   B. any changes in vision.
   C. any history of color blindness.
   D. the use of any eye medications.

_____ **7.** The delicate tissues of the eye can be burned by _____, often causing permanent damage.
   A. chemicals
   B. heat
   C. light rays
   D. all of the above

_____ **8.** To care for thermal burns of the eyes and eyelids, you should:
   A. cover both eyes with a sterile dressing moistened with sterile saline.
   B. apply eye shields over the dressing.
   C. transport promptly.
   D. all of the above

_____ 9. Retinal injuries that are caused by exposure to extremes of light are generally not _____ but may result in permanent damage to vision.

   A. deep

   B. painful

   C. lacerated

   D. none of the above

_____ 10. Superficial burns of the eyes can result from:

   A. light from prolonged exposure to a sunlamp.

   B. reflected light from a bright snow-covered area.

   C. ultraviolet rays from an arc welding unit.

   D. all of the above

_____ 11. Lacerations of the eyelids may cause heavy bleeding that can usually be controlled by:

   A. firm pressure.

   B. gentle pressure.

   C. pressure dressings.

   D. flushing the eyes.

_____ 12. If there is a laceration of the globe itself, you should apply:

   A. firm pressure.

   B. gentle pressure.

   C. no pressure.

   D. pressure dressings.

_____ 13. Important guidelines in treating penetrating injuries of the eye include all of the following, except:

   A. Never exert pressure on or manipulate the globe in any way.

   B. If part of the eyeball is exposed, gently apply a moist, sterile dressing to prevent drying.

   C. Cover the injured eye with a protective metal eye shield or sterile dressing.

   D. Gently replace the eyeball if it is displaced out of its socket.

_____ 14. A "black eye" is a result of:

   A. bleeding into tissue around the orbit.

   B. a fracture of the orbit.

   C. a torn retina.

   D. none of the above

_____ 15. Bleeding into the anterior chamber of the eye that obscures part or all of the iris is called:

   A. hematemesis.

   B. hyphema.

   C. hyphoma.

   D. hemoptysis.

_____ 16. Fracture of the orbit, particularly of the bones that form its floor and support of the globe is known as a:

   A. blowout fracture.

   B. retinal detachment.

   C. hyphema.

   D. black eye.

_____ **17.** Eye findings that should alert you to the possibility of a head injury include:

    A. one pupil larger than the other.

    B. bleeding under the conjunctiva.

    C. protrusion or bulging of one eye.

    D. all of the above

_____ **18.** The only time that contact lenses should be removed immediately in the field is in the case of a _____ the eye.

    A. blowout fracture of

    B. retinal detachment of

    C. chemical burn in

    D. broken contact lens in

Vocabulary

Define the following terms using the space provided.

1. Blowout fracture:

_____

_____

_____

2. Hyphema:

_____

_____

_____

3. Conjunctivitis:

_____

_____

_____

Fill-in

Read each item carefully, then complete the statement by filling in the missing word(s).

1. The glands that produce fluids to keep the eye moist are called _____

_____ .

2. A cranial nerve that transmits visual information to the brain is called an

_____ _____ .

3. The eye works like a _____ , with the iris and pupil making adjustments to light and the retina acting like film.

5

4. Never remove contact lenses from an injured eye unless the injury is a

_____ _____.

5. The _____ is composed of the adjacent bones of the face and skull.

6. When performing an examination, you are looking for specific _____ or conditions that may suggest the nature of the problem.

7. Large objects are prevented from penetrating the eye by the protective

_____ that surrounds it.

8. _____ is inflammation and redness of the conjunctiva.

True/False

If you believe the statement to be more true than false, write the letter "T" in the space provided. If you believe the statement to be more false than true, write the letter "F."

1. _____ Objects impaled in the eye should be removed before applying dressing.

2. _____ Vitreous humor can be replaced.

3. _____ Aqueous humor can be replaced.

4. _____ Lacrimal glands help keep the eye dry.

5. _____ Contact lenses should always be removed in the field.

6. _____ Foreign objects stuck to the cornea should be removed prior to transport.

7. _____ Bleeding soon after irritation or injury can result in a bright yellow conjunctiva.

8. _____ In a normal, uninjured eye the entire circle of the iris is visible.

9. _____ If a small foreign object is lying on the surface of the patient's eye, you should use a dextrose solution to gently irrigate the eye.

Short Answer

Complete this section with short written answers using the space provided.

1. Describe a retinal detachment, including common signs and symptoms.

_____

_____

_____

2. List the three guidelines for treating penetrating eye injuries.

_____

_____

_____

3. List assessment findings in the eye that may indicate head injury.

_____

_____

_____

Word Fun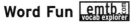

The following crossword puzzle is an activity provided to reinforce correct spelling and understanding of medical terminology associated with emergency care and the EMT-B. Use the clues in the column to complete the puzzle.

Across

2. White portion of eye

5. Membrane that lines eyelids and surface of eye

6. Eye socket

7. Bleeding into anterior chamber of eye

8. Transmits visual sensations to the brain

9. Muscle regulating light into eye

Down

1. Break in the bones of the orbit

3. Tear producer

4. Transparent tissue in front of pupil

Ambulance Calls

The following real case scenarios provide an opportunity to explore the concerns associated with patient management. Read each scenario, then answer each question in detail.

1. You are dispatched to a 12-year-old male with a pencil impaled in his right eye. The patient is alert and very upset. His mother is hysterical. You manage to learn from her that he has no prior medical history.

How would you best manage this patient?

_____

_____

_____

2. You are called to an industrial site where a 32-year-old male had acid splashed in his eyes. Coworkers have tried pouring water into his eyes, but he is still unable to see and is in severe pain.

How would you best manage this patient?

3. You are dispatched to a baseball field where a 15-year-old male was hit in the right eye with a baseball. He is sitting on the bench complaining of double vision and presents with swelling and bruising around his right orbit.

How would you best manage this patient?

Skill Drills

Test your knowledge of skill drills by filling in the correct words in the photo captions.

Skill Drill 25-1: Removing a Foreign Object from Under the Upper Eyelid

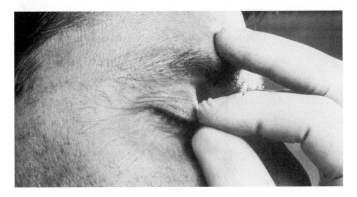

1. Have the patient look _____ , grasp the _____ _____, and gently pull the lid away from the eye.

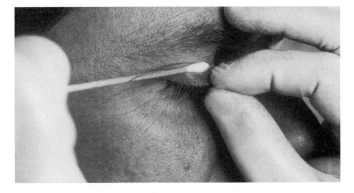

2. Place a _____ _____ on the upper lid.

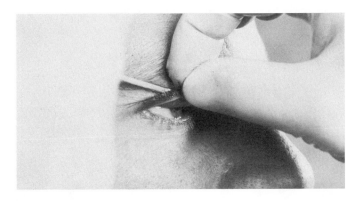

3. Pull the lid _____ and _____, folding it back over the applicator.

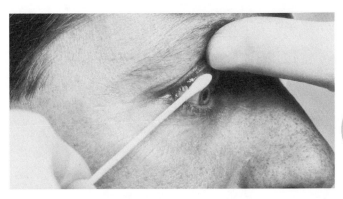

4. Gently remove the foreign object with a _____, _____ applicator.

5

Skill Drill 25-2: Stabilizing a Foreign Object Impaled in the Eye

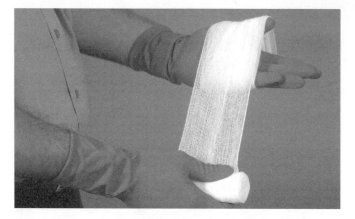

1. To prepare a doughnut ring, wrap a _____ roll around your fingers and thumb _____ times. Adjust the diameter by _____ your fingers.

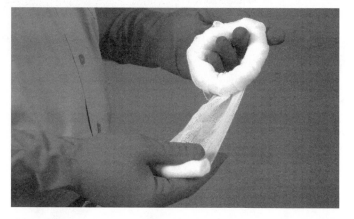

2. Wrap the remainder of the roll, . . .

3. . . . working around the ring.

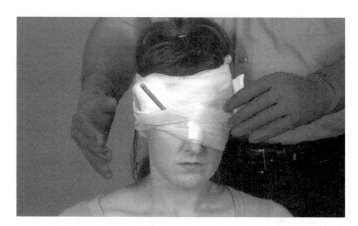

4. Place the dressing over the _____ to hold the impaled object in place, then _____ it with a _____ dressing.

Notes

Workbook Activities

The following activities have been designed to help you. Your instructor may require you to complete some or all of these activities as a regular part of your EMT-B training program. You are encouraged to complete any activity that your instructor does not assign as a way to enhance your learning in the classroom.

Chapter Review

The following exercises provide an opportunity to refresh your knowledge of this chapter.

Matching

Match each of the terms in the left column to the appropriate definition in the right column.

_____ **1.** Cranium   **A.** upper jaw

_____ **2.** Occiput   **B.** posterior cranium

_____ **3.** Pinna   **C.** contains the brain

_____ **4.** Zygomas   **D.** pull/tear away

_____ **5.** Mandible   **E.** visible part of ear

_____ **6.** Avulse   **F.** cheekbones

_____ **7.** Maxilla   **G.** lower jawbone

Multiple Choice

Read each item carefully, then select the best response.

_____ **1.** As an EMT-B, your objective is to:
A. prevent further injury.
B. manage any acute airway problems.
C. control bleeding.
D. all of the above

_____ **2.** The head is divided into two parts: the cranium and the:
A. brain.
B. face.
C. skull.
D. medulla oblongata.

_____ **3.** The brain connects to the spinal cord through a large opening at the base of the skull known as the:
A. eustachian tube.
B. spinous process.
C. foramen magnum.
D. vertebral foramina.

C H A P T E R 26

Face and Throat Injuries

_____ **4.** Approximately _____ of the nose is composed of bone. The remainder is composed of cartilage.

A. 9/10

B. 2/3

C. 3/4

D. 1/3

_____ **5.** Motion of the mandible occurs at the:

A. temporomandibular joint.

B. mastoid process.

C. chin.

D. mandibular angle.

_____ **6.** The _____ may be found on either side of the trachea, along with the jugular veins and several nerves.

A. hypothalamus

B. subclavian arteries

C. cricoid cartilage

D. carotid arteries

_____ **7.** The _____ connects the cricoid cartilage and thyroid cartilage.

A. larynx

B. cricoid membrane

C. cricothyroid membrane

D. thyroid membrane

_____ **8.** Upper airway obstructions caused by facial injuries may be due to:

A. heavy bleeding.

B. loosened teeth or dentures.

C. soft-tissue swelling.

D. all of the above

_____ 9. If you find portions of avulsed skin that have become separated, you should:

    A. wrap the skin in a moist, sterile dressing/place in a plastic bag in ice water/transport with patient.

    B. place the skin in plastic "biohazard" bag and dispose of properly.

    C. place the skin in a plastic bag filled with ice and transport to the ER.

    D. leave it at the scene to be disposed of later.

_____ 10. The nasal cavity is divided into two chambers by the:

    A. frontal sinus.

    B. middle turbinate.

    C. zygoma.

    D. nasal septum.

_____ 11. You may be able to control bleeding from the nose by:

    A. applying a sterile dressing.

    B. pinching the nostrils together.

    C. putting the patient in a supine position.

    D. having the patient hold ice in his or her mouth.

_____ 12. The middle ear is connected to the nasal cavity by the:

    A. frontal sinus.

    B. zygomatic process.

    C. eustachian tube.

    D. superior trachea.

_____ 13. A basilar skull fracture may present with:

    A. leakage of CSF from the ears and nose.

    B. distended neck veins.

    C. Battle's signs.

    D. periorbital ecchymosis.

_____ 14. Signs of a possible facial fracture include:

    A. bleeding in the mouth.

    B. absent or loose teeth.

    C. loose and/or moveable bone fragments.

    D. all of the above

_____ 15. The presence of air in the soft tissues of the neck that produces a crackling sensation is called:

    A. the "Rice Krispy" effect.

    B. a pneumothorax.

    C. rales.

    D. subcutaneous emphysema.

_____ 16. Most bleeding from the neck can be controlled by:

    A. direct pressure.

    B. a pressure point.

    C. elevation.

    D. a tourniquet.

Labeling [emtb vocab explorer]

Label the following diagrams with the correct terms.

1. Face/Skull

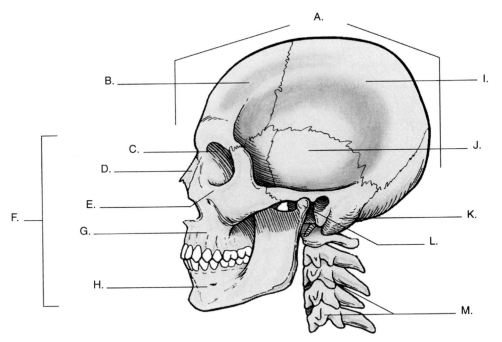

A. _____

B. _____

C. _____

D. _____

E. _____

F. _____

G. _____

H. _____

I. _____

J. _____

K. _____

L. _____

M. _____

Vocabulary [emtb vocab explorer]

Define the following terms using the space provided.

1. Air embolism:

_____

_____

_____

2. Hematoma:

_____

_____

_____

3. Sternocleidomastoid muscle:

_____

_____

_____

NOTES

5

4. Subcutaneous emphysema:

_____

_____

_____

5. Temporomandibular joint (TMJ):

_____

_____

_____

Fill-in

Read each item carefully, then complete the statement by filling in the missing word(s).

1. Pulsations in the neck are felt in the _____ vessels.

2. The _____ vertebrae are in the neck.

3. The _____ lobes of the cranium are located on the lateral portion of the head.

4. The _____ is located in the anterior portion of the neck.

5. The rings of the trachea are made of _____.

6. The Adam's Apple is more prominent in _____ than in _____.

7. The _____ _____ is a large opening at the base of the skull.

8. The _____ is the upper part of the jaw.

9. The _____ lobes lie laterally between the temporal and occipital lobes.

10. The _____ connects the larynx with the main air passage.

True/False

If you believe the statement to be more true than false, write the letter "T" in the space provided. If you believe the statement to be more false than true, write the letter "F."

1. _____ Injuries to the face often lead to airway problems.

2. _____ Care for facial injuries begins with BSI precautions and the ABCs.

3. _____ Exposed eye or brain injuries are covered with a dry dressing.

4. _____ Clear fluid in the outer ear is normal.

5. _____ Any crushing injury of the upper part of the neck likely involves the larynx or the trachea.

6. _____ Soft-tissue injuries to the face are common.

Short Answer

Complete this section with short written answers using the space provided.

1. Describe bleeding control methods for facial injuries.

_____

_____

_____

2. Describe bleeding control methods for lacerations to veins or arteries in the neck.

_____

_____

_____

Word Fun

The following crossword puzzle is a good way to review correct spelling and meaning of medical terminology associated with emergency care and the EMT-B. Use the clues in the column to complete the puzzle.

C L U E S

Across

1. Posterior portion of skull

4. Upper jawbone

5. Skull

8. Lower jawbone

10. Connects middle ear to nasal cavity

11. To pull or tear away

12. Blood accumulated in soft tissue

Down

2. Layers of bones in nasal cavity

3. Bony mass about 1" posterior to ear

6. Air in the veins

7. Fleshy bulge anterior to ear canal

9. Large opening in base of skull

5

Ambulance Calls

The following real case scenarios provide an opportunity to explore the concerns associated with patient management. Read each scenario, then answer each question in detail.

1. You are called to a local school playground for a 7-year-old female complaining of moderate epistaxis. She has no history and her mother has been called. The child is sitting in the nurse's office holding her face over a trash can.

   How would you best manage this patient?

   _____

   _____

   _____

   _____

   _____

   _____

   _____

   _____

2. You are dispatched to a 37-year-old man with a large laceration to the right side of his neck. Bleeding is dark and heavy. He is alert, but weak.

   How would you best manage this patient?

   _____

   _____

   _____

   _____

   _____

   _____

   _____

   _____

3. You are called to a lumberyard where a 25-year-old male had his ear cut completely off with a saw. Bleeding is controlled with direct pressure. He is alert and oriented.

How would you best manage this patient?

_____

_____

_____

_____

_____

_____

_____

_____

_____

_____

Skill Drills

Test your knowledge of skill drills by filling in the correct words in the photo captions.

Skill Drill 26-1: Controlling Bleeding from a Neck Injury

5

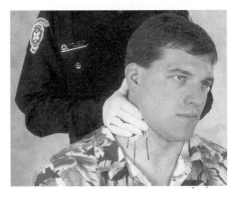  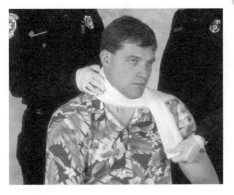

1. Apply _____ _____ to control bleeding.

2. Use a _____ _____ to secure a dressing in place.

3. Wrap the bandage around and under the patient's _____.

Workbook Activities

The following activities have been designed to help you. Your instructor may require you to complete some or all of these activities as a regular part of your EMT-B training program. You are encouraged to complete any activity that your instructor does not assign as a way to enhance your learning in the classroom.

Chapter Review

The following exercises provide an opportunity to refresh your knowledge of this chapter.

NOTES

Matching

Match each of the terms in the left column to the appropriate definition in the right column.

_____ **1.** Thoracic cage   **A.** chest rises

_____ **2.** Diaphragm   **B.** chest

_____ **3.** Exhalation   **C.** chest falls

_____ **4.** Inhalation   **D.** separates chest from abdomen

_____ **5.** Aorta   **E.** major artery in the chest

_____ **6.** Closed chest injury   **F.** penetrating wound

_____ **7.** Hemoptysis   **G.** rapid respirations

_____ **8.** Pericardium   **H.** unusually blunt trauma

_____ **9.** Open chest injury   **I.** coughing up blood

_____ **10.** Tachypnea   **J.** sac around the heart

Multiple Choice

Read each item carefully, then select the best response.

_____ **1.** Air is supplied to the lungs via the:
   A. esophagus.
   B. trachea.
   C. nares.
   D. oropharnyx.

_____ **2.** The _____ separates the thoracic cavity from the abdominal cavity.
   A. diaphragm
   B. mediastinum
   C. xyphoid process
   D. inferior border of the ribs

Chest Injuries

_____ **3.** On inhalation, all of the following occur, except:

    A. the intercostal muscles contract, elevating the rib cage.

    B. the diaphragm contracts.

    C. the pressure inside the chest increases.

    D. air enters through the nose and mouth.

_____ **4.** Blunt trauma to the chest may:

    A. bruise the lungs and heart.

    B. fracture whole areas of the chest wall.

    C. damage the aorta.

    D. all of the above

_____ **5.** Symptoms of chest injury include:

    A. cyanosis around the lips or fingertips.

    B. rapid, weak pulse.

    C. hemoptysis.

    D. pain at the site of injury.

_____ **6.** Common causes of dyspnea include:

    A. airway obstruction.

    B. lung compression.

    C. damage to the chest wall.

    D. all of the above

_____ **7.** _____ occurs when the chest wall does not expand on each side when the patient inhales.

    A. Flail segment

    B. Paradoxical motion

    C. Pneumothorax

    D. Hemoptysis

_____ **8.** The principle reason for concern about a patient who has a chest injury is:

    A. hemoptysis.

    B. cyanosis.

    C. that the body has no means of storing oxygen.

    D. a rapid, weak pulse and low blood pressure.

NOTES

_____ **9.** A _____ results when an injury allows air to enter through a hole in the chest wall or the surface of the lung as the patient attempts to breathe, causing the lung on that side to collapse.

   A. tension pneumothorax

   B. hemothorax

   C. hemopneumothorax

   D. pneumothorax

_____ **10.** A sucking chest wound should be treated:

   A. after assessing ABCs.

   B. after confirming mental status.

   C. immediately by covering with a gloved hand, then an occlusive dressing.

   D. by using a stack of gauze dressings.

_____ **11.** A spontaneous pneumothorax:

   A. presents with a sudden sharp chest pain.

   B. presents with increasing difficulty breathing.

   C. should be treated the same as a traumatic pneumothorax.

   D. all of the above

_____ **12.** As a pneumothorax develops tension:

   A. air gradually increases the pressure in the chest.

   B. it causes the complete collapse of the affected lung.

   C. it prevents blood from returning through the venae cavae to the heart.

   D. all of the above

_____ **13.** Common signs and symptoms of tension pneumothorax include:

   A. increasing respiratory distress.

   B. distended neck veins.

   C. tracheal deviation away from the injured site.

   D. all of the above

_____ **14.** A hemothorax results from blood collecting in the pleural space from:

   A. a bleeding rib cage.

   B. a bleeding lung.

   C. a bleeding great vessel.

   D. all of the above

_____ **15.** A fractured rib that penetrates into the pleural space may lacerate the surface of the lung, causing a:

   A. tension pneumothorax.

   B. hemothorax.

   C. hemopneumothorax.

   D. all of the above

_____ **16.** In what is called paradoxical movement, the detached portion of the chest wall:

   A. moves opposite of normal.

   B. moves out instead of in during inhalation.

   C. moves in instead of out during expiration.

   D. all of the above

_____ **17.** Traumatic asphyxia:

    A. is bruising of the lung.

    B. occurs when three or more adjacent ribs are fractured in two or more places.

    C. is a sudden, severe compression of the chest.

    D. all of the above

_____ **18.** Traumatic asphyxia results in a very characteristic appearance, including:

    A. distended neck veins.

    B. cyanosis.

    C. hemorrhage into the sclera of the eye.

    D. all of the above

_____ **19.** Signs and symptoms of a pericardial tamponade include:

    A. low blood pressure.

    B. a weak pulse.

    C. muffled heart tones.

    D. all of the above

_____ **20.** Large blood vessels in the chest that can result in massive hemorrhaging include all of the following, except:

    A. the pulmonary arteries.

    B. the femoral arteries.

    C. the aorta.

    D. four main pulmonary veins.

Labeling emtb.com anatomy review

Label the following diagrams with the correct terms.

    1. Anterior Aspect of the Chest

A. _____

B. _____

C. _____

D. _____

E. _____

E. _____

F. _____

5

2. The Ribs

A. _____

B. _____

C. _____

D. _____

E. _____

F. _____

G. _____

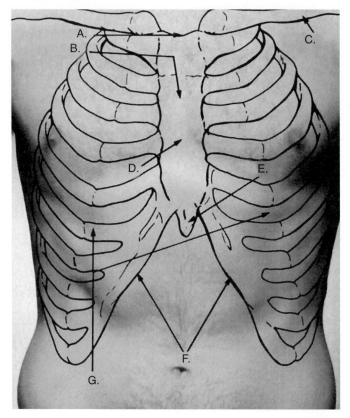

Vocabulary

Define the following terms using the space provided.

1. Flail chest:

_____

_____

_____

2. Paradoxical motion:

_____

_____

_____

3. Pericardial tamponade:

_____

_____

_____

4. Spontaneous pneumothorax:

_____

_____

_____

5. Sucking chest wound:

_____

_____

_____

6. Tension pneumothorax:

_____

_____

_____

Fill-in

Read each item carefully, then complete the statement by filling in the missing word.

1. The esophagus is located in the _____ of the chest.

2. During inhalation, the pressure in the chest _____.

3. In the anterior chest, ribs connect to the _____.

4. The trachea divides into the right and left main stem _____.

5. The _____ nerves supply the diaphragm.

6. Contents of the chest are protected by the _____.

7. The chest extends from the lower end of the neck to the _____.

8. _____ line the area between the lungs and chest wall.

9. The great vessel located in the chest is the _____.

10. During inhalation, the diaphragm _____.

True/False

If you believe the statement to be more true than false, write the letter "T" in the space provided. If you believe the statement to be more false than true, write the letter "F."

1. _____ Dyspnea is difficulty with breathing.

2. _____ Tachypnea is slow respirations.

3. _____ Distended neck veins may be a sign of a tension pneumothorax.

4. _____ Rib fractures are common in children.

5. _____ Narrowing pulse pressure is related to spontaneous pneumothorax.

6. _____ Laceration of the large blood vessels in the chest can cause minimal hemorrhage.

7. _____ The thoracic cage extends from the lower end of the neck to the umbilicus.

8. _____ Patients with spinal cord injuries at C3 or above can lose their ability to breathe entirely.

9. _____ Almost one third of people who are killed immediately in car crashes die as a result of traumatic rupture of the myocardium.

Short Answer

Complete this section with short written answers using the space provided.

1. List the signs and symptoms associated with a chest injury.

_____

_____

_____

2. Describe the two methods for sealing a sucking chest wound.

_____

_____

_____

3. Describe the method(s) for immobilizing a flail chest wall segment.

_____

_____

_____

4. Define traumatic asphyxia and describe its signs.

_____

_____

_____

Word Fun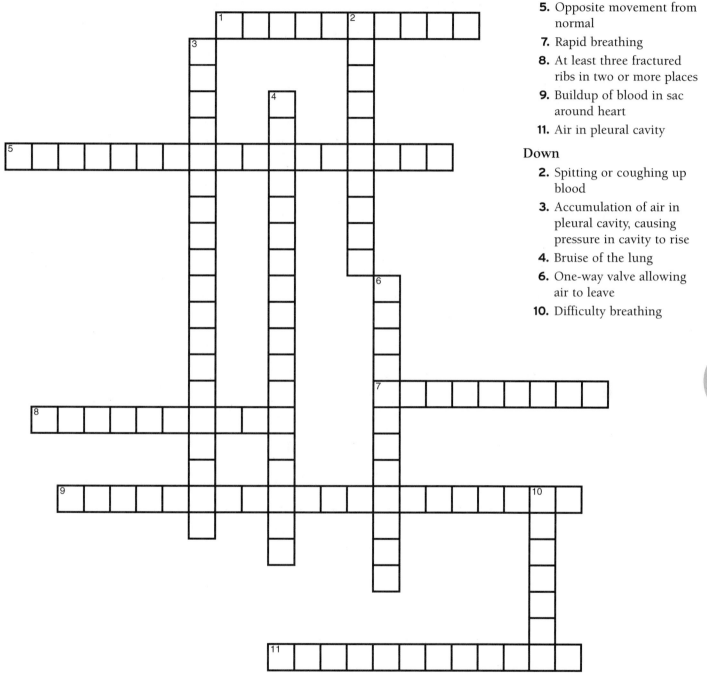

The following crossword puzzle is an activity provided to reinforce correct spelling and understanding of medical terminology associated with emergency care and the EMT-B. Use the clues in the column to complete the puzzle.

■ CLUES ■

Across

1. Blood in the pleural cavity
5. Opposite movement from normal
7. Rapid breathing
8. At least three fractured ribs in two or more places
9. Buildup of blood in sac around heart
11. Air in pleural cavity

Down

2. Spitting or coughing up blood
3. Accumulation of air in pleural cavity, causing pressure in cavity to rise
4. Bruise of the lung
6. One-way valve allowing air to leave
10. Difficulty breathing

5

Ambulance Calls

The following real case scenarios provide an opportunity to explore the concerns associated with patient management. Read each scenario, then answer each question in detail.

1. You are dispatched to a head-on motor vehicle crash. Your patient, a 67-year-old male, unrestrained driver, is unresponsive and has a very weak pulse and decreasing respirations as you pull him from the vehicle. Once you have him immobilized on the long backboard, he becomes apneic and pulseless.

How would you best manage this patient?

_____

_____

_____

_____

_____

_____

_____

2. Your patient, a 22-year-old male, is complaining of a sudden onset of right-sided chest pain with a sudden onset of difficulty breathing. He tells you he was out running when it started.

How would you best manage this patient?

_____

_____

_____

_____

_____

_____

_____

_____

_____

3. You are dispatched to a lumberyard where a 27-year-old male was crushed by a piece of heavy equipment. Coworkers pulled the equipment off the patient. He presents with distended neck veins, cyanosis, and bloodshot eyes.

How would you best manage this patient?

_____

_____

_____

_____

_____

_____

_____

_____

_____

5

Workbook Activities

The following activities have been designed to help you. Your instructor may require you to complete some or all of these activities as a regular part of your EMT-B training program. You are encouraged to complete any activity that your instructor does not assign as a way to enhance your learning in the classroom.

Chapter Review

The following exercises provide an opportunity to refresh your knowledge of this chapter.

■ NOTES ■

Matching

Match each of the terms in the left column to the appropriate definition in the right column.

_____ 1. Hollow organs   **A.** blood in urine

_____ 2. Solid organs   **B.** organs outside of the body

_____ 3. Peritonitis   **C.** abdominal lining inflammation

_____ 4. Genitourinary system   **D.** kidneys, liver, spleen

_____ 5. Filtering system   **E.** kidneys

_____ 6. Evisceration   **F.** abdomen

_____ 7. Hematuria   **G.** stomach, bladder, ureters

_____ 8. Peritoneal cavity   **H.** controls reproductive functions and the waste discharge system

Multiple Choice

Read each item carefully, then select the best response.

_____ 1. The abdomen contains several organs that make up the:

    A. digestive system.

    B. urinary system.

    C. genitourinary system.

    D. all of the above

_____ 2. Hollow organs of the abdomen include the:

    A. stomach.

    B. ureters.

    C. bladder.

    D. all of the above

Abdomen and Genitalia Injuries

_____ **3.** Solid organs of the abdomen include all of the following, except the:

    A. liver.

    B. spleen.

    C. gallbladder.

    D. pancreas.

_____ **4.** The first signs of peritonitis include:

    A. severe abdominal pain.

    B. tenderness.

    C. muscular spasm.

    D. all of the above

_____ **5.** Late signs of peritonitis may include:

    A. a soft abdomen.

    B. nausea.

    C. normal bowel sounds.

    D. all of the above

_____ **6.** _____ takes place in the solid organs.

    A. Digestion

    B. Excretion

    C. Energy production

    D. all of the above

_____ **7.** Because solid organs have a rich supply of blood, any injury can result in major:

    A. hemorrhaging.

    B. damage.

    C. pain.

    D. guarding.

_____ **8.** A patient who has abdominal bleeding may experience all of the following, except:

   A. pain or tenderness.

   B. rigidity.

   C. urticaria.

   D. distention.

_____ **9.** The major soft-tissue landmark is (are) the _____, which overlie(s) the fourth lumbar vertebra.

   A. iliac crests

   B. umbilicus

   C. pubic symphysis

   D. anterior iliac spines

_____ **10.** The abdomen is divided into four:

   A. quadrants.

   B. planes.

   C. sections.

   D. angles.

_____ **11.** Injuries to the abdomen may involve:

   A. hollow organs.

   B. open injuries.

   C. solid organs.

   D. all of the above

_____ **12.** Open abdominal injuries are also known as:

   A. blunt injuries.

   B. eviscerations.

   C. penetrating injuries.

   D. peritoneal injuries.

_____ **13.** Closed abdominal injuries may result from:

   A. a stab wound.

   B. seat belts.

   C. a gunshot wound.

   D. all of the above

_____ **14.** The major complaint of patients with abdominal injury is:

   A. pain.

   B. tachycardia.

   C. rigidity.

   D. swelling.

_____ **15.** The most common sign of significant abdominal injury is:

   A. pain.

   B. tachycardia.

   C. rigidity.

   D. distention.

_____ **16.** Late signs of abdominal injury include all of the following, except:

   A. distention.

   B. increased blood pressure.

   C. rigidity.

   D. shallow respirations.

_____ **17.** Your primary concern when dealing with an unresponsive patient with an open abdominal injury is:

    A. covering the wound with a moist dressing.

    B. maintaining the airway.

    C. controlling the bleeding.

    D. monitoring vital signs.

_____ **18.** A patient with blunt abdominal trauma may present with:

    A. severe bruises on the abdominal wall.

    B. laceration of the liver or spleen.

    C. rupture of the intestine.

    D. all of the above

_____ **19.** It is imperative that a patient who has received severe blunt abdominal trauma be:

    A. log rolled onto a backboard.

    B. transported rapidly.

    C. given oxygen.

    D. all of the above

_____ **20.** When used alone, diagonal shoulder safety belts can cause:

    A. a bruised chest.

    B. a lacerated liver.

    C. decapitation.

    D. all of the above

_____ **21.** When evaluating a patient involved in a motor vehicle crash where airbags have deployed, it is important to look for:

    A. debris inside the vehicle.

    B. condition of the tires.

    C. damage to the steering column underneath the airbag.

    D. all of the above

_____ **22.** Patients with penetrating abdominal injuries often complain of:

    A. pain.

    B. nausea.

    C. vomiting.

    D. all of the above

_____ **23.** When caring for a patient with a penetrating abdominal injury, you should assume that the object:

    A. has penetrated the peritoneum.

    B. entered the abdominal cavity.

    C. possibly injured one or more organs.

    D. all of the above

_____ **24.** When treating a patient with an evisceration, you should:

    A. attempt to replace the abdominal contents.

    B. cover the protruding organs with a dry, sterile dressing.

    C. cover the protruding organs with moist, adherent dressings.

    D. cover the protruding contents with moist, sterile gauze compresses.

5

_____ **25.** The solid organs of the urinary system include the:

   A. kidneys.

   B. ureters.

   C. bladder.

   D. urethra.

_____ **26.** All of the male genitalia lie outside the pelvic cavity with the exception of the:

   A. urethra.

   B. penis.

   C. seminal vesicles.

   D. testes.

_____ **27.** Suspect kidney damage if the patient has a history or physical evidence of:

   A. an abrasion, laceration, or contusion in the flank.

   B. a penetrating wound in the region of the lower rib cage or the upper abdomen.

   C. fractures on either side of the lower rib cage.

   D. all of the above

_____ **28.** Signs of injury to the kidney may include:

   A. bruises or lacerations on the overlying skin.

   B. shock.

   C. hematuria.

   D. all of the above

_____ **29.** Suspect a possible injury of the urinary bladder in all of the following findings, except:

   A. bruising to the left upper quadrant.

   B. blood at the urethral opening.

   C. blood at the tip of the penis or a stain on the patient's underwear.

   D. physical signs of trauma on the lower abdomen, pelvis, or perineum.

_____ **30.** When treating a patient with an amputation of the penile shaft, your top priority is:

   A. locating the amputated part.

   B. controlling bleeding.

   C. keeping the remaining tissue dry.

   D. delaying transport until bleeding is controlled.

_____ **31.** Treatment of injuries involving the external male genitalia includes:

   A. making the patient as comfortable as possible.

   B. using sterile, moist compresses to cover areas that have been stripped of skin.

   C. applying direct pressure with dry, sterile gauze dressings to control bleeding.

   D. all of the above

_____ **32.** In cases of sexual assault, you must treat the medical injuries but also provide:

   A. privacy.

   B. support.

   C. reassurance.

   D. all of the above

Labeling

Label the following diagrams with the correct terms.

1. Hollow Organs

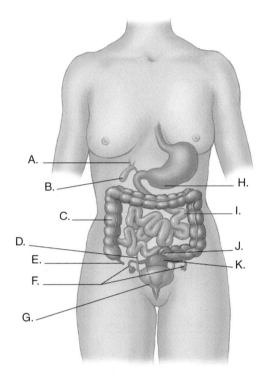

A. _____

B. _____

C. _____

D. _____

E. _____

F. _____

G. _____

H. _____

I. _____

J. _____

K. _____

2. Solid Organs

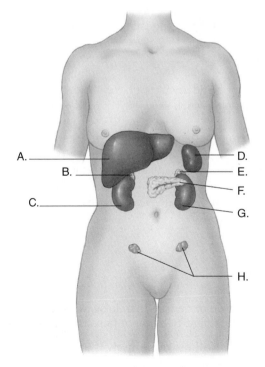

A. _____

B. _____

C. _____

D. _____

E. _____

F. _____

G. _____

H. _____

Vocabulary

Define the following terms using the space provided.

1. Closed abdominal injury:

_____

_____

_____

2. Open abdominal injury:

_____

_____

_____

3. Guarding:

_____

_____

_____

Fill-in

Read each item carefully, then complete the statement by filling in the missing word.

1. Severe bleeding may occur with injury to _____ organs.

2. The _____ system is responsible for filtering waste.

3. Kidneys are located in the _____ space.

4. Injuries to the kidneys or bladder will not have obvious _____, but there are usually more subtle clues such as lower rib pain or a possible pelvic fracture.

5. When ruptured, the organs of the abdominal cavity can spill their contents into the peritoneal cavity, causing an intense inflammatory reaction called _____.

6. Blood within the peritoneal cavity does not provoke a(n) _____, and may not cause pain or tenderness.

7. Closed abdominal injuries are also known as _____.

True/False

If you believe the statement to be more true than false, write the letter "T" in the space provided. If you believe the statement to be more false than true, write the letter "F."

1. _____ Hollow organs will bleed profusely if injured.

2. _____ The most common sign of an abdominal injury is an elevated heart rate.

3. _____ Patients with abdominal injuries should be kept supine with head elevated.

4. _____ Peritoneal irritation is in response to hollow organ injury.

5. _____ Eviscerated organs should be covered with a dry dressing.

6. _____ Injuries to the kidneys usually occur in isolation.

Short Answer

Complete this section with short written answers using the space provided.

1. List the hollow organs of the abdomen and urinary system.

_____

_____

_____

2. List the solid organs of the abdomen and urinary system.

_____

_____

_____

3. List the signs and symptoms of an abdominal injury.

_____

_____

_____

4. List the steps to care for a penetrating abdominal injury.

_____

_____

_____

5. List the steps to care for an open abdominal wound with exposed organs.

_____

_____

_____

6. List the major history or physical findings associated with possible kidney damage.

_____

_____

_____

Word Fun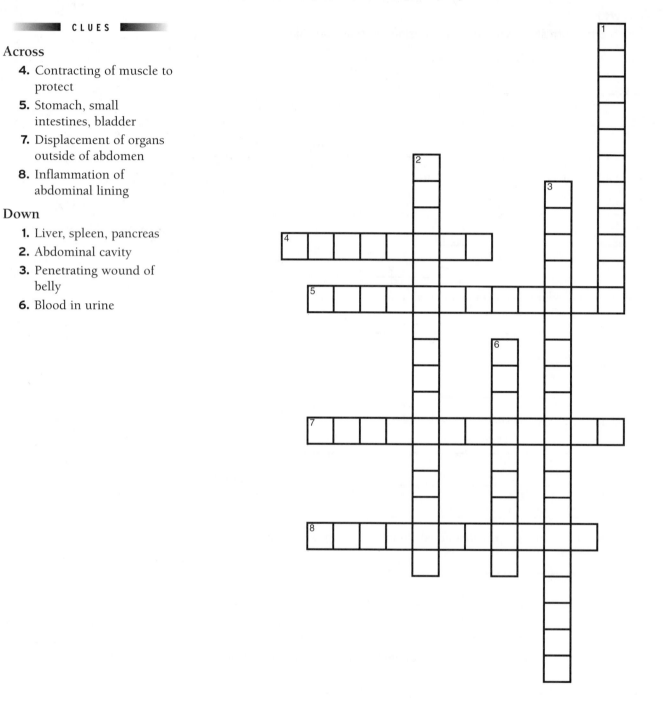

The following crossword puzzle is an activity provided to reinforce correct spelling and understanding of medical terminology associated with emergency care and the EMT-B. Use the clues in the column to complete the puzzle.

■ CLUES ■

Across

4. Contracting of muscle to protect
5. Stomach, small intestines, bladder
7. Displacement of organs outside of abdomen
8. Inflammation of abdominal lining

Down

1. Liver, spleen, pancreas
2. Abdominal cavity
3. Penetrating wound of belly
6. Blood in urine

Ambulance Calls

The following real case scenarios provide an opportunity to explore the concerns associated with patient management. Read each scenario, then answer each question in detail.

1. You are dispatched to a local bar where your patient, a 26-year-old male, was involved in an altercation. He has several superficial lacerations to his arms, and a knife is impaled in his right upper quadrant. He is lying supine on the floor. He is alert and bar patrons tell you that he did not fall; they helped him to the floor.
   How would you best manage this patient?

   _____

   _____

   _____

   _____

   _____

   _____

   _____

   _____

   _____

2. You are called to a significant motor vehicle crash where a 21-year-old female restrained driver is complaining of pain to the abdomen, but no other injury. She believes that the pain is just from the seat belt. She is alert and oriented and vital signs are all within normal limits. You explain that due to the mechanism of injury and her complaint of abdominal pain, she should be seen at the emergency room for evaluation. She agrees to go with you.
   How would you best manage this patient?

   _____

   _____

   _____

   _____

   _____

   _____

   _____

   _____

3. You are dispatched to a fall at a residence where you find a 35-year-old male who tripped and fell onto a barbed wire fence. He has a bowel evisceration and is in extreme pain. His neighbor tells you that there was no trauma involved; he slipped and fell from a standing position onto the wire, tearing open his abdomen in the process. He lowered himself to the ground before passing out. He is voice responsive and tachycardic. His breathing is 24 and deep.

How would you best manage this patient?

_____

_____

_____

_____

_____

_____

_____

_____

_____

Notes

Workbook Activities

The following activities have been designed to help you. Your instructor may require you to complete some or all of these activities as a regular part of your EMT-B training program. You are encouraged to complete any activity that your instructor does not assign as a way to enhance your learning in the classroom.

Chapter Review

The following exercises provide an opportunity to refresh your knowledge of this chapter.

Matching

Match each of the terms in the left column to the appropriate definition in the right column.

_____ **1.** Striated

_____ **2.** Tendons

_____ **3.** Smooth

_____ **4.** Joint

_____ **5.** Ligaments

_____ **6.** Closed fracture

_____ **7.** Point tenderness

_____ **8.** Displaced fracture

_____ **9.** Articular cartilage

_____ **10.** Open fracture

_____ **11.** Traction

A. Any injury that makes the limb appear in an unnatural position

B. Any fracture in which the skin has not been broken

C. A thin layer of cartilage, covering the articular surface of bones in synovial joints

D. Involuntary muscle

E. Any break in the bone in which the overlying skin has been damaged as well

F. Hold joints together

G. Skeletal muscle

H. The act of exerting a pulling force on a structure

I. Where two bones contact

J. Attach muscle to bone

K. Tenderness sharply located at the site of an injury

Multiple Choice

Read each item carefully, then select the best response.

_____ **1.** Blood in the urine is known as a:

A. hematuria.

B. hemotysis.

C. hematocrit.

D. hemoglobin.

CHAPTER 29

Musculoskeletal Care

_____ **2.** Smooth muscle is found in the:

   A. back.

   B. blood vessels.

   C. heart.

   D. all of the above

_____ **3.** The bones in the skeleton produce _____ in the bone marrow.

   A. red blood cells

   B. minerals

   C. electrolytes

   D. white blood cells

_____ **4.** _____ are held together in a tough fibrous structure known as a capsule.

   A. Tendons

   B. Joints

   C. Ligaments

   D. Bones

_____ **5.** Joints are bathed and lubricated by _____ fluid.

   A. cartilaginous

   B. articular

   C. synovial

   D. cerebrospinal

_____ **6.** A _____ is a disruption of a joint in which the bone ends are no longer in contact.

   A. torn ligament

   B. dislocation

   C. fracture dislocation

   D. sprain

 NOTES

_____ **7.** A _____ is a joint injury in which there is both some partial or temporary dislocation of the bone ends and partial stretching or tearing of the supporting ligaments.

    A. dislocation

    B. strain

    C. sprain

    D. torn ligament

_____ **8.** A _____ is a stretching or tearing of the muscle.

    A. strain

    B. sprain

    C. torn ligament

    D. split

_____ **9.** The zone of injury includes the:

    A. adjacent nerves.

    B. adjacent blood vessels.

    C. surrounding soft tissue.

    D. all of the above

_____ **10.** A(n) _____ fractures the bone at the point of impact.

    A. direct blow

    B. indirect force

    C. twisting force

    D. high-energy injury

_____ **11.** A(n) _____ may cause a fracture or dislocation at a distant point.

    A. direct blow

    B. indirect force

    C. twisting force

    D. high-energy injury

_____ **12.** When caring for patients who have fallen, you must identify the _____ and the mechanism of injury so that you will not overlook associated injuries.

    A. site of injury

    B. height of fall

    C. point of contact

    D. twisting forces

_____ **13.** _____ produce severe damage to the skeleton, surrounding soft tissues, and vital internal organs.

    A. Direct blows

    B. Indirect forces

    C. Twisting forces

    D. High-energy injuries

_____ **14.** Regardless of the extent and severity of the damage to the skin, you should treat any injury that breaks the skin as a possible:

    A. closed fracture.

    B. open fracture.

    C. nondisplaced fracture.

    D. displaced fracture.

_____ **15.** A _____ is also known as a hairline fracture.

    A. closed fracture

    B. open fracture

    C. nondisplaced fracture

    D. displaced fracture

_____ **16.** A _____ produces actual deformity, or distortion, of the limb by shortening, rotating, or angulating it.

    A. closed fracture

    B. open fracture

    C. nondisplaced fracture

    D. displaced fracture

_____ **17.** When examining an injured extremity, you should compare the injured limb to:

    A. the opposite uninjured limb.

    B. one of your limbs or one of your partner's limbs.

    C. an injury chart.

    D. none of the above

_____ **18.** _____ is the most reliable indicator of an underlying fracture.

    A. Crepitus

    B. Deformity

    C. Point tenderness

    D. Absence of distal pulse

_____ **19.** A(n) _____ is a fracture that occurs in a growth section of a child's bone, which may prematurely stop growth if not properly treated.

    A. greenstick fracture

    B. comminuted fracture

    C. pathological fracture

    D. epiphyseal fracture

_____ **20.** A(n) _____ is an incomplete fracture that passes only partway through the shaft of a bone but may still cause severe angulation.

    A. greenstick fracture

    B. comminuted fracture

    C. pathological fracture

    D. epiphyseal fracture

_____ **21.** A(n) _____ is a fracture of a weakened or diseased bone, seen in patients with osteoporosis or cancer.

    A. greenstick fracture

    B. comminuted fracture

    C. pathological fracture

    D. epiphyseal fracture

_____ **22.** A(n) _____ is a fracture in which the bone is broken into two or more fragments.

    A. greenstick fracture

    B. comminuted fracture

    C. pathological fracture

    D. epiphyseal fracture

5

23. Rapid swelling usually indicates _____ from a fracture site and is typically followed by severe pain.
    A. bleeding
    B. laceration
    C. locked joint
    D. compartment syndrome

24. Fractures are almost always associated with _____ of the surrounding soft tissue.
    A. laceration
    B. crepitus
    C. ecchymosis
    D. swelling

25. Signs and symptoms of a dislocated joint include:
    A. marked deformity.
    B. tenderness or palpation.
    C. locked joint.
    D. all of the above

26. Signs and symptoms of sprains include all of the following, except:
    A. point tenderness.
    B. pain prevents the patient from moving or using the limb normally.
    C. marked deformity.
    D. instability of the joint is indicated by increased motion.

27. Assessment of patients with musculoskeletal injuries must include:
    A. initial assessment followed by a focused physical exam.
    B. evaluation of neurovascular function.
    C. applying oxygen as needed.
    D. all of the above

28. Compartment syndrome:
    A. occurs within 6 to 12 hours after injury.
    B. usually is a result of excessive bleeding, a severely crushed extremity, or the rapid return of blood to an ischemic limb.
    C. is characterized by pain that is out of proportion to the injury.
    D. all of the above

29. Always check neurovascular function:
    A. after any manipulation of the limb.
    B. before applying a splint.
    C. after applying a splint.
    D. all of the above

30. Splinting will help to prevent:
    A. excessive bleeding of the tissues at the injury site caused by broken bone ends.
    B. laceration of the skin by broken bone ends.
    C. increased pain from movement of bone ends.
    D. all of the above

_____ **31.** In-line _____ is the act of exerting a pulling force on a body structure in the direction of its normal alignment.

    A. stabilization

    B. immobilization

    C. traction

    D. direction

_____ **32.** Basic types of splints include:

    A. rigid.

    B. formable.

    C. traction.

    D. all of the above

_____ **33.** Do not use traction splints for any of the following conditions, except:

    A. injuries of the pelvis.

    B. an isolated femur fracture.

    C. partial amputation or avulsions with bone separation.

    D. lower leg or ankle injury.

_____ **34.** Hazards associated with improper application of splints include:

    A. compression of nerves, tissues, and blood vessels.

    B. delay in transport of a patient with a life-threatening injury.

    C. reduction of distal circulation if the splint is too tight.

    D. all of the above

_____ **35.** The _____ is one of the most commonly fractured bones in the body.

    A. scapula

    B. clavicle

    C. humerus

    D. radius

_____ **36.** Indications that blood vessels have likely been injured include:

    A. a cold, pale hand.

    B. weak or absent pulse.

    C. poor capillary refill.

    D. all of the above

_____ **37.** Signs and symptoms associated with hip dislocation include:

    A. severe pain in the hip.

    B. lateral and posterior aspects of the hip region will be tender on palpation.

    C. you may be able to palpate the femoral head deep within the muscles of the buttock.

    D. all of the above

_____ **38.** There is always a significant amount of blood loss, as much as _____ mL, after a fracture of the shaft of the femur.

    A. 100 to 250

    B. 250 to 500

    C. 500 to 1,000

    D. 100 to 1,500

5

_____39. The knee is especially susceptible to _____ injuries, which occur when abnormal bending or twisting forces are applied to the joint.

    A. tendon

    B. ligament

    C. dislocation

    D. fracture-dislocation

_____40. Signs and symptoms of knee ligament injury include:

    A. swelling.

    B. point tenderness.

    C. joint effusion.

    D. all of the above

_____41. Although substantial ligament damage always occurs with a knee dislocation, the more urgent injury is to the _____ artery, which is often lacerated or compressed by the displaced tibia.

    A. tibial

    B. femoral

    C. popliteal

    D. dorsalis pedis

_____42. Because of local tenderness and swelling, it is easy to confuse a nondisplaced or minimally displaced fracture about the knee with a:

    A. tendon injury.

    B. ligament injury.

    C. dislocation.

    D. fracture-dislocation.

_____43. Fracture of the tibia and fibula are often associated with _____ as a result of the distorted positions of the limb following injury.

    A. vascular injury

    B. muscular injury

    C. tendon injury

    D. ligament injury

_____44. The _____ is the most commonly injured joint.

    A. knee

    B. elbow

    C. ankle

    D. hip

Labeling

Label the following diagram with the correct terms.

1. The Human Skeleton

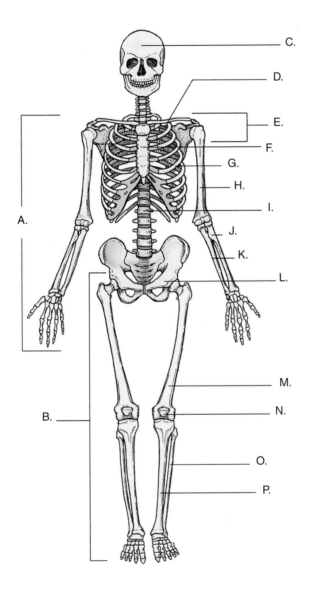

A. _____

B. _____

C. _____

D. _____

E. _____

F. _____

G. _____

H. _____

I. _____

J. _____

K. _____

L. _____

M. _____

N. _____

O. _____

P. _____

5

Vocabulary

Define the following terms using the space provided.

1. Acromioclavicular (A/C) joint:

_____

_____

_____

2. Compartment syndrome:

_____

_____

_____

3. Dislocation:

_____

_____

_____

4. Nondisplaced fracture:

_____

_____

_____

5. Position of function:

_____

_____

_____

6. Sling:

_____

_____

_____

7. Swathe:

_____

_____

_____

Fill-in

Read each item carefully, then complete the statement by filling in the missing word.

1. Atrophy is the _____ of muscle tissue.

2. Bone marrow produces _____ blood cells.

3. The knee and elbow are _____ and socket joints.

4. The _____ is one of the most commonly fractured bones in the body.

5. Always carefully assess the _____ to try to determine the amount of kinetic energy that an injured limb has absorbed.

6. Penetrating injury should alert you to the possibility of a(n) _____.

7. The _____ is the most important nerve in the lower extremity; it controls the activity of muscles in the thigh and below the knee.

8. The _____ is the longest and largest bone in the body.

9. A grating or grinding sensation known as _____ can be felt and sometimes even heard when fractured bone ends rub together.

10. A dislocated joint sometimes will spontaneously _____, or return to its normal position.

11. If you suspect that a patient has compartment syndrome, splint the affected limb, keeping it at the level of the heart, and provide immediate transport, checking

_____ frequently during transport.

True/False

If you believe the statement to be more true than false, write the letter "T" in the space provided. If you believe the statement to be more false than true, write the letter "F."

1. _____ All extremity injuries should be splinted before moving a patient unless the patient's life is in immediate danger.

2. _____ Splinting reduces pain and prevents the motion of bone fragments.

3. _____ You should use traction to reduce a fracture and force all bone fragments back into alignment.

4. _____ When applying traction, the direction of pull is always along the axis of the limb.

5. _____ Cover wounds with a dry, sterile dressing before applying a splint.

6. _____ When splinting a fracture, you should be careful to immobilize only the joint above the injury site.

7. _____ One of the steps of the neurological examination is to palpate the pulse distal to the point of injury.

8. _____ Assessment of neurovascular function should be repeated every 5 to 10 minutes until the patient arrives at the hospital.

9. _____ A patient's ability to sense light touch in the fingers and toes distal to the injury site is a good indication that the nerve supply is intact.

5

Short Answer

Complete this section with short written answers using the space provided.

1. List the four types of forces that may cause injury to a limb.

_____

_____

_____

_____

2. List five of the signs associated with a possible fracture.

_____

_____

_____

_____

_____

3. List the four items to check when assessing neurovascular function.

_____

_____

_____

_____

4. List the general principles of splinting.

_____

_____

_____

5. What are the three goals of in-line traction?

_____

_____

_____

Word Fun

The following crossword puzzle is an activity provided to reinforce correct spelling and understanding of medical terminology associated with emergency care and the EMT-B. Use the clues in the column to complete the puzzle.

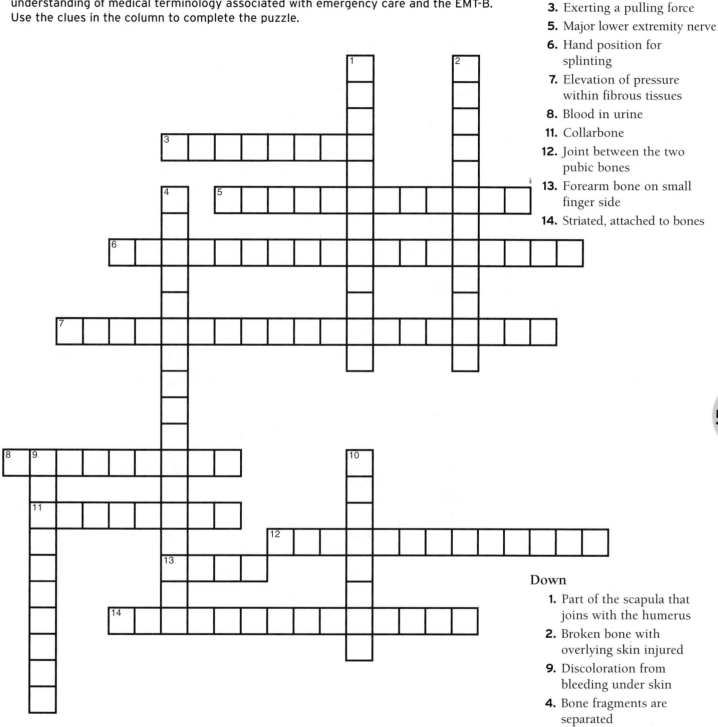

CLUES

Across

3. Exerting a pulling force
5. Major lower extremity nerve
6. Hand position for splinting
7. Elevation of pressure within fibrous tissues
8. Blood in urine
11. Collarbone
12. Joint between the two pubic bones
13. Forearm bone on small finger side
14. Striated, attached to bones

Down

1. Part of the scapula that joins with the humerus
2. Broken bone with overlying skin injured
9. Discoloration from bleeding under skin
4. Bone fragments are separated
10. Grating sound of bone ends

5

Ambulance Calls

The following real case scenarios provide an opportunity to explore the concerns associated with patient management. Read each scenario, then answer each question in detail.

1. You are dispatched to a construction site for a 27-year-old male complaining of severe thoracic pain posteriorly. Coworkers tell you he was hit in the upper back by the bucket of a backhoe. He is alert and oriented and closer inspection reveals bruising and deformity over the left scapula with pain and crepitus on palpation. How would you best manage this patient?

_____

_____

_____

_____

_____

_____

_____

_____

2. You are called to a local park where an 11-year-old girl fell off the parallel bars onto her right elbow. She is cradling the arm to her chest. She has obvious swelling and deformity in the area. She has good pulse, motor, and sensation at the wrist. ABCs are normal. How would you best manage this patient?

_____

_____

_____

_____

_____

_____

_____

_____

3. You are dispatched to a rollover motor vehicle crash. Your patient is a 29-year-old female unrestrained driver complaining of severe lower back pain. She is slightly tachycardic, but the ABCs are all normal. While performing a rapid trauma assessment, you find crepitus and an unstable pelvis. She is becoming less responsive.

How would you best manage this patient?

_____

_____

_____

_____

_____

_____

_____

_____

_____

5

Skill Drills

Skill Drill 29-1: Assessing Neurovascular Pulse
Test your knowledge of this skill drill by filling in the correct words in the photo captions.

1. Palpate the _____ pulse in the upper extremity.

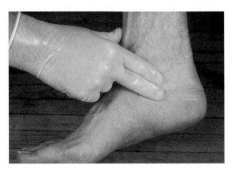

2. Palpate the _____ _____ pulse in the lower extremity.

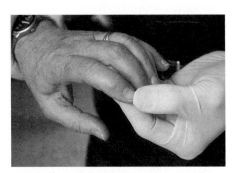

3. Assess capillary refill by blanching a fingernail or _____.

4. Assess sensation on the flesh near the _____ of the _____ finger.

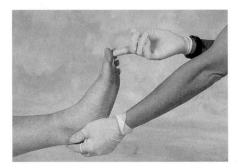

5. On the foot, first check sensation on the flesh near the _____ of the _____ _____.

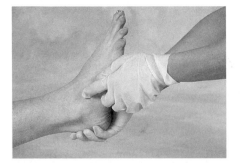

6. Also check foot sensation on the _____ _____.

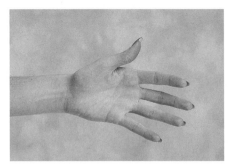

7. Evaluate motor function by asking the patient to _____ the hand. (Perform motor tests only if the hand or foot is not _____. _____ a test if it causes pain.)

8. Also ask the patient to _____ _____ _____.

(continued)

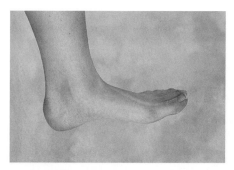
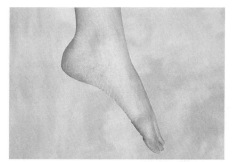

9. To evaluate motor function in the foot, ask the patient to _____ the foot.

10. Also have the patient _____ the foot and _____ the toes.

Skill Drill 29-2: Caring for Musculoskeletal Injuries
Test your knowledge of this skill drill by filling in the correct words in the photo captions.

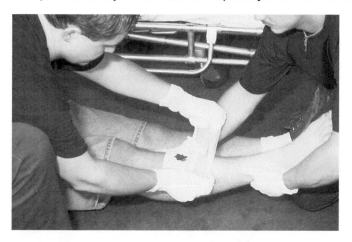

1. Cover open wounds with a _____, _____ dressing, and _____ _____ to control bleeding.

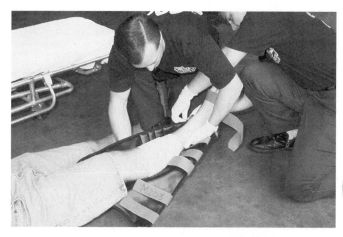

2. Apply a splint and elevate the extremity about _____ inch(es) (slightly above the level of the _____).

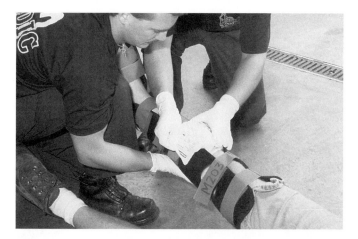

3. Apply _____ _____ if there is swelling, but do not place them _____ on the skin.

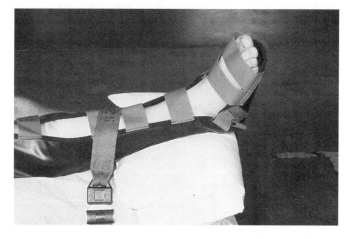

4. _____ the patient for transport and _____ the injured area.

5

Skill Drill 29-3: Applying a Rigid Splint
Test your knowledge of this skill drill by filling in the correct words in the photo captions.

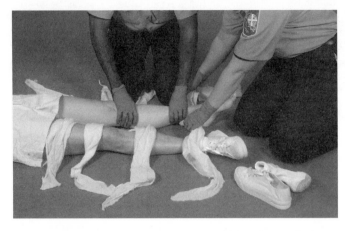

1. Provide gentle _____ and _____ _____ of the limb.

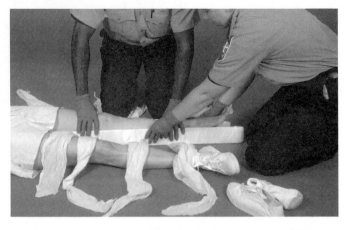

2. Second EMT-B places the splint _____ or _____ the limb.

_____ between the limb and the splint as needed to ensure even pressure and contact.

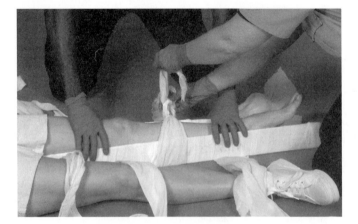

3. Secure the splint to the limb with _____.

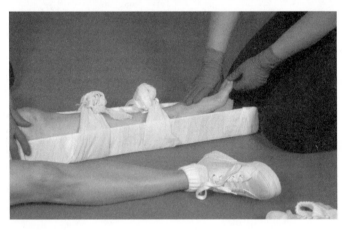

4. Assess and record _____ _____ function.

Skill Drill 29-4: Applying a Zippered Air Splint
Test your knowledge of this skill drill by filling in the correct words in the photo captions.

1. Support the injured limb and apply gentle _____ as your partner applies the open, deflated splint.

2. Zip up the splint, inflate it by _____ or by _____, and test the _____. Check and record _____ _____ function.

Skill Drill 29-5: Applying an Unzipped Air Splint
Test your knowledge of this skill drill by filling in the correct words in the photo captions.

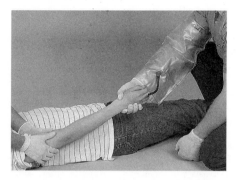 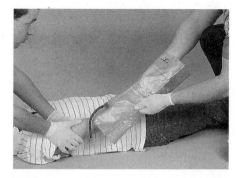 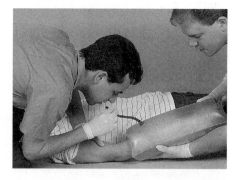

1. _____ the injured limb. Have your partner place his or her arm through the splint to grasp the patient's _____ or _____.

2. Apply gentle _____ while sliding the splint onto the injured limb.

3. _____ the splint.

Skill Drill 29-6: Applying a Vacuum Splint
Test your knowledge of this skill drill by filling in the correct words in the photo captions.

1. _____ and _____ the injury.

2. Place the splint and _____ it around the limb.

3. _____ the air _____ _____ the splint and _____ the valve.

5

Skill Drill 29-7: Applying a Hare Traction Splint

Test your knowledge of this skill drill by placing the photos below in the correct order. Number the first step with a "1," the second step with a "2," etc.

_____

Slide the splint into position under the injured limb.

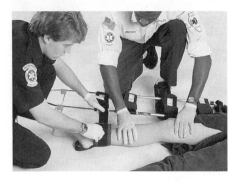

_____

Support the injured limb as your partner fastens the ankle hitch about the foot and ankle.

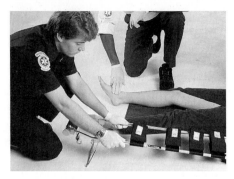

_____

Expose the injured limb and check pulse, motor, and sensory function.

Place the splint beside the uninjured limb, adjust the splint to proper length, and prepare the straps.

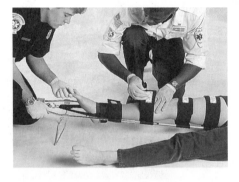

_____

Secure and check support straps. Assess pulse, motor, and sensory functions.

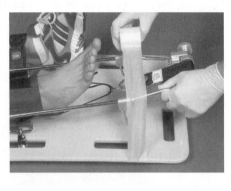

_____

Secure the patient and splint to the backboard in a way that will prevent movement of the splint during patient movement and transport.

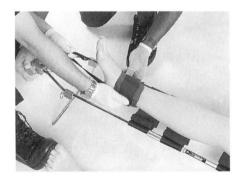

_____

Connect the loops of the ankle hitch to the end of the splint as your partner continues to maintain traction. Carefully tighten the ratchet to the point that the splint holds adequate traction.

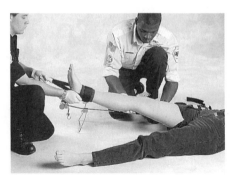

_____

Continue to support the limb as your partner applies gentle in-line traction to the ankle hitch and foot.

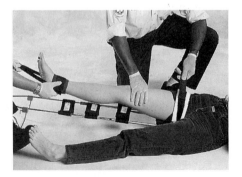

_____

Pad the groin and fasten the ischial strap.

Skill Drill 29-8: Applying a Sager Traction Splint

Test your knowledge of this skill drill by placing the photos below in the correct order. Number the first step with a "1," the second step with a "2," etc.

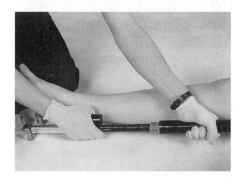

_____

Estimate the proper length of the splint by placing it next to the injured limb.

Fit the ankle pads to the ankle.

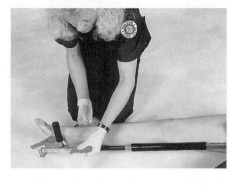

_____

Tighten the ankle harness just above the malleoli.

Snug the cable ring against the bottom of the foot.

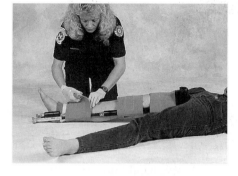

_____

Secure the splint with elasticized cravats.

_____

Extend the splint's inner shaft to apply traction of about 10% of body weight.

_____

Place the splint at the inner thigh, apply the thigh strap at the upper thigh, and secure snugly.

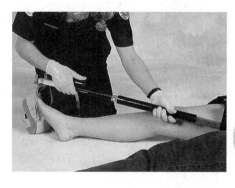

_____

After exposing the injured area, check the patient's pulse and motor and sensory function.

Adjust the thigh strap so that it lies anteriorly when secured.

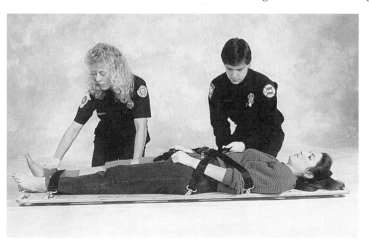

_____

Secure the patient to a long spine board.

Check pulse, motor and sensory functions.

Skill Drill 29-9: Splinting the Hand and Wrist

Test your knowledge of this skill drill by filling in the correct words in the photo captions.

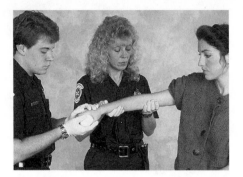

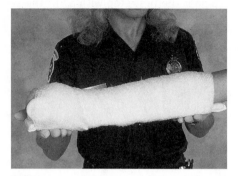

1. Move the hand into the _____ _____ _____. Place a soft _____ _____ in the palm.

2. Apply a _____ _____ splint on the _____ side with fingers _____.

3. Secure the splint with a _____ _____.

Notes

Workbook Activities

The following activities have been designed to help you. Your instructor may require you to complete some or all of these activities as a regular part of your EMT-B training program. You are encouraged to complete any activity that your instructor does not assign as a way to enhance your learning in the classroom.

Chapter Review

The following exercises provide an opportunity to refresh your knowledge of this chapter.

NOTES

Matching

Match each of the terms in the left column to the appropriate definition in the right column.

_____ 1. Cerebellum

_____ 2. Brain stem

_____ 3. Somatic nervous system

_____ 4. Autonomic nervous system

_____ 5. Spinal column

_____ 6. Central nervous system

_____ 7. Cerebral edema

_____ 8. Connecting nerves

_____ 9. Intervertebral disk

_____ 10. Meninges

A. consists of 33 bones

B. swelling of the brain

C. the brain and spinal cord

D. controls movement

E. the part of the central nervous system that controls virtually all the functions that are absolutely necessary for life

F. three distinct layers of tissue that surround and protect the brain and spinal cord within the skull and spinal cord

G. the part of the nervous system that regulates involuntary functions

H. the part of the nervous system that regulates voluntary activities

I. located in the brain and spinal cord, these connect the motor and sensory nerves

J. cushion that lies between the vertebrae

Multiple Choice

Read each item carefully, then select the best response.

_____ 1. The nervous system includes:

A. the brain.

B. the spinal cord.

C. billions of nerve fibers.

D. all of the above

CHAPTER 30

Head and Spine Injuries

_____ **2.** The nervous system is divided into two parts: the central nervous system and the:

   A. autonomic nervous system.

   B. peripheral nervous system.

   C. sympathetic nervous system.

   D. somatic nervous system.

_____ **3.** The brain is divided into three major areas: the cerebrum, the cerebellum, and the:

   A. foramen magnum.

   B. meninges.

   C. brain stem.

   D. spinal column.

_____ **4.** Injury to the head and neck may indicate injury to the:

   A. thoracic spine.

   B. lumbar spine.

   C. cervical spine.

   D. sacral spine.

_____ **5.** The _____ is composed of three layers of tissue that surround the brain and spinal cord within the skull and spinal canal.

   A. meninges

   B. dura mater

   C. pia mater

   D. arachnoid space

_____ **6.** The skull is divided into two large structures: the cranium and the:

   A. occipital.

   B. face.

   C. parietal.

   D. foramen magnum.

_____ **7.** Peripheral nerves include:

   A. connecting nerves.

   B. sensory nerves.

   C. motor nerves.

   D. all of the above

_____ **8.** The brain and spinal cord float in cerebrospinal fluid (CSF), which:

   A. acts as a shock absorber.

   B. bathes the brain and spinal cord.

   C. buffers them from injury.

   D. all of the above

_____ **9.** The autonomic nervous system is composed of two parts: the sympathetic nervous system and the:

   A. peripheral nervous system.

   B. central nervous system.

   C. parasympathetic nervous system.

   D. somatic nervous system.

_____ **10.** The most prominent and the most easily palpable spinous process is at the _____ cervical vertebra at the base of the neck.

   A. 7th

   B. 6th

   C. 5th

   D. 4th

_____ **11.** When identifying the mechanism of injury of an unresponsive patient, _____ may have helpful information.

   A. first responders

   B. family members

   C. bystanders

   D. all of the above

_____ **12.** Emergency medical care of a patient with a possible spinal injury begins with:

   A. opening the airway.

   B. level of consciousness.

   C. scene of safety.

   D. BSI precautions.

_____ **13.** The _____ is a tunnel running the length of the spine, which encloses and protects the spinal cord.

   A. foramen magnum

   B. spinal canal

   C. foramen foranina

   D. meninges

_____ **14.** Once the head and neck are manually stabilized, you should assess:

   A. pulse.

   B. motor function.

   C. sensation.

   D. all of the above

_____ **15.** You must maintain manual stabilization of the head until:

   A. the patient's head and torso are in line.

   B. the patient is secured to a backboard with the head immobilized.

   C. the rigid cervical collar is in place.

   D. the patient arrives at the hospital.

_____ **16.** The ideal procedure for moving a patient from the ground to the backboard is the:

   A. four-person log roll.

   B. lateral slide.

   C. four-person lift.

   D. push and pull maneuver.

_____ **17.** You can almost always control bleeding from a scalp laceration by:

   A. direct pressure.

   B. elevation.

   C. pressure point.

   D. tourniquet.

_____ **18.** Exceptions to using a short spinal extrication device include all of the following, except:

   A. you or the patient is in danger.

   B. the patient is conscious and complaining of lumbar pain.

   C. you need to gain immediate access to other patients.

   D. the patient's injuries justify immediate removal.

_____ **19.** Applying excessive pressure to an open wound with a skull fracture could:

   A. increase intracranial pressure.

   B. push bone fragments into the brain.

   C. increase the size of the soft-tissue injury.

   D. all of the above

_____ **20.** A _____ is a temporary loss or alteration of a part or all of the brain's abilities to function without actual physical damage to the brain:

   A. contusion

   B. concussion

   C. hematoma

   D. subdural hematoma

_____ **21.** Symptoms of a concussion include:

   A. dizziness.

   B. weakness.

   C. visual changes.

   D. all of the above

_____ **22.** Intracranial bleeding outside of the dura and under the skull is known as a(n):

   A. concussion.

   B. intracerebral hemorrhage.

   C. subdural hematoma.

   D. epidural hematoma.

5

_____23. The difference in signs and symptoms of traumatic vs. nontraumatic brain injuries is the:

    A. lack of altered mental status.

    B. lack of mechanism of injury.

    C. lack of swelling.

    D. increase in blood pressure.

_____24. _____ is the most reliable sign of a closed head injury.

    A. Vomiting

    B. Decreased LOC

    C. Seizures

    D. Numbness and tingling in extremities

_____25. _____ is one of the most common, and one of the most serious, complications of a head injury.

    A. Cyanosis

    B. Hypoxia

    C. Vomiting

    D. Cerebral edema

_____26. Common causes of head injuries include all of the following, except:

    A. direct blows.

    B. motor vehicle crashes.

    C. seizure activity.

    D. sports injuries.

_____27. Assessment of mental status is accomplished through the use of the mnemonic:

    A. SAMPLE.

    B. OPQRST.

    C. AVPU.

    D. AEIOU-TIPS.

_____28. Unequal pupil size may indicate:

    A. increased intracranial pressure.

    B. a congenital problem.

    C. damage to the nerves that control dilation and constriction.

    D. all of the above

_____29. Patients with head injuries often have injuries to the _____ as well.

    A. face

    B. torso

    C. cervical spine

    D. extremities

_____30. Proper order of treatment for traumatic head injuries includes:

    A. scene safety, airway, LOC with c-spine control, breathing, circulation.

    B. LOC with c-spine control, airway, breathing, circulation.

    C. LOC, airway, breathing, circulation, c-spine.

    D. BSI, ABCs, LOC, c-spine control.

_____31. A cervical collar should be applied to a patient with a possible spinal injury based on:

    A. the mechanism of injury.

    B. the history.

    C. signs and symptoms.

    D. all of the above

_____**32.** Helmets must be removed in all of the following cases, except:

    A. cardiac arrest.

    B. when the helmet allows for excessive movement.

    C. when there are no impending airway or breathing problems.

    D. when a shield cannot be removed.

_____**33.** Your best choice of action for a child involved in a motor vehicle crash and found in their car seat is to:

    A. immobilize the child in the car seat.

    B. rule out spinal injury and place the child with a parent.

    C. pad sides of car seat but leave space to allow for lateral movement.

    D. move the child to a pediatric immobilization device.

Labeling

Label the following diagrams with the correct terms.

1. The Brain

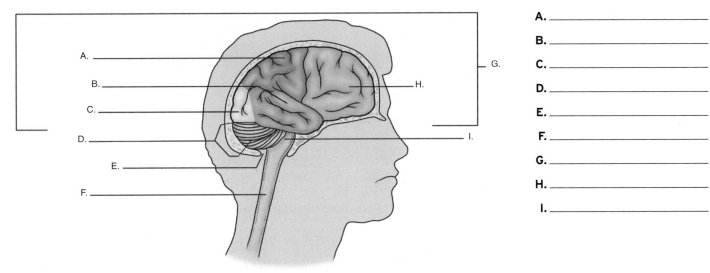

A. _____

B. _____

C. _____

D. _____

E. _____

F. _____

G. _____

H. _____

I. _____

2. The Connecting Nerves in the Spinal Cord

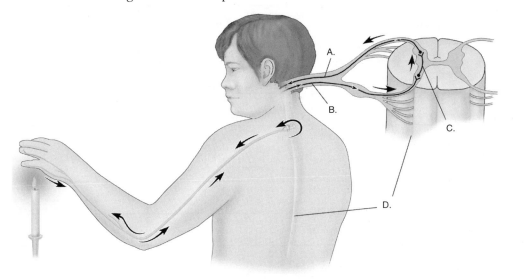

A. _____

B. _____

C. _____

D. _____

3. The Spinal Column

A. _____

B. _____

C. _____

D. _____

E. _____

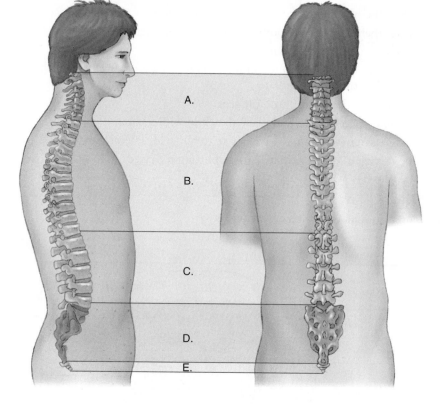

Vocabulary

Define the following terms using the space provided.

1. Retrograde amnesia:

_____

_____

_____

2. Anterograde (posttraumatic) amnesia:

_____

_____

_____

3. Closed head injury

_____

_____

_____

4. Eyes-forward position:

_____

_____

_____

5. Open head injury:

_____

_____

_____

Fill-in
Read each item carefully, then complete the statement by filling in the missing word(s).

1. The _____ nerves carry information to the muscles.

2. The dura mater, arachnoid, and pia mater are layers of _____ within the skull and spinal canal.

3. The brain and spinal cord are part of the _____ nervous system.

4. Within the peripheral nervous system, there are _____ pairs of spinal nerves.

5. The _____ nerves pass through holes in the skull and transmit sensations directly to the brain.

6. Vertebrae are separated by cushions called _____ .

7. The skull has two large structures of bone, the _____ and the

_____ .

8. The _____ and _____ produce cerebrospinal fluid (CSF).

9. The _____ nervous system reacts to stress.

10. The _____ nervous system causes the body to relax.

True/False
If you believe the statement to be more true than false, write the letter "T" in the space provided. If you believe the statement to be more false than true, write the letter "F."

1. _____   A distracted spine has been moved laterally.

2. _____   If a sensory nerve in the reflex arc detects an irritating stimulus, it will bypass the motor nerve and send a message directly to the brain.

3. _____   Voluntary activities are those actions we perform unconsciously.

4. _____   The autonomic nervous system is composed of the sympathetic nervous system and the parasympathetic nervous system.

5. _____ The parasympathetic nervous system reacts to stress with the fight-or-flight response whenever it is confronted with a threatening situation.

6. _____ All patients with suspected head and/or spine injuries should have their head realigned to an in-line neutral position.

7. _____ When assessing a patient for possible spinal injury, you should begin with a focused history and physical exam.

Short Answer

Complete this section with short written answers using the space provided.

1. List the five basic questions to ask a conscious patient when conducting an assessment of a head or head and spine injury.

_____

_____

_____

_____

_____

2. List the reasons for not placing the head/spine injury patient's head into a neutral in-line position.

_____

_____

_____

3. List the three major types of brain injuries.

_____

_____

_____

4. List at least five signs and symptoms of a head injury.

_____

_____

_____

_____

_____

5. List the three general principles for treating a head injury.

_____

_____

_____

6. List the six questions to ask yourself when deciding whether or not to remove a helmet.

_____

_____

_____

_____

_____

_____

5

Word Fun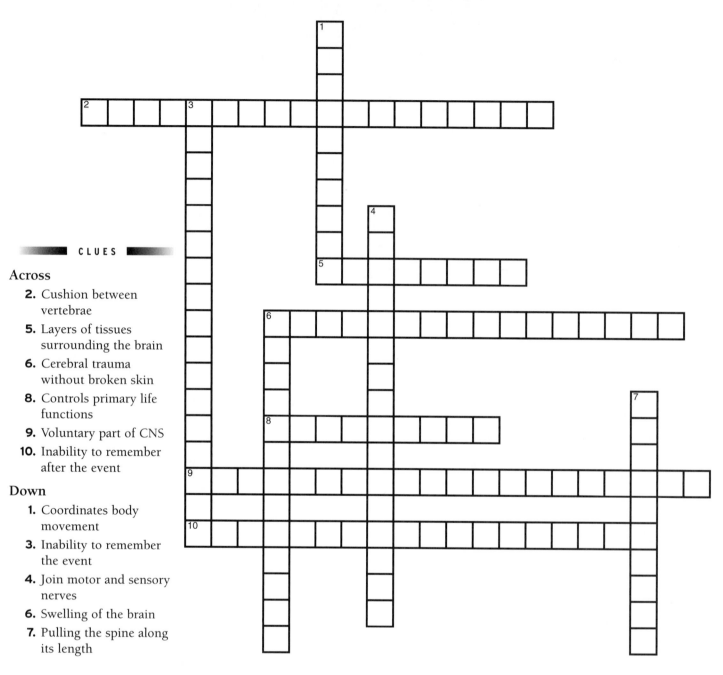

The following crossword puzzle is an activity provided to reinforce correct spelling and understanding of medical terminology associated with emergency care and the EMT-B. Use the clues in the column to complete the puzzle.

◼ CLUES ◼

Across

2. Cushion between vertebrae
5. Layers of tissues surrounding the brain
6. Cerebral trauma without broken skin
8. Controls primary life functions
9. Voluntary part of CNS
10. Inability to remember after the event

Down

1. Coordinates body movement
3. Inability to remember the event
4. Join motor and sensory nerves
6. Swelling of the brain
7. Pulling the spine along its length

Ambulance Calls

The following real case scenarios provide an opportunity to explore the concerns associated with patient management. Read each scenario, then answer each question in detail.

1. You are dispatched to a moderate damage motor vehicle crash. Your patient is a 52-year-old female, restrained driver, hit from the rear, complaining of neck and back pain. She is alert and oriented and her vital signs are within the normal limits. Pulse, motor, and sensory functions are intact in all extremities.

   How would you best manage this patient?

   _____

   _____

   _____

   _____

   _____

   _____

   _____

   _____

   _____

2. You are called to the scene of a baseball game where a 10-year-old boy was accidentally hit with a baseball bat on the left side of his head. He has a depression in the left temporal region and severe vomiting. He is pain responsive and bleeding is minimal.

   How would you best manage this patient?

   _____

   _____

   _____

   _____

   _____

   _____

   _____

   _____

5

3. You are dispatched to a motor vehicle crash with major damage to the patient compartment. Your patient, an 18-month-old male, is still in his car seat in the center of the back seat. He responds appropriately and there is no damage to his seat. He has no visible injuries, but a front seat passenger was killed.

How would you best manage this patient?

_____

_____

_____

_____

_____

_____

_____

_____

Skill Drills

Skill Drill 30-1: Performing Manual In-Line Stabilization
Test your knowledge of this skill drill by filling in the correct words in the photo captions.

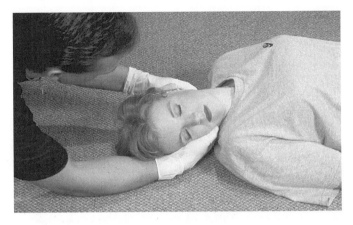

1. Kneel behind the patient and place your hands firmly around the _____ of the _____ on either _____.

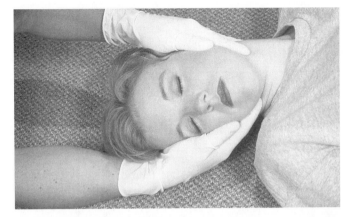

2. Support the lower jaw with your _____ and _____ fingers, and the head with your _____.

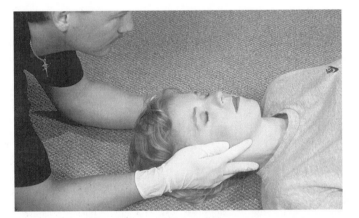

3. Gently _____ the head into a _____, _____ position, aligned with the torso. Do not _____ the head or neck excessively.

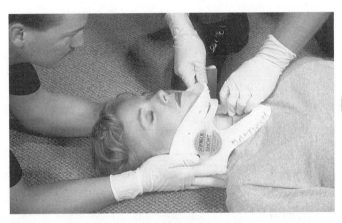

4. Continue to _____ the head manually while your partner places a rigid _____ _____ around the neck. Maintain _____ _____ until you have the patient secured to a backboard.

5

Skill Drill 30-2: Immobilizing a Patient to a Long Backboard

Test your knowledge of this skill drill by placing the photos below in the correct order. Number the first step with a "1," the second step with a "2," etc. Also, fill in the correct words in the photo captions.

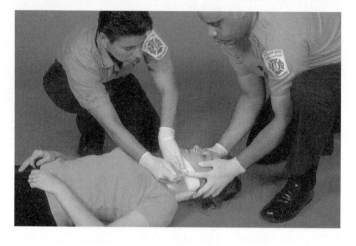

_____

Apply a_____ _____.

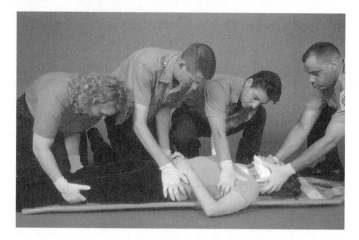

_____

_____ the patient on the board.

_____

Place _____ across the patient's forehead.

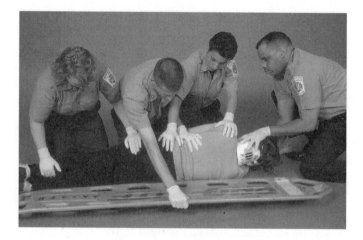

_____

On command, rescuers _____ the patient toward themselves, quickly examine the _____, slide the backboard under the patient, and roll the patient onto the board.

(continued)

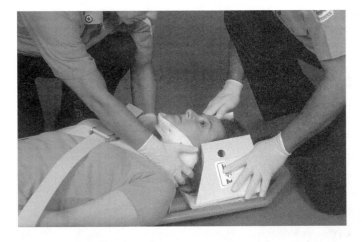

Begin to secure the patient's head using a commercial immobilization device or _____ _____.

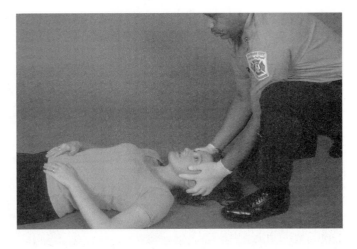

Apply and maintain _____ _____.
Assess _____ _____ in all extremities.

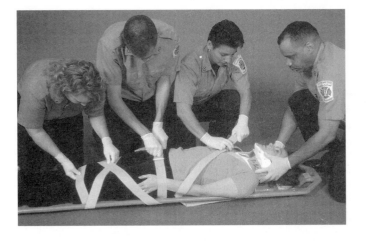

Secure the _____, _____, and _____ _____.

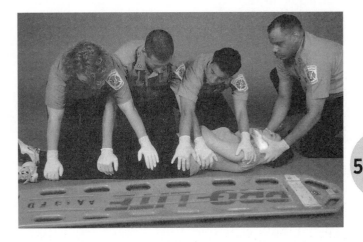

Rescuers _____ on one side of the patient and place _____ on the far side of the patient.

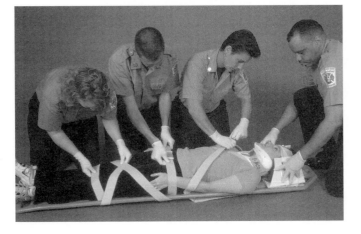

Check all _____ and readjust as needed.
Reassess _____ _____ in all extremities.

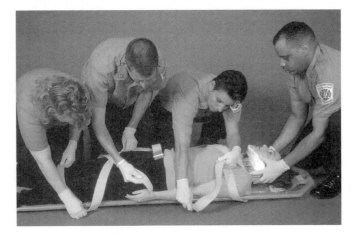

Secure the _____ _____ first.

Skill Drill 30-3: Immobilizing a Patient Found in a Sitting Position
Test your knowledge of this skill drill by placing the photos below in the correct order.
Number the first step with a "1," the second step with a "2," etc.

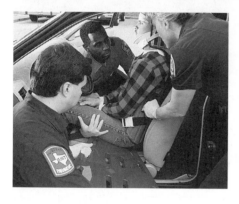

Wedge a long backboard next to the patient's buttocks.

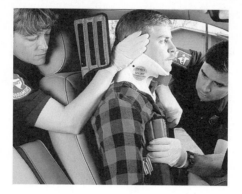

Open the side flaps, and position them around the patient's torso, snug around the armpits.

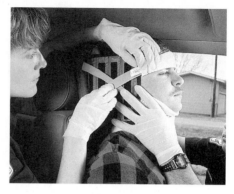

Pad between the head and the device as needed.
Secure the forehead strap and fasten the lower head strap around the collar.

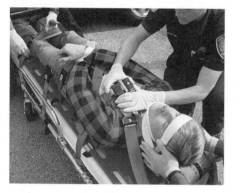

Secure the immobilization devices to each other.
Reassess pulse, motor, and sensory functions in each extremity.

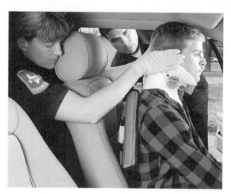

Insert a short spine immobilization device between the patient's upper back and the seat.

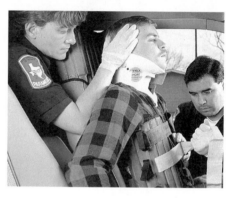

Secure the upper torso flaps, then the midtorso flaps.

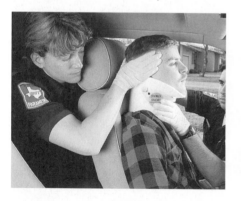

Stabilize the head and neck in a neutral, in-line position.
Assess pulse, motor, and sensory function in each extremity.
Apply a cervical collar.

Secure the groin (leg) straps. Check and adjust torso straps.

Turn and lower the patient onto the long board.
Lift the patient, and slip the long board under the spine device.

Skill Drill 30-4: Immobilizing a Patient Found in a Standing Position
Test your knowledge of this skill drill by filling in the correct words in the photo captions.

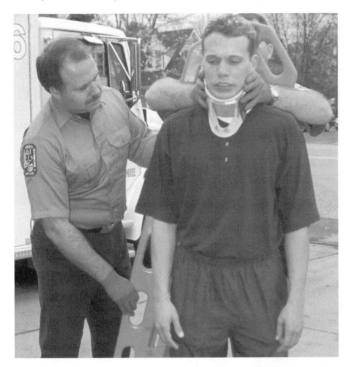

1. After _____ stabilizing the head and neck, apply a _____ _____. Position the board _____ the patient.

2. Position EMT-Bs at _____ and _____ the patient. Side EMT-Bs reach under patient's _____ and grasp _____ at or slightly above _____ level.

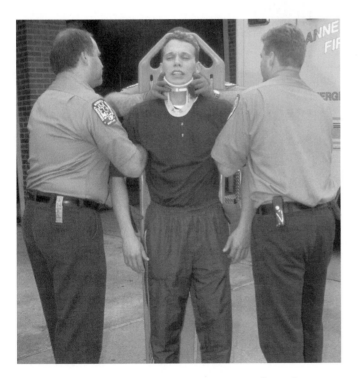

3. Prepare to lower the patient. EMT-Bs on the sides should be _____ the EMT-B at the head and _____ for his or her _____.

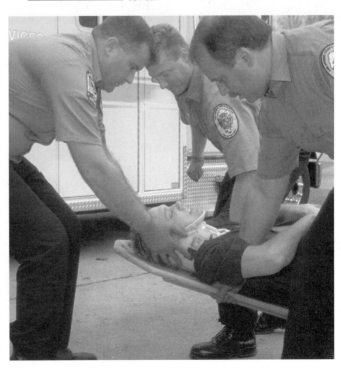

4. On command, _____ the backboard to the ground.

5

Skill Drill 30-5: Application of a Cervical Collar

Test your knowledge of this skill drill by filling in the correct words in the photo captions.

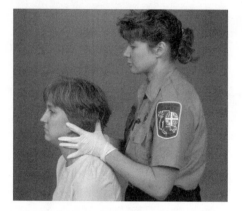

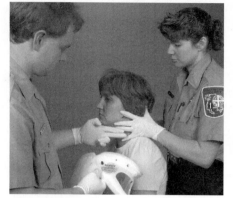

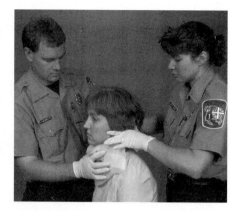

1. Apply _____ stabilization.

2. Measure the proper _____ _____.

3. Place the _____ _____ first.

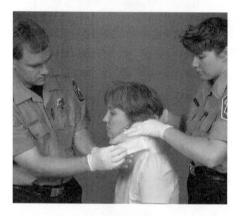

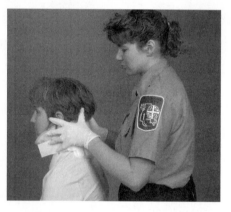

4. _____ the collar around the neck and _____ the collar.

5. Assure proper _____ and maintain _____, _____ stabilization.

Skill Drill 30-6: Removing a Helmet

Test your knowledge of this skill drill by placing the photos below in the correct order. Number the first step with a "1," the second step with a "2," etc. Also, fill in the correct words in the photo captions.

_____

Prevent head movement by placing your _____ on either side of the helmet and fingers on the _____ _____. Have your partner _____ the strap.

_____

Kneel down at the patient's head with your _____ at one side. Open the face shield to assess _____ and _____. Remove _____ if present.

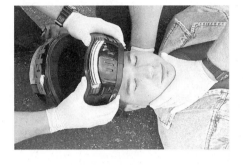

_____

Have your partner slide the hand from the _____ to the _____ of the head to prevent it from snapping back.

_____

Gently slip the helmet about _____ off, then stop.

_____

Remove the helmet and _____ the cervical spine. Apply a _____ _____ and secure the patient to a _____ _____. _____ as needed to prevent neck flexion or extension.

_____

Have your partner place one hand at the _____ of the _____ _____ and the other at the _____.

5

Workbook Activities

The following activities have been designed to help you. Your instructor may require you to complete some or all of these activities as a regular part of your EMT-B training program. You are encouraged to complete any activity that your instructor does not assign as a way to enhance your learning in the classroom.

Chapter Review

The following exercises provide an opportunity to refresh your knowledge of this chapter.

■ NOTES ■

Matching

Match each of the terms in the left column to the appropriate definition in the right column.

_____ **1.** Gastrostomy tube   **A.** ages 12 to 18 years

_____ **2.** Shunt   **B.** ages 3 to 6 years

_____ **3.** Tracheostomy tube   **C.** soft openings within the skull of an infant

_____ **4.** Toddler   **D.** specialized medical practice devoted to the care of children

_____ **5.** Preschool-age children   **E.** used for breathing

_____ **6.** Adolescents   **F.** diverts excess cerebrospinal fluid

_____ **7.** Infancy   **G.** first month after birth

_____ **8.** Neonate   **H.** the first year of life

_____ **9.** Fontanels   **I.** used for feeding

_____ **10.** Pediatrics   **J.** after infancy, until about 3 years of age

Multiple Choice

Read each item carefully, then select the best response.

_____ **1.** In addition to the tongue, the _____ help(s) to produce a smaller opening to move air easily.

    A. tonsils

    B. adenoids

    C. soft pallet

    D. all of the above

_____ **2.** A respiratory rate of _____ breaths/min is normal for the newborn.

    A. 12 to 20

    B. 20 to 40

    C. 30 to 50

    D. 40 to 60

Pediatric Assessment

_____ **3.** Breathing requires the use of the _____ and diaphragm.

    A. chest muscles

    B. neck muscles

    C. subclavian muscles

    D. abdominal muscles

_____ **4.** Anything that puts pressure on the abdomen of a young child can block the movement of the _____ and cause respiratory compromise.

    A. thorax

    B. air

    C. diaphragm

    D. lungs

_____ **5.** The primary method for the body to compensate for decreased oxygenation is to:

    A. increase the respiratory rate.

    B. increase the heart rate.

    C. increase the blood pressure.

    D. increase diaphragm contractions.

_____ **6.** Signs of vasoconstriction can include:

    A. weak peripheral pulses.

    B. delayed capillary refill.

    C. cool hands or feet.

    D. all of the above

_____ **7.** Infants respond mainly to _____ stimuli.

    A. social

    B. mental

    C. physical

    D. all of the above

_____ **8.** Look for _____ when doing your initial assessment of an infant from across the room.

A. work of breathing

B. skin color and alertness

C. level of activity

D. all of the above

_____ **9.** Injuries in the _____ age group are more frequent.

A. infant

B. toddler

C. preschool

D. school

_____ **10.** At the _____ age, children are easily distracted with counting games and small toys.

A. infant

B. toddler

C. preschool

D. adolescent

_____ **11.** As the work-of-breathing increases, you may see:

A. faster breathing.

B. retractions along the chest wall.

C. the child sitting in a position to allow for more chest expansion.

D. all of the above

_____ **12.** Assessment of the _____ will give clues to the amount of oxygen reaching the end-organs of the body.

A. heart rate

B. respiratory rate

C. level of responsiveness

D. skin color

_____ **13.** When evaluating the respiratory rate in children younger than 3 years and in infants, you should count the number of times the:

A. chest rises in 15 seconds.

B. chest rises in 30 seconds.

C. abdomen rises in 15 seconds.

D. abdomen rises in 30 seconds.

_____ **14.** When assessing a special needs child, you must first determine the child's:

A. baseline vital signs.

B. normal baseline status.

C. previous injuries.

D. over-the-counter medications.

_____ **15.** Your first priority in treating a special needs child includes:

A. obtaining an extensive history.

B. determining mode of transportation.

C. assessing the airway.

D. obtaining the patient's medications to take to the hospital.

_____ **16.** Potential problems associated with tracheostomy tubes include:

A. obstruction of the tube by mucous plugs.

B. bleeding.

C. air leakage around the tube.

D. all of the above

_____ **17.** Tubes that extend from the brain to the abdomen to drain excess cerebrospinal fluid that may accumulate near the brain are called:

A. shunts.

B. central lines.

C. G-tubes.

D. tracheostomy tubes.

Vocabulary emtb. vocab explorer

Define the following terms using the space provided.

1. Central IV lines:

_____

_____

_____

2. Fluid reservoir:

_____

_____

_____

3. Work-of-breathing (WOB):

_____

_____

_____

4. Mucous plugs:

_____

_____

_____

Fill-in

Read each item carefully, then complete the statement by filling in the missing word.

1. The specialized medical practice devoted to the care of the young is called

_____ .

2. The _____ is longer and more rounded compared to the size of the mandible, or lower jaw, in younger children.

3. In a child, the _____ is softer and narrower.

4. An infant's heart rate can become as high as _____ beats or more per minute if the body needs to compensate for injury or illness.

6

5. _____ is an early sign that the child may be compensating for decreased perfusion.

6. Infants have two soft openings within the skull called _____.

7. Most _____ are able to think abstractly and can participate in decision-making.

True/False

If you believe the statement to be more true than false, write the letter "T" in the space provided. If you believe the statement to be more false than true, write the letter "F."

1. _____ Toddlers often resist separation and demonstrate stranger anxiety.

2. _____ The skeletal system contains growth plates at the ends of long bones, which enable these bones to grow during childhood.

3. _____ Adulthood begins at age 18.

4. _____ Infants are usually afraid of strangers, because they are the center of attention in most families.

5. _____ Preschool-age children have a rich fantasy life, which can make them particularly fearful of pain and change involving their bodies.

6. _____ Normal respirations are a common sign of illness or injury in children.

7. _____ In infants, feel for a pulse over the brachial or femoral area.

8. _____ When checking capillary refill, color should return in less than 3 seconds after you let go.

9. _____ The parent or caregiver of a special needs child will be an important part of your assessment.

10. _____ Children with diabetes who receive insulin and tube feedings may become hyperglycemic quickly if tube feedings are discontinued.

11. _____ If a shunt becomes clogged due to infection, changes in mental status and respiratory arrest may occur.

12. _____ Infants and small children are not very susceptible to temperature changes.

13. _____ Be sure to keep the newborn's face covered, and make sure the ambulance is warm.

14. _____ When transporting children, do not allow the parent to hold the child during the actual transport.

Short Answer

Complete this section with short written answers using the space provided.

1. Discuss developmental considerations for infancy and approach for caregivers.

_____

_____

_____

2. Give four examples of special needs children.

_____

_____

_____

_____

3. Discuss developmental considerations for toddlers and approach for caregivers.

_____

_____

_____

4. Discuss developmental considerations for the school-age child and approach for caregivers.

_____

_____

_____

5. Discuss developmental considerations for the adolescent and approach for caregivers.

_____

_____

_____

6

Across

4. Rapid breathing, chest wall retractions
5. Tube diverting cerebrospinal fluid to abdomen
6. Feeding tube placed through abdominal wall
8. Specialized medical practice devoted to care of children
9. First month after birth

Down

1. First year of life
2. Inserted into neck to aid with breathing
3. Two soft openings within the skull
7. Terrible twos stage

Word Fun

The following crossword puzzle is an activity provided to reinforce correct spelling and understanding of medical terminology associated with emergency care and the EMT-B. Use the clues in the column to complete the puzzle.

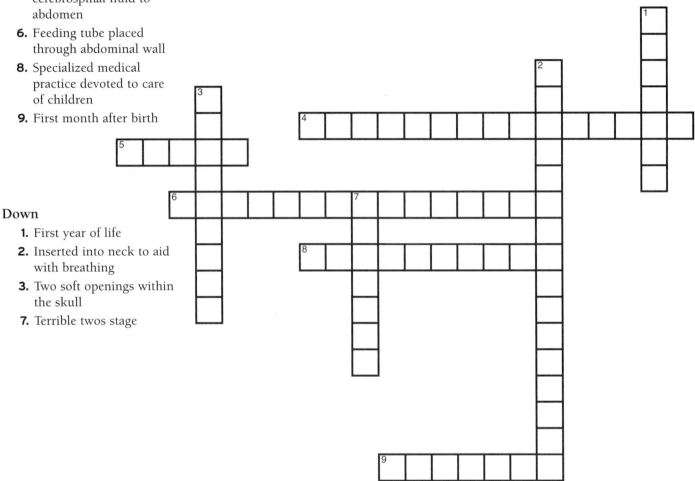

Ambulance Calls

The following real case scenarios provide an opportunity to explore the concerns associated with patient management. Read each scenario, then answer each question in detail.

1. You are dispatched to a residence for an 8-year-old girl complaining of a possible tibia fracture. Her mother tells you that she tripped when she jumped out of a swing and twisted her leg when she landed. The child is lying on the ground and is fairly calm. You see obvious deformity to the lower left leg.

How would you best manage this patient?

_____

_____

_____

_____

_____

_____

_____

_____

_____

2. You are dispatched to a residence of a 3-year-old child with a history of lung problems. The child, a very small boy, is cyanotic and lethargic. He is pain responsive. He has copious mucous secretions in his airway. The grandmother, who was sitting with the child, is hysterical.

How would you best manage this patient?

_____

_____

_____

_____

_____

_____

_____

_____

_____

3. You are called to a residence for a 5-year-old child with a G-tube complaining of difficulty breathing. The child is more lethargic than normal, but is alert. He has retractions and is gasping with a respiratory rate of 48.

How would you best manage this patient?

_____

_____

_____

_____

_____

_____

_____

_____

Workbook Activities

The following activities have been designed to help you. Your instructor may require you to complete some or all of these activities as a regular part of your EMT-B training program. You are encouraged to complete any activity that your instructor does not assign as a way to enhance your learning in the classroom.

Chapter Review

The following exercises provide an opportunity to refresh your knowledge of this chapter.

■ NOTES ■

Matching

Match each of the terms in the left column to the appropriate definition in the right column.

_____ **1.** Croup

_____ **2.** Wheezing

_____ **3.** Tonic

_____ **4.** Epiglottis

_____ **5.** Rales

_____ **6.** Tonic-clonic movements

_____ **7.** *Neisseria meningitidis*

_____ **8.** Stridor

_____ **9.** Septum

_____ **10.** Rigor mortis

_____ **11.** Shock

_____ **12.** Apnea

_____ **13.** Blanching

_____ **14.** Meningitis

_____ **15.** Xiphoid process

_____ **16.** Febrile seizure

_____ **17.** Tripod position

A. leaning forward on two arms stretched forward

B. rigid extremity that cannot be made to relax

C. seizure relating to a fever

D. inflammation of the meninges

E. rhythmic back and forth movement of an extremity with body stiffness

F. causes rash characterized by small cherry-red spots or purple-black color

G. stiffening of the body after death

H. absence of breathing

I. infection of the airway below the level of the vocal chords

J. turning white

K. whistling sound made from air moving through narrowed bronchioles

L. the lower tip of the sternum

M. infection of soft tissue in the area above the vocal chords

N. the central divider in the nose

O. crackling sound caused by flow of air through liquid

P. insufficient blood to body organs

Q. high-pitched sound caused by swelling around vocal chords

Pediatric Airway and Medical Emergencies

Multiple Choice

Read each item carefully, then select the best response.

_____ **1.** Anatomic differences between adults and children include all of the following, except:

    A. the heart is higher in the child's chest.

    B. the child's lungs are smaller.

    C. the adult's opening to the trachea is higher in the neck.

    D. the child's neck is shorter.

_____ **2.** Because of the smaller diameter of the trachea in infants, which is about the same diameter as a drinking straw, their airway is easily obstructed by:

    A. secretions.

    B. blood.

    C. swelling.

    D. all of the above

_____ **3.** Since intercostal muscles are not well developed in children, movement of the _____ dictates the amount of air they inspire.

    A. diaphragm

    B. ribs

    C. abdomen

    D. lungs

_____ **4.** Positioning the airway in a neutral sniffing position:

    A. keeps the trachea from kinking when the neck is hyperextended.

    B. keeps the trachea from kinking when the neck is flexed.

    C. maintains the proper alignment if you have to immobilize the spine.

    D. all of the above

_____ **5.** Benefits of using a nasopharyngeal airway include all of the following, except:

    A. It is usually well tolerated.

    B. It may be used in the presence of head trauma.

    C. It is not as likely as the oropharyngeal airway to cause vomiting.

    D. It is used for conscious patients or those with altered levels of consciousness.

_____ **6.** Indications for assisting ventilations in a child include:
  A. respiratory rate of less than 12 breaths/min.
  B. respiratory rate of greater than 60 breaths/min.
  C. inadequate tidal volume.
  D. all of the above

_____ **7.** Errors in technique, when providing ventilations with a BVM device, that may result in gastric distention include:
  A. providing too much volume.
  B. squeezing the bag too forcefully.
  C. ventilating too fast.
  D. all of the above

_____ **8.** _____ is an infection of the airway below the level of the vocal cords, usually caused by a virus.
  A. Croup
  B. Tonsillitis
  C. Epiglottitis
  D. Pharyngitis

_____ **9.** Signs of complete airway obstruction include:
  A. inability to speak or cry.
  B. increasing respiratory difficulty, with stridor.
  C. cyanosis.
  D. all of the above

_____ **10.** Early signs of respiratory distress include all of the following, except:
  A. combativeness.
  B. anxiety.
  C. cyanosis.
  D. restlessness.

_____ **11.** Signs of increased work of breathing include:
  A. nasal flaring.
  B. wheezing, stridor, or other abnormal airway sounds.
  C. accessory muscle use.
  D. all of the above

_____ **12.** _____ is a continuous seizure, or multiple seizures without a return to consciousness, for 30 minutes or more.
  A. Status epilepticus
  B. Grand mal seizure
  C. Absence seizure
  D. Focal motor seizure

_____ **13.** During the postictal period, the patient may appear:
  A. sleepy.
  B. confused.
  C. unresponsive.
  D. all of the above

_____ **14.** Most pediatric seizures are due to _____, which is why they are called febrile seizures.
  A. infection
  B. fever
  C. ingestion
  D. trauma

_____ **15.** Signs that a patient is not breathing adequately include:

    A. very slow respirations.

    B. very shallow breaths.

    C. cyanosis or pale lips.

    D. all of the above

_____ **16.** Care of the actively seizing child includes all of the following, except:

    A. assessing and managing the ABCs.

    B. noting the type of movement and position of the eyes.

    C. cooling the patient with alcohol if there is fever.

    D. making sure the patient is protected from hitting anything.

_____ **17.** Nonverbal infants may demonstrate consciousness by:

    A. tracking.

    B. babbling and cooing.

    C. crying.

    D. all of the above

_____ **18.** In the mnemonic AEIOU-TIPS, the "E" stands for:

    A. epilepsy.

    B. endocrine.

    C. electrolyte abnormalities.

    D. all of the above

_____ **19.** Signs and symptoms of poisoning vary widely, depending on:

    A. the substance.

    B. age.

    C. weight.

    D. all of the above

_____ **20.** Common sources of poisoning in children include all of the following, except:

    A. street drugs.

    B. baking soda.

    C. house plants.

    D. vitamins.

_____ **21.** Common causes of high temperature in a child include all of the following, except:

    A. drug ingestion.

    B. infection.

    C. going from one temperature extreme to another.

    D. cancer.

_____ **22.** _____ is an increase in body temperature caused by an inability of the body to cool itself.

    A. Fever

    B. Hyperthermia

    C. Hypothermia

    D. Thermoregulation

_____ **23.** Meningitis is an infection caused by:

    A. bacteria.

    B. a virus.

    C. fungi.

    D. all of the above

6

NOTES

24. Signs of meningitis in children and infants include all of the following, except:
    A. bulging fontanels.
    B. neck pain.
    C. altered levels of consciousness.
    D. fever.

25. Children with *N meningitidis* are at serious risk of:
    A. sepsis.
    B. shock.
    C. death.
    D. all of the above

26. The first step in caring for a child with suspected meningitis is:
    A. to obtain a thorough history including onset and changes in behavior.
    B. to maintain c-spine control so as not to further irritate the meninges.
    C. to take BSI precautions.
    D. to secure the airway and give high-flow oxygen.

27. Common causes of shock in children include all of the following, except:
    A. heart attack.
    B. head trauma.
    C. dehydration.
    D. pneumothorax.

28. Pediatric patients respond initially to fluid loss by:
    A. decreasing heart rate.
    B. increasing respirations.
    C. showing signs of pink or red skin.
    D. decreasing blood pressure.

29. Life-threatening dehydration can overcome an infant in a matter of:
    A. minutes.
    B. hours.
    C. days.
    D. weeks.

30. At birth, most infants only need stimulation to:
    A. cry.
    B. wake up.
    C. breathe.
    D. move.

31. Pulse rate in the newborn should be palpated at the brachial artery or:
    A. carotid artery.
    B. radial artery.
    C. femoral artery.
    D. the base of the umbilical cord.

32. Respiratory problems leading to cardiopulmonary arrest in children may be caused by:
    A. a foreign body in the airway.
    B. near drowning.
    C. sudden infant death syndrome.
    D. all of the above

_____**33.** If you find an unresponsive child while not on duty, who is apneic and pulseless, you should:

    A. roll the patient to the side and call EMS immediately.

    B. open the airway, go call EMS, and provide BLS.

    C. provide BLS for approximately 1 minute, then call EMS.

    D. call EMS before touching the patient, then return and provide BLS.

_____**34.** _____ will decrease the risk of gastric distention and aspiration of vomitus by pushing the larynx back to compress and close off the esophagus.

    A. Using a BVM device

    B. The Sellick maneuver

    C. Abdominal compression

    D. none of the above

_____**35.** While checking a pulse, you may observe other signs of circulation including:

    A. breathing.

    B. coughing.

    C. movement.

    D. all of the above

_____**36.** _____ is the leading cause of death in infants younger than 1 year.

    A. SIDS

    B. Congenital heart disease

    C. Respiratory arrest

    D. Foreign body airway obstruction

_____**37.** When dealing with the death of an infant, your assessment of the scene should include:

    A. the site where the infant was discovered.

    B. the general condition of the house.

    C. family interaction.

    D. all of the above

_____**38.** A classic apparent life-threatening event is characterized by:

    A. a distinct change in muscle tone.

    B. choking or gagging.

    C. a child that may appear healthy after the event.

    D. all of the above

_____**39.** Signs of posttraumatic stress include:

    A. nightmares.

    B. difficulty sleeping.

    C. lack of appetite.

    D. all of the above

_____**40.** In dealing with the family after the death of a child, you should:

    A. use the child's name.

    B. acknowledge their feelings.

    C. keep any instructions short and simple.

    D. all of the above

6

Vocabulary emtb vocab explorer

Define the following terms using the space provided.

1. Meningeal irritation:

_____

_____

_____

2. Sudden infant death syndrome (SIDS):

_____

_____

_____

3. Sellick maneuver:

_____

_____

_____

4. Pediatric resuscitation tape measure:

_____

_____

_____

5. Altered level of consciousness:

_____

_____

_____

6. Apparent life-threatening event (ALTE):

_____

_____

_____

7. Dependent lividity:

_____

_____

_____

8. Epiglottitis:

_____

_____

_____

9. Meconium:

_____

_____

_____

Fill-in

Read each item carefully, then complete the statement by filling in the missing word(s).

1. The term _____ is used to describe a continuous seizure, or multiple seizures without a return to consciousness, for 30 minutes or more.

2. Because a young child might not be able to speak, your assessment of his or her condition must be based in large part on what you can _____ and

_____.

3. The child's _____ is larger relative to the small mandible and can easily block the airway.

4. _____ can interfere with movement of the diaphragm and lead to hypoventilation.

5. _____ occurs when fluid losses are greater than fluid intake.

6. _____ are devices that help to maintain the airway or assist in providing artificial ventilation.

7. An oropharyngeal airway should be used in neither conscious patients nor those who have a decreased level of consciousness, as both will have a _____.

8. _____ indicates the amount of oxygen getting to the organs of the body.

9. The _____ is the amount of air that is delivered to the lungs and airways in one inhalation.

10. _____ should be considered as a possible cause of airway obstruction if a child has congestion, fever, drooling, and cold symptoms.

6

11. A _____ is the result of disorganized electrical activity in the brain.

12. _____ is an increase in body temperature, usually in response to an infection.

True/False

If you believe the statement to be more true than false, write the letter "T" in the space provided. If you believe the statement to be more false than true, write the letter "F."

1. _____ You must always assist ventilations in all pediatric patients who have respiratory rates greater than 60 breaths/min.

2. _____ Febrile seizures are self-limiting and do not need transport unless they recur.

3. _____ Partial seizures may present as eye deviation only.

4. _____ Febrile seizures may not be accompanied by a postictal phase.

5. _____ Alcohol applied to skin is a recommended method of cooling a patient.

6. _____ Increasing irritability, especially with handling, may be a sign of meningitis in infants.

Short Answer

Complete this section with short written answers using the space provided.

1. How is urine output assessed in infants?

_____

_____

_____

2. List ten common causes of altered level of consciousness in pediatric patients.

_____   _____

_____   _____

_____   _____

_____   _____

_____   _____

3. List four signs of increased work of breathing in children.

_____

_____

_____

_____

4. List four environmental factors you should assess when responding to a situation in which an infant has died.

_____

_____

_____

_____

5. List three things you can do for a family of a SIDS baby. List four things you should NOT say to the family of a SIDS baby.

_____

_____

_____

_____

Word Fun

The following crossword puzzle is an activity provided to reinforce correct spelling and understanding of medical terminology associated with emergency care and the EMT-B. Use the clues in the column to complete the puzzle.

Across

4. Inadequate tissue perfusion

5. Pooling of blood after death

6. Stiffening of the body after death

7. Turning white

8. External openings of nasal passages

10. Central divider of the nose

Down

1. Lower tip of sternum

2. Dark green material in amniotic fluid

3. Inflammation of the meninges

9. Absence of breathing

6

Ambulance Calls

The following real case scenarios will give you an opportunity to explore the concerns associated with patient management. Read each scenario, then answer each question in detail.

1. You are called to a child care center for a 3-year-old male with difficulty breathing. The patient is still alert, but gasping for breath when you arrive. His respirations are 52 and shallow.

   How would you best manage this patient?

   _____

   _____

   _____

   _____

   _____

   _____

   _____

   _____

   _____

2. You are called to a residence for a 2-year-old child with difficulty breathing. The little girl has stridor and expiratory wheezes, as well as intercostal retractions. She is very upset by your arrival and clings to her mother. Her breathing worsens with agitation. Her mother tells you she is currently taking medication for an upper respiratory infection and has spent much of her life in and out of hospitals with respiratory problems.

   How would you best manage this patient?

   _____

   _____

   _____

   _____

   _____

   _____

   _____

   _____

3. You are dispatched to a residence for seizures in progress. You arrive on the scene to find a 5-year-old male playing on the floor. He is flushed and very warm to touch. The parents tell you that he was "shaking with his eyes rolled back" for over 5 minutes. He has no history of seizures, but woke up with a runny nose this morning.

How would you best manage this patient?

_____

_____

_____

_____

_____

_____

_____

_____

_____

_____

Skill Drills

Skill Drill 32-1: Positioning the Airway in a Child
Test your knowledge of this skill drill by filling in the correct words in the photo captions.

 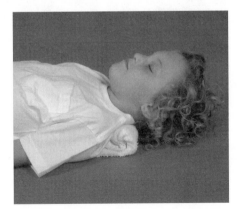 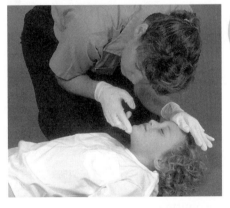

1. Position the child on a _____ surface.

2. Place a _____ towel about _____ inch(es) thick under the _____ and _____.

3. _____ the forehead to limit _____.

Skill Drill 32-2: Inserting an Oropharyngeal Airway in a Child
Test your knowledge of this skill drill by filling in the correct words in the photo captions.

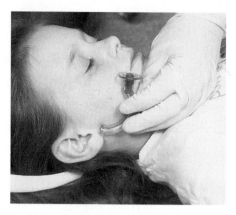

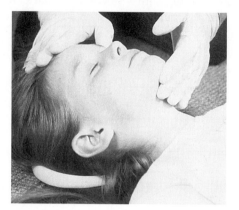

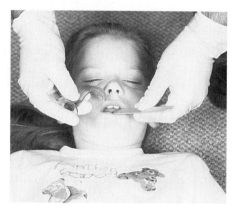

1. Determine the _____ _____ airway. Confirm the correct size _____, next to the patient's _____.

2. Position the patient's _____ with the appropriate method.

3. Open the mouth. Insert the airway until the _____ rests against the _____. _____ the airway.

Skill Drill 32-3: Inserting a Nasopharyngeal Airway in a Child
Test your knowledge of this skill drill by filling in the correct words in the photo captions.

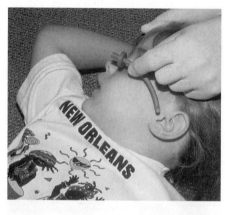

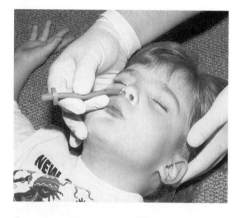

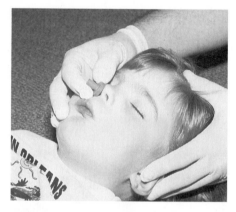

1. Determine the correct airway size by comparing its _____ to the opening of the _____ (naris).
Place the airway next to the patient's _____ to confirm correct _____.
_____ the airway.

2. _____ the airway. Insert the _____ into the right naris with the bevel pointing toward the _____.

3. Carefully move the tip forward until the _____ rests against the _____ of the nostril. Reassess the _____.

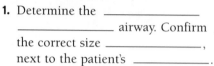

Skill Drill 32-4: One-Person BVM Ventilation on a Child
Test your knowledge of this skill drill by placing the photos below in the correct order.
Number the first step with a "1," the second step with a "2," etc.

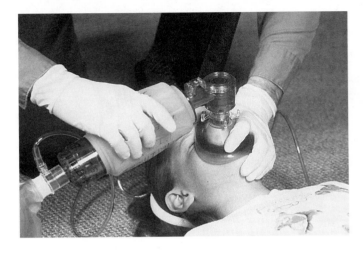

Hold the mask on the patient's face with a one-handed
head tilt-chin lift technique ("E-C grip").
Ensure a good mask-face seal while maintaining the airway.

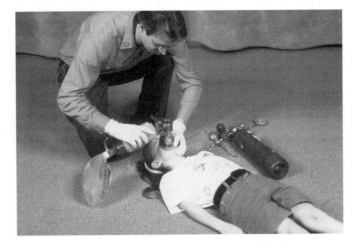

Assess effectiveness of ventilation by watching bilateral
rise and fall of the chest.

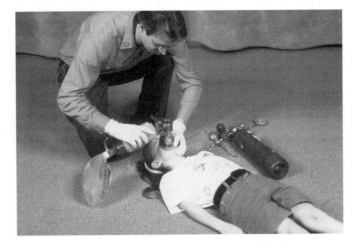

Open the airway and insert the appropriate airway
adjunct.

Squeeze the bag 20 times/min for a child, or 30 times/min
for an infant. Allow adequate time for exhalation.

Skill Drill 32-5: Removing a Foreign Body Airway Obstruction in an Unconscious Child
Test your knowledge of this skill drill by placing the photos below in the correct order.
Number the first step with a "1," the second step with a "2," etc.

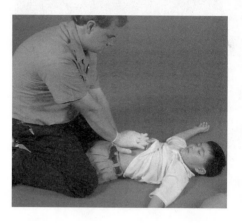

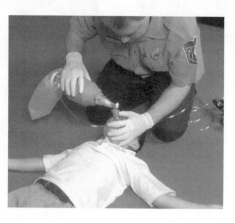

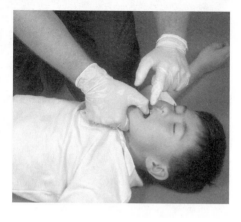

If ventilation is unsuccessful, position your hands on the abdomen above the navel and well below the chest cage. Give five abdominal thrusts.

Attempt rescue breathing. If unsuccessful, reposition the head and try again.

Only try to remove the obstruction if you can see it.
Attempt rescue breathing. If unsuccessful, reposition the head and try again.
Repeat abdominal thrusts if obstruction persists.

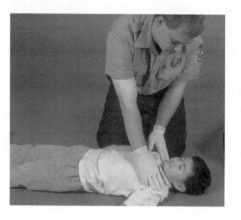

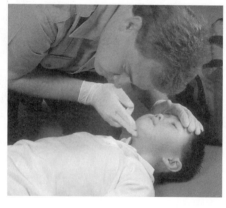

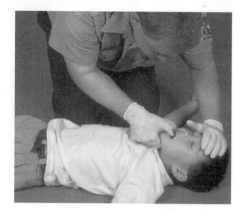

Position the child on a firm, flat surface.

Inspect the airway. Remove any visible foreign object, if you can see it.

Open the airway again to try to see the object.

Skill Drill 32-6: Caring for a Child in Shock
Test your knowledge of this skill drill by placing the photos below in the correct order.
Number the first step with a "1," the second step with a "2," etc.

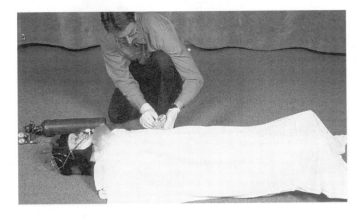

_____

Keep the patient warm with blankets.
Transport immediately.
Continue to monitor vital signs.
Consider ALS backup.
Allow a caregiver to accompany the child if possible.

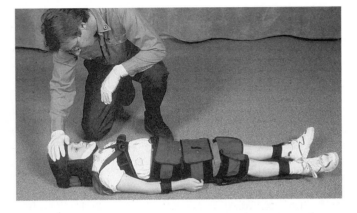

_____

Open the airway.
Be prepared to ventilate.
Control any bleeding.

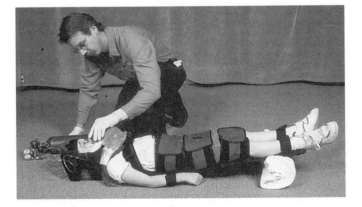

_____

Begin supplemental oxygen.

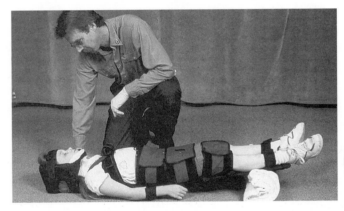

_____

Position the patient with the head lower than the feet.

6

Skill Drill 32-7: Performing Infant Chest Compressions
Test your knowledge of this skill drill by filling in the correct words in the photo captions.

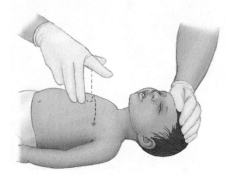

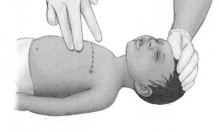

1. Position the infant on a _____ surface while _____ the airway.
Place two _____ in the _____ of the sternum just below a line between the _____.

2. Use two fingers to _____ the chest about _____ inch(es) to _____ inch(es) at a rate of _____ times/min.
Allow the sternum to return _____ to its _____ position between compressions.

3. Coordinate rapid _____ and _____ in a _____ ratio.
Check for return of _____ and _____ after _____ minute(s), then every _____ minute(s).

Skill Drill 32-8: Performing CPR on a Child
Test your knowledge of this skill drill by filling in the correct words in the photo captions.

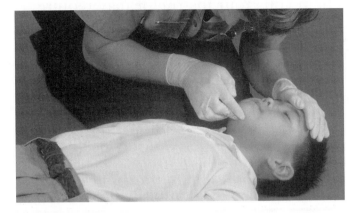

1. Place the child on a _____ surface and maintain the airway with one _____.

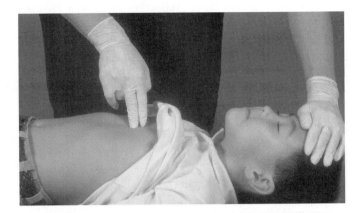

2. Place the _____ of your other hand over the lower half of the _____, avoiding the _____.

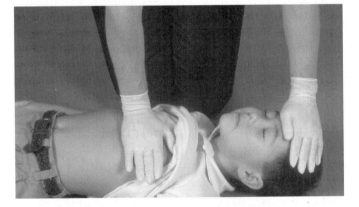

3. Compress the chest about _____ inch(es) to _____ inch(es) at a rate of _____ /min (about _____ /min with pauses for ventilations). Coordinate compressions with ventilations in a _____ ratio, pausing for _____.

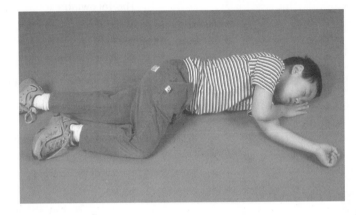

4 Reassess for _____ and _____ after about _____ minute(s), and then every _____ minute(s). If the child resumes _____ breathing, place him or her in the _____ position.

6

Workbook Activities

The following activities have been designed to help you. Your instructor may require you to complete some or all of these activities as a regular part of your EMT-B training program. You are encouraged to complete any activity that your instructor does not assign as a way to enhance your learning in the classroom.

Chapter Review

The following exercises provide an opportunity to refresh your knowledge of this chapter.

NOTES

Matching

Match each of the terms in the left column to the appropriate definition in the right column.

_____ 1. Child abuse   **A.** mnemonic for assessing possible child abuse

_____ 2. Hyperventilation   **B.** the principal injury from submersion

_____ 3. Trauma   **C.** fluid within the lungs

_____ 4. Child safety seats   **D.** weak spots in the bone often injured as a result of trauma

_____ 5. Jaw-thrust maneuver   **E.** effective in decreasing the severity of injuries in motor vehicle crashes

_____ 6. ABCs   **F.** number one killer of children in the United States

_____ 7. Growth plates   **G.** abnormal airway sound made by turbulent airflow

_____ 8. Hypoxia   **H.** improper or excessive action that injures or harms a child or infant

_____ 9. Pulmonary edema   **I.** used to open the airway with suspected spinal injuries

_____ 10. Stridor   **J.** first priority for all patients

_____ 11. CHILD ABUSE   **K.** rapid ventilation

Multiple Choice

Read each item carefully, then select the best response.

_____ 1. Children differ from adults in that they:

    A. have less circulating blood volume.

    B. have a larger body surface area in relation to body mass.

    C. have more flexible and elastic bones.

    D. all of the above

CHAPTER 33

Pediatric Trauma

_____ **2.** When approaching a child, you should look for all of the following, except:
  A. work of breathing.
  B. pulse.
  C. level of activity.
  D. skin color.

_____ **3.** Because children are less mature psychologically, they are often injured because of their underdeveloped judgment and:
  A. lack of experience.
  B. need for car seats.
  C. participation in organized sports.
  D. all of the above

_____ **4.** Children who are not restrained in child safety seats are at greater risk for:
  A. head injuries.
  B. neck injuries.
  C. spinal injuries.
  D. all of the above

_____ **5.** When caring for children with sports-related injuries, you should remember to:
  A. elevate the extremities.
  B. assist ventilations.
  C. immobilize the cervical spine.
  D. remove all helmets.

_____ **6.** Your single most important step in caring for a child with a head injury is to:
  A. immobilize the cervical spine.
  B. bandage all wounds.
  C. ensure an open airway.
  D. obtain a SAMPLE history.

_____ **7.** _____ can occur as a result of head injuries in children.
  A. Cervical spine injuries
  B. Nausea and vomiting
  C. Respiratory arrest
  D. all of the above

_____ **8.** Signs of shock in children include all of the following, except:

   A. tachycardia.

   B. hypotension.

   C. poor capillary refill.

   D. mental status changes.

_____ **9.** Children's bones bend more easily than adult's bones and as a result, incomplete or _____ fractures can occur.

   A. spiral

   B. comminuted

   C. greenstick

   D. compound

_____ **10.** Indications for the use of the PASG in children include all of the following, except:

   A. pelvic instability.

   B. obvious abdominal trauma.

   C. obvious lower extremity trauma.

   D. clear signs and symptoms of decompensated shock.

_____ **11.** The most common cause(s) of burns in children is (are):

   A. exposure to hot substances.

   B. hot items on a stove.

   C. exposure to caustic substances.

   D. all of the above

_____ **12.** In submersion situations, your first priority is to always take steps to:

   A. stay properly protected from body fluids.

   B. maintain the airway.

   C. ensure your own safety.

   D. retrieve the patient from the water.

_____ **13.** The principal injury from submersion is:

   A. lack of oxygen.

   B. neck and spinal cord injuries.

   C. drowning.

   D. hypothermia.

_____ **14.** Secondary drowning develops minutes or hours later from:

   A. infection.

   B. bacteria.

   C. pulmonary edema.

   D. submersion.

_____ **15.** For children who have had traumatic injuries, use a child-sized BVM device at a rate of one breath every _____ seconds.

   A. 2

   B. 3

   C. 4

   D. 5

_____ **16.** Signs and symptoms of brain swelling associated with head trauma include:

   A. one dilated, nonreactive pupil.

   B. decorticate or decerebrate posturing.

   C. altered breathing patterns.

   D. all of the above

_____ **17.** If _____—an abnormal airway sound made by turbulent airflow—is present, use the jaw-thrust maneuver to keep the airway open.

    A. rales

    B. rhonchi

    C. stridor

    D. wheezing

_____ **18.** All patients who have unstable or potentially unstable injuries require immediate transportation to the closest _____ facility.

    A. emergency

    B. trauma

    C. pediatric

    D. most appropriate

_____ **19.** Child abuse may include:

    A. physical abuse.

    B. neglect.

    C. emotional abuse.

    D. all of the above

_____ **20.** The "H" in the CHILD ABUSE mnemonic stands for:

    A. history of present injury.

    B. history inconsistent with injury.

    C. history of previous injuries.

    D. history of sibling injuries.

_____ **21.** Fractures that may be associated with a fall from a bed include all of the following, except:

    A. wrist.

    B. finger.

    C. femur.

    D. none of the above

_____ **22.** The "U" in the CHILD ABUSE mnemonic stands for:

    A. unusual injury patterns.

    B. unusual circumstances.

    C. unusual parental concerns.

    D. unusual history.

_____ **23.** An abused child may appear:

    A. withdrawn.

    B. fearful.

    C. hostile.

    D. all of the above

_____ **24.** The child abuser may be:

    A. a friend of the family.

    B. a parent.

    C. a relative.

    D. any of the above

_____ **25.** After difficult incidents involving children, _____ is helpful in working through the stress and trauma.

    A. having a drink

    B. talking with the family

    C. debriefing

    D. putting the incident out of your mind

6

Vocabulary emtb vocab explorer

Define the following terms using the space provided.

1. Shaken baby syndrome:

_____

_____

_____

2. Child neglect:

_____

_____

_____

3. Decorticate posturing:

_____

_____

_____

4. Decerebrate posturing:

_____

_____

_____

Fill-in

Read each item carefully, then complete the statement by filling in the missing word.

1. _____ is the number one killer of children in the United States.

2. A child's _____ are softer and more flexible than an adult's and may compress the heart and lungs, causing serious injury with no obvious external damage.

3. Because a child's _____ is proportionately larger than an adult's, it exerts greater stress on the neck structures during a deceleration injury.

4. Although child safety seats are effective in decreasing the severity of injuries, children may still sustain _____ and lower spinal injuries as the result of trauma caused by improperly used passenger restraints.

5. When struck by a vehicle, the exact area that is struck depends on the child's height and the final position of the _____ at the time of impact.

6. You should suspect a serious _____ _____ in any child who experiences nausea and vomiting after a traumatic event.

7. Children can lose a greater proportion of their blood volume than adults can before signs and symptoms of _____ develop.

8. _____ _____ are potential weak spots in the bone and are often injured as a result of trauma.

NOTES

True/False

If you believe the statement to be more true than false, write the letter "T" in the space provided. If you believe the statement to be more false than true, write the letter "F. "

1. _____ Children are simply little adults.
2. _____ Children may experience significant internal injuries with little or no obvious outside signs.
3. _____ Child safety seats are effective in decreasing the severity of injuries.
4. _____ Head injuries are uncommon in children.
5. _____ Children can lose a greater proportion of blood than adults before showing signs or symptoms of shock.
6. _____ Children have soft and flexible ribs.
7. _____ The intentional injury of a child is rare in our society.
8. _____ EMT-Bs must report all cases of suspected child abuse.
9. _____ Child protection agencies are mandated to investigate all reported child abuse.

Short Answer

Complete this section with short written answers using the space provided.

1. List the severity and body area involved for the three categories of burns in children.

_____

_____

_____

2. List four questions to ask yourself when you suspect physical or sexual abuse of a child.

_____

_____

_____

_____

3. What are the most common causes of burns in a child?

_____

_____

_____

6

Word Fun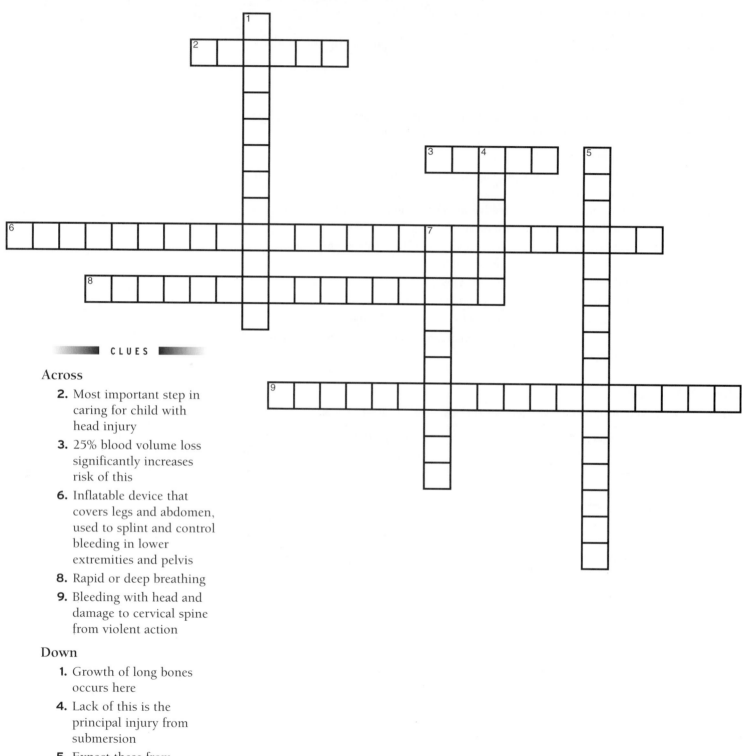

The following crossword puzzle is an activity provided to reinforce correct spelling and understanding of medical terminology associated with emergency care and the EMT-B. Use the clues in the column to complete the puzzle.

CLUES

Across

2. Most important step in caring for child with head injury

3. 25% blood volume loss significantly increases risk of this

6. Inflatable device that covers legs and abdomen, used to splint and control bleeding in lower extremities and pelvis

8. Rapid or deep breathing

9. Bleeding with head and damage to cervical spine from violent action

Down

1. Growth of long bones occurs here

4. Lack of this is the principal injury from submersion

5. Expect these from chemical ingestion

7. Improper, excessive action against child

Ambulance Calls

The following real case scenarios provide an opportunity to explore the concerns associated with patient management. Read each scenario, then answer each question in detail.

1. You are dispatched to a residence for a possible drowning of a 7-year-old. When you arrive, the child is lying at the side of the pool where his father pulled him out. He is breathing shallowly at 8 breaths/min and has faint radial pulses.

   How would you best manage this patient?

   _____

   _____

   _____

   _____

   _____

   _____

   _____

   _____

   _____

2. You are called to a residence for a baby "not acting right." The infant, 6 months old, has unequal pupils, is unresponsive, and is posturing. He is breathing irregularly at about 32 breaths/min. The parents tell you he would not stop crying and then he started acting this way.

   How would you best manage this patient?

   _____

   _____

   _____

   _____

   _____

   _____

   _____

   _____

   _____

6

3. You respond to a possible rape of a 13-year-old girl. Upon arrival, you find the police already on scene and the parents are very upset and belligerent. The girl is cowering in a chair in the corner of the room. The parents tell you that the perpetrator is a 19-year-old neighbor who attacked the girl and held her at knifepoint while he raped her. The police are handling the report and tell you they will follow you to the hospital to gather the girl's clothing for evidence. The mother also tells you that the girl has some moderate vaginal bleeding.

How would you best manage this patient?

_____

_____

_____

_____

_____

_____

_____

_____

_____

_____

Skill Drills emt-b video clips

Skill Drill 33-1: Immobilizing a Child

Test your knowledge of this skill drill by filling in the correct words in the photo captions.

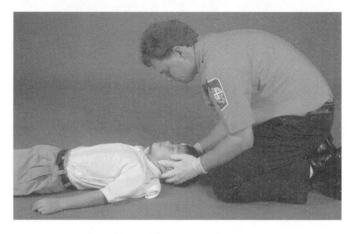

1. Use a towel under the _____ to maintain the head in a _____ position.

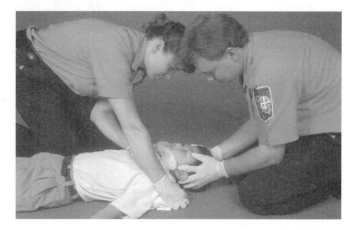

2. Apply an appropriately sized _____.

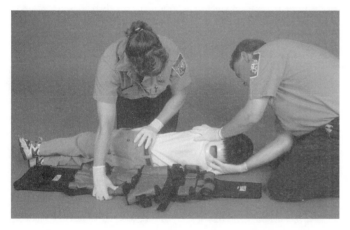

3. _____ _____ the child onto the _____ device.

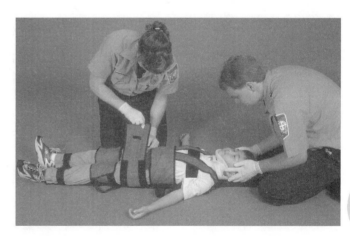

4. Secure the _____ first.

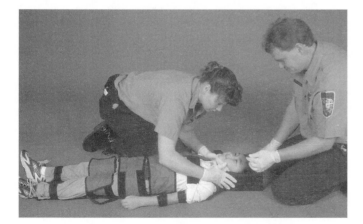

5. Secure the _____.

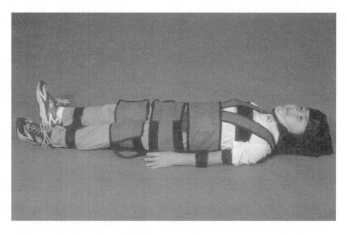

6. Ensure that the child is _____ properly.

6

Skill Drill 33-2: Immobilization of an Infant
Test your knowledge of this skill drill by placing the photos below in the correct order.
Number the first step with a "1," the second step with a "2," etc.

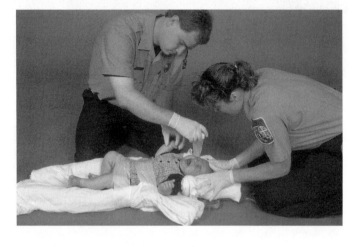

Secure the head.

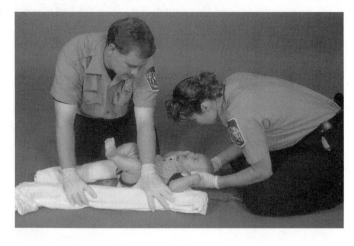

Slide the infant onto the board.

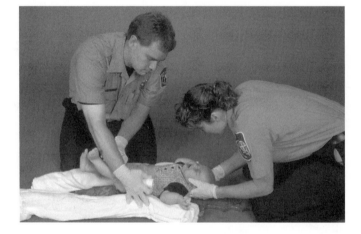

Place a towel under the shoulders to ensure neutral head
position.

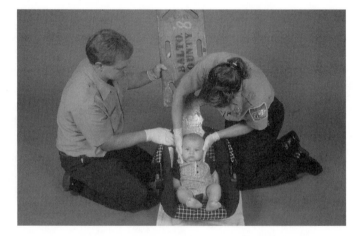

Stabilize the head in neutral position.

(continued)

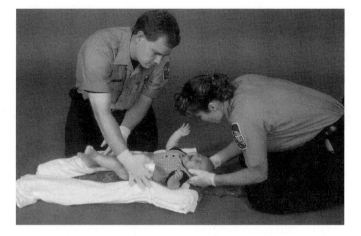

Secure the torso first; pad any voids.

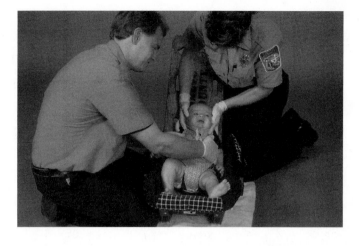

Place an immobilization device between the patient and the surface he or she is resting on.

6

Workbook Activities

The following activities have been designed to help you. Your instructor may require you to complete some or all of these activities as a regular part of your EMT-B training program. You are encouraged to complete any activity that your instructor does not assign as a way to enhance your learning in the classroom.

Chapter Review

The following exercises provide an opportunity to refresh your knowledge of this chapter.

Matching

Match each of the terms in the left column to the appropriate definition in the right column.

_____ 1. Aneurysm
    A. a protein that is the chief component of connective tissue and bones

_____ 2. Cataract
    B. narrowing of a blood vessel

_____ 3. Delirium
    C. clouding of the lens of the eye

_____ 4. Dementia
    D. early shock

_____ 5. Syncope
    E. abnormal blood-filled dilation of a blood vessel

_____ 6. Dyspnea
    F. difficulty breathing

_____ 7. Compensated shock
    G. inability to focus, think logically, or maintain attention

_____ 8. Decompensated shock
    H. late shock

_____ 9. Collagen
    I. slow onset of progressive disorientation

_____ 10. Vasodilation
    J. widening of a blood vessel

_____ 11. Vasoconstriction
    K. fainting

Multiple Choice

Read each item carefully, then select the best response.

_____ 1. Leading causes of death in the elderly include all of the following, except:
    A. heart disease.
    B. AIDS.
    C. cancer.
    D. diabetes.

_____ 2. Lifesaving interventions for geriatric patients may include:
    A. reviewing the home environment.
    B. providing information on preventing falls.
    C. making referrals to appropriate social services agencies.
    D. all of the above

CHAPTER 34

Geriatric Assessment

_____ **3.** Simple preventive measures can help the elderly to avoid:
  A. further injury.
  B. costly medical treatment.
  C. death.
  D. all of the above

_____ **4.** Acute illness and trauma are more likely to involve _____ beyond those initially involved.
  A. organ systems
  B. bones
  C. fractures
  D. vessels

_____ **5.** Risk factors that affect mortality in elderly patients include all of the following, except:
  A. living alone.
  B. unsound mind.
  C. regular exercise.
  D. recent hospitalization.

_____ **6.** Loss of collagen makes the skin:
  A. wrinkled.
  B. thinner.
  C. more susceptible to injury.
  D. all of the above

_____ **7.** Driving and walking become more hazardous because the pupils of the eyes begin to lose the ability to:
  A. dilate.
  B. handle changes in light.
  C. constrict.
  D. detect color.

_____ 8. Problems with balance are usually related to changes in the:

   A. blood pressure.

   B. vision.

   C. inner eyes.

   D. cardiovascular system.

_____ 9. Although the alveoli became enlarged, their elasticity decreases, resulting in a decreased ability to:

   A. cough and thereby increasing the chance of infection.

   B. exchange oxygen and carbon dioxide.

   C. monitor the changes in oxygen and carbon dioxide.

   D. force carbon dioxide out of the lungs.

_____ 10. Compensation for an increased demand on the cardiovascular system is accomplished by:

   A. increasing heart rate.

   B. increasing contraction of the heart.

   C. constricting the blood vessels to nonvital organs.

   D. all of the above

_____ 11. Aging decreases a person's ability to _____ because of stiffer vessels.

   A. vasoconstrict

   B. vasodilate

   C. circulate blood

   D. exchange oxygen

_____ 12. An accumulation of fatty materials in the arteries is known as:

   A. myocardial infarction.

   B. stroke.

   C. atherosclerosis.

   D. aneurysm.

_____ 13. With a decrease in renal function, levels of _____ may rise, creating the impression of an overdose.

   A. medications

   B. toxins

   C. acid

   D. alkali

_____ 14. By age 85, a 10% reduction in brain weight can result in:

   A. increased risk of head trauma.

   B. short-term memory impairment.

   C. slower reflex times.

   D. all of the above

_____ 15. As a person ages, fractures are more likely to occur because of a decrease in bone:

   A. cartilage.

   B. density.

   C. length.

   D. tissue.

_____ 16. Scene clues that can provide important information include:

   A. the general condition of the home.

   B. the number and type of pill bottles around.

   C. any hazards that could cause a fall.

   D. all of the above

_____ 17. The best rule of thumb when assessing mental status is to always compare the patient's current level of consciousness or ability to function with the level or ability:

 A. of another adult in the household.

 B. before the problem began.

 C. of a person of the same age.

 D. none of the above

_____ 18. The _____ is usually the key in helping to assess the elderly patient's problem.

 A. history

 B. medication

 C. environment

 D. all of the above

_____ 19. An elderly patient's diminished _____ may hamper communication.

 A. sight

 B. hearing

 C. speaking ability

 D. all of the above

_____ 20. The term applied to prescribing multiple medications is:

 A. hypermedicating.

 B. hyperpharmacy.

 C. polypharmacy.

 D. overmedicating.

_____ 21. The sensation of pain may be _____ in an elderly patient, leading to "silent" heart attacks.

 A. enhanced

 B. diminished

 C. overstated

 D. false

_____ 22. In order to understand a patient's baseline condition and how today's behavior differs from it, you should ask the nursing home staff questions concerning the patient's:

 A. mobility.

 B. activities of daily living.

 C. ability to speak.

 D. all of the above

_____ 23. You must consider the body's decreasing ability to _____ simple trauma when you are assessing and caring for an elderly patient.

 A. isolate

 B. separate

 C. heal

 D. recognize

_____ 24. An isolated hip fracture in an 85-year-old patient can produce a systemic impact that results in:

 A. deterioration.

 B. shock.

 C. life-threatening conditions.

 D. all of the above

6

_____25. When assessing an elderly patient who has fallen, it is important to determine why the fall occurred because it may have been the result of a medical problem such as:

   A. fainting.

   B. a cardiac rhythm disturbance.

   C. a medication interaction.

   D. all of the above

_____26. Your assessment of the patient's condition and stability must include past medical conditions, even if they are not currently acute or:

   A. symptomatic.

   B. asymptomatic.

   C. complaining.

   D. on medication.

_____27. A common complaint from the patient experiencing an abdominal aortic aneurysm (AAA) is pain in the:

   A. abdomen.

   B. back.

   C. leg with decreased blood flow.

   D. all of the above

_____28. All of the following are true of delirium, except:

   A. It may have metabolic causes.

   B. The patient may be hypoglycemic.

   C. It develops slowly over a period of years.

   D. The memory remains mostly intact.

_____29. For a DNR order to be valid, it must:

   A. be signed by the patient or legal guardian.

   B. be signed by one or more physicians.

   C. be dated within the preceding 12 months.

   D. all of the above

_____30. When in doubt about whether an advance directive is valid, or if one is in place, your best course of action is to:

   A. call medical control to see if an order is needed.

   B. take resuscitation action that is appropriate to the situation.

   C. wait for the family or caregivers to produce the appropriate document.

   D. none of the above

_____31. Signs and symptoms of possible abuse include all of the following, except:

   A. chronic pain.

   B. no history of repeated visits to the emergency department or clinic.

   C. depression or lack of energy.

   D. self-destructive behavior.

_____32. Signs of neglect include:

   A. lack of hygiene.

   B. poor dental hygiene.

   C. lack of reasonable amenities in the home.

   D. all of the above

Vocabulary emtb. vocab explorer

Define the following terms using the space provided.

1. Advance directives:

_____

_____

_____

2. Atherosclerosis:

_____

_____

_____

3. Arteriosclerosis:

_____

_____

_____

4. Elder abuse:

_____

_____

_____

5. Osteoporosis:

_____

_____

_____

Fill-in

Read each item carefully, then complete the statement by filling in the missing word.

1. Geriatric or elderly patients are individuals who are older than _____ years.

2. The aging body of the geriatric person may _____ serious medical conditions.

3. Common _____ about the elderly include the presence of mental confusion, illness, a sedentary lifestyle, and immobility.

4. Older skin feels dry due to fewer _____.

5. _____ is a measure of the workload of the heart.

6. An _____ is an abnormal blood-filled dilation of the wall of a blood vessel.

7. Flexion at the neck and a forward curling of the shoulders produce a condition

called _____.

True/False

If you believe the statement to be more true than false, write the letter "T" in the space provided. If you believe the statement to be more false than true, write the letter "F."

1. _____ Vasodilation is a narrowing of a blood vessel.

2. _____ Cardiovascular disease is one of the leading causes of death in the elderly.

3. _____ Mental confusion and immobility are common stereotypes about the elderly.

4. _____ Assessment of an elderly patient usually takes less time than a middle-aged person.

5. _____ The sensation of pain in an elderly patient may be diminished.

6. _____ Elderly people are more prone to hypothermia than younger people.

7. _____ Falls are the leading cause of trauma, death, and disability in the elderly.

8. _____ 20% to 30% of elderly have "silent" heart attacks.

9. _____ Elderly abuse is on the decline in the United States.

Short Answer

Complete this section with short written answers using the space provided.

1. List the three major categories of elder abuse.

_____

_____

_____

2. Briefly describe the three possible causes of syncope in an elderly patient.

_____

_____

_____

3. List at least five informational items that may be important in assessing possible elder abuse.

_____

_____

_____

_____

_____

4. Name three common signs and symptoms of a heart attack in an elderly patient.

_____

_____

_____

Word Fun

The following crossword puzzle is an activity provided to reinforce correct spelling and understanding of medical terminology associated with emergency care and the EMT-B. Use the clues in the column to complete the puzzle.

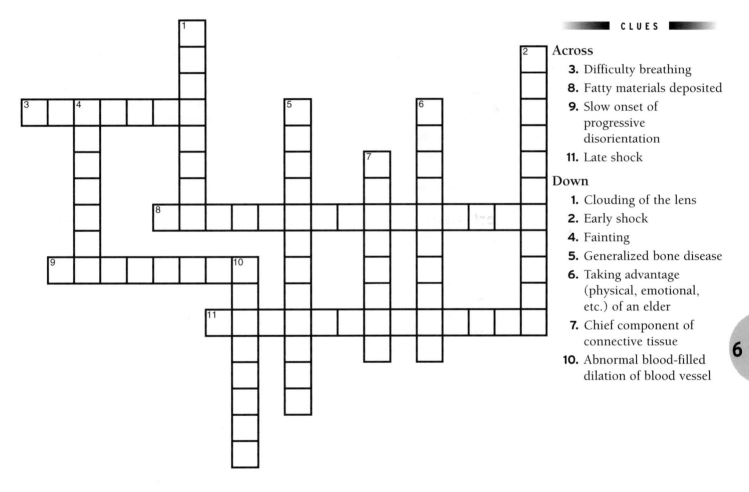

CLUES

Across

3. Difficulty breathing

8. Fatty materials deposited

9. Slow onset of progressive disorientation

11. Late shock

Down

1. Clouding of the lens

2. Early shock

4. Fainting

5. Generalized bone disease

6. Taking advantage (physical, emotional, etc.) of an elder

7. Chief component of connective tissue

10. Abnormal blood-filled dilation of blood vessel

Ambulance Calls

The following real case scenarios provide an opportunity to explore the concerns associated with patient management. Read each scenario, then answer each question in detail.

1. You are called to the residence of an 87-year-old female who is "not acting right." Family members tell you they think she may have taken her medication twice this morning. She is lethargic, confused, and hypotensive.

How would you best manage this patient?

_____

_____

_____

_____

_____

_____

_____

_____

_____

_____

2. You are dispatched to a single vehicle motor vehicle crash with minor damage to the front where the car rolled into a culvert at a low speed. Bystanders tell you that the driver, an 82-year-old male, slumped over the steering wheel before the vehicle veered off the road. The patient is now alert and says he does not remember what happened. He has a 2-inch laceration above his right eye with moderate bleeding.

How would you best manage this patient?

_____

_____

_____

_____

_____

_____

_____

_____

_____

3. You are dispatched to a residence for a fall. Upon arrival, you find an 87-year-old female complaining of severe pain in her left hip. Her daughter tells you that she falls a lot. You notice that the patient appears malnourished and lethargic. In the process of immobilizing the patient on a long spine board, you see what appears to be small circular burns to the backs of her thighs. Her daughter tells you she has a history of dementia and accidentally sat on a heater grate. The pattern is incompatible with such an injury.

How would you best manage this patient?

_____

_____

_____

_____

_____

_____

_____

_____

6

Workbook Activities

The following activities have been designed to help you. Your instructor may require you to complete some or all of these activities as a regular part of your EMT-B training program. You are encouraged to complete any activity that your instructor does not assign as a way to enhance your learning in the classroom.

Chapter Review

The following exercises provide an opportunity to refresh your knowledge of this chapter.

Matching

Match each of the terms in the left column to the appropriate definition in the right column.

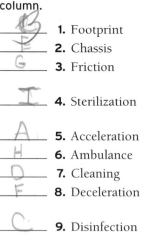

B	**1.** Footprint	**A.** the process of increasing speed
E	**2.** Chassis	**B.** area of contact between tire and road surface
G	**3.** Friction	**C.** the killing of pathogenic agents by direct application of chemicals
I	**4.** Sterilization	**D.** the process of removing dirt, dust, blood, or other visible contaminants
A	**5.** Acceleration	**E.** vehicle frame
H	**6.** Ambulance	**F.** the process of slowing down
D	**7.** Cleaning	**G.** resistance to motion
F	**8.** Deceleration	**H.** specialized vehicle for treating and transporting sick and injured patients
C	**9.** Disinfection	**I.** removes microbial contamination

Multiple Choice

Read each item carefully, then select the best response.

_____ **1.** Ambulances today are designed according to strict government regulations based on ____C____ standards.

   A. local

   B. state

   C. national

   D. individual

C H A P T E R 35

Ambulance Operations

_____ A _____ **2.** Features of the modern ambulance include all of the following, except:

    A. self-contained breathing apparatus.

    B. a patient compartment.

    C. two-way radio communication.

    D. a driver's compartment.

_____ A _____ **3.** During the _____ phase, make sure that equipment and supplies are in their proper place.

    A. preparation

    B. dispatch

    C. arrival at scene

    D. transport

_____ B _____ **4.** The Type _____ ambulance is a standard van with forward-control integral cab-body.

    A. I

    B. II

    C. III

    D. IV

_____ D _____ **5.** Items needed to care for life-threatening conditions include:

    A. equipment for airway management.

    B. equipment for artificial ventilation.

    C. oxygen delivery devices.

    D. all of the above

_____ D _____ **6.** Oropharyngeal airways can be used for:

    A. adults.

    B. children.

    C. infants.

    D. all of the above

_____ A _____ **7.** A BVM device, when attached to oxygen supply with the oxygen reservoir in place, is able to supply almost _____ oxygen.

A. 100%

B. 95%

C. 90%

D. 85%

_____ D _____ **8.** BVM devices should be transparent so that you can:

A. monitor the patient's respirations.

B. notice any color changes in the patient.

C. detect vomiting.

D. all of the above

_____ D _____ **9.** Oxygen masks, with and without nonbreathing bags, should be transparent and disposable and in sizes for:

A. adults.

B. children.

C. infants.

D. all of the above

_____ A B _____ **10.** Basic wound care supplies include all of the following, except:

A. sterile sheets.

B. an OB kit.

C. an assortment of band-aids.

D. large safety pins.

_____ D _____ **11.** Think of the jump kit as containing anything you might need in the first _____ minutes with the patient except for the semiautomated external defibrillator and possibly the oxygen cylinder and portable suctioning unit.

A. 2

B. 3

C. 4

D. 5

_____ A B _____ **12.** Stretchers must be easy to move, store, clean, and:

A. fold.

B. disinfect.

C. lift.

D. wash.

_____ A _____ **13.** Deceleration straps over the shoulders prevent the patient from continuing to move _____ in case the ambulance suddenly slows or stops.

A. forward

B. backwards

C. laterally

D. down

_____ D _____ **14.** The ambulance inspection should include checks of:

A. fuel levels.

B. brake fluid.

C. wheels and tires.

D. all of the above

_____ **15.** All medical equipment and supplies should be checked at least:

   A. after every call.

   B. after every emergency transport.

   C. every 12 hours.

   D. every day.

_____ **16.** For every emergency request, the dispatcher should gather and record all of the following, except:

   A. the nature of the call.

   B. the location of the patient(s).

   C. medications that the patient is currently taking.

   D. the number of patients and possible severity of their condition.

_____ **17.** During the _____ phase, the team should review dispatch information and assign specific initial duties and scene management tasks to each team member.

   A. preparation

   B. dispatch

   C. en route

   D. transport

_____ **18.** Basic requirements for the driver to safely operate an ambulance include:

   A. physical fitness.

   B. emotional fitness.

   C. proper attitude.

   D. all of the above

_____ **19.** The _____ phase may be the most dangerous part of the call.

   A. preparation

   B. en route

   C. transport

   D. on scene

_____ **20.** In order to operate an emergency vehicle safely, you must know how it responds to _____ under various conditions.

   A. steering

   B. braking

   C. acceleration

   D. all of the above

_____ **21.** You must always drive:

   A. offensively.

   B. defensively.

   C. under the speed limit.

   D. all of the above

_____ **22.** When driving with lights and siren, you are _____ for drivers to yield right of way.

   A. requesting

   B. demanding

   C. offering

   D. none of the above

23. The _____ is a measure of the tire's grip on the road.

A. coefficient of friction

B. friction

C. footprint

D. centrifugal force

24. Steering technique includes:

A. the way you hold the steering wheel.

B. the way it moves.

C. the timing of the movements.

D. all of the above

25. _____ is the transfer of weight to different points on the chassis.

A. Chassis set

B. Acceleration

C. Deceleration

D. Sliding

26. Vehicle size and _____ will greatly influence braking and stopping distances.

A. length

B. height

C. weight

D. width

27. When on an emergency call, before proceeding past a stopped school bus with its lights flashing, you should stop before reaching the bus and wait for the driver to:

A. make sure the children are safe.

B. close the bus door.

C. turn off the warning lights.

D. all of the above

28. The _____ is probably the most overused piece of equipment on an ambulance.

A. stethoscope

B. siren

C. cardiac monitor

D. stretcher

29. The _____ is the most visible, effective warning device for clearing traffic in front of the vehicle.

A. front light bar

B. rear light bar

C. high-beam flasher unit

D. standard headlight

30. If you are involved in a motor vehicle crash while operating an emergency vehicle and are found to be at fault, you may be charged:

A. civilly.

B. criminally.

C. both civilly and criminally.

D. neither civilly nor criminally.

_B_ **31.** _____ crashes are the most common and usually the most serious type of collision in which ambulances are involved.

    A. T-bone

    B. Intersection

    C. Lateral

    D. Rollover

_D_ **32.** Guidelines for sizing up the scene include:

    A. looking for safety hazards.

    B. evaluating the need for additional units or other assistance.

    C. evaluating the need to stabilize the spine.

    D. all of the above

_D_ **33.** The main objectives in directing traffic include:

    A. warning other drivers.

    B. preventing additional crashes.

    C. keeping vehicles moving in an orderly fashion.

    D. all of the above

_C_ **34.** Transferring the patient to receiving staff member occurs during the _____ phase.

    A. arrival

    B. transport

    C. delivery

    D. postrun

_D_ **35.** Cleaning the vehicle inside and out, refueling the vehicle, disposing of contaminated waste, and replacing equipment and supplies all are accomplished during the _____ phase.

    A. preparation

    B. transport

    C. delivery

    D. postrun

_D_ **36.** Air medical unit crews may include:

    A. EMTs.

    B. paramedics.

    C. physicians.

    D. all of the above

_A_ **37.** The proper approach area for a helicopter is between the _____ o'clock and _____ o'clock positions as the pilot faces forward.

    A. 2 and 10

    B. 11 and 5

    C. 1 and 11

    D. 3 and 9

_D_ **38.** When clearing a landing site for an approaching helicopter, look for:

    A. loose debris.

    B. electric or telephone wires.

    C. poles.

    D. all of the above

Vocabulary

Define the following terms using the space provided.

1. Air ambulances:

_____

_____

_____

2. Coefficient of friction:

_____

_____

_____

3. Decontaminate:

_____

_____

_____

4. Hydroplaning:

_____

_____

_____

Fill-in

Read each item carefully, then complete the statement by filling in the missing word.

1. A _____ is a portable kit containing items that are used in the initial care of the patient.

2. The six-pointed star that identifies vehicles that meet federal specifications as

licensed or certified ambulances is known as the _____.

3. For many decades after 1906, a _____ was the vehicle that was most often used as an ambulance.

4. _____ respond initially to the scene with personnel and equipment to treat the sick and injured until an ambulance can arrive.

5. An ambulance call has _____ phases.

6. Devices should be either disposable or easy to clean and _____, which means to remove radiation, chemical, or other hazardous materials.

7. Suction tubing must reach the patient's _____ , regardless of the patient's position.

8. A _____ provides a firm surface under the patient's torso so that you can give effective chest compressions.

9. _____ is resistance to the motion of one body against another.

True/False

If you believe the statement to be more true than false, write the letter "T" in the space provided. If you believe the answer to be more false than true, write the letter "F."

1. _____ Equipment and supplies should be placed in the unit according to their relative importance and frequent use.

2. _____ A CPR board is a pocket-sized reminder that the EMT-B carries to help recall CPR procedures.

3. _____ Having the ability to exchange equipment between units or between your unit and the emergency department decreases the time that you and your unit must stay at the hospital.

4. _____ In most instances, if the patient is properly assessed and stabilized at the scene, speeding during transport is unnecessary, undesirable, and dangerous.

5. _____ The en route or response phase of the emergency call is the least dangerous for the EMT-B.

6. _____ Controlled acceleration is the use of acceleration to control the vehicle.

7. _____ When the siren is on, you can speed up and assume that you have the right-of-way.

8. _____ Use the "4-second-rule" to help you maintain a safe following distance.

9. _____ Always approach a helicopter from the front.

10. _____ Fixed-wing air ambulances are generally used for short-haul patient transfers.

11. _____ A clear landing zone of 50' by 50' is recommended for EMS helicopters.

Short Answer

Complete this section with short written answers using the space provided.

1. Describe the three basic ambulance designs.

_____

_____

_____

2. List the phases of an ambulance call.

_____

_____

_____

7

3. List the five factors that contribute to the use of excessive speed.

_____

_____

_____

_____

_____

4. Describe the three basic principles that govern the use of warning lights and sirens.

_____

_____

_____

5. List four guidelines for safe ambulance driving.

_____

_____

_____

_____

6. Describe the correct technique for approaching a helicopter that is "hot" (rotors turning).

_____

_____

_____

7. List the general considerations used for selecting a helicopter landing site.

_____

_____

_____

Word Fun

The following crossword puzzle is an activity provided to reinforce correct spelling and understanding of medical terminology associated with emergency care and the EMT-B. Use the clues in the column to complete the puzzle.

CLUES

Across

1. Device for use under patient's torso when performing chest compressions
6. Portable set-up of first needed gear
7. Resistance of one body moving against another
8. Vehicle frame
9. Used to remove microbial contaminates
10. Stretcher-type device

Down

2. Remove or neutralize
3. Tires lifted off street because of water build-up
4. Area of contact between road and tire
5. Opposite of ground transportation for injured

Ambulance Calls

The following real case scenarios provide an opportunity to explore the concerns associated with patient management. Read each scenario, then answer each question in detail.

NOTES

1. You are dispatched to a call on the opposite side of town. The dispatcher tells you that the bridge over Second Street is still blocked from an earlier crash. This was the shortest, most direct route to your destination. You do not know of another route.

   How would you best manage this situation?

_____

_____

_____

_____

_____

7

2. You are called to the scene of a motor vehicle crash. En route to the scene, in heavy rain, your ambulance begins to hydroplane. You are on a straightaway and no one is in the oncoming lane.

How would you best manage this situation?

3. You are called to the scene of a motor vehicle crash. The car is situated in a curve and traffic is heavy. Police are not on the scene. Your patient is alert and looking around, but stuck in the vehicle due to traffic. You see blood smeared across her face, but it appears to be minimal.

How would you best manage this situation?

Notes

Workbook Activities

The following activities have been designed to help you. Your instructor may require you to complete some or all of these activities as a regular part of your EMT-B training program. You are encouraged to complete any activity that your instructor does not assign as a way to enhance your learning in the classroom.

Chapter Review

The following exercises provide an opportunity to refresh your knowledge of this chapter.

NOTES

Matching

Match each of the terms in the left column to the appropriate definition in the right column.

_____ **1.** Extrication
_____ **2.** Simple access
_____ **3.** Complex access
_____ **4.** Access
_____ **5.** Incident commander
_____ **6.** Structure fire
_____ **7.** Hazardous material

A. access requiring no special tools and training

B. individual who has overall command of the scene in the field

C. a substance that causes injury or death with exposure

D. access requiring special tools and training

E. removal from entrapment or a dangerous situation or position

F. gaining entry to an enclosed area to reach a patient

G. fire in a house, apartment building, or other building

Multiple Choice

Read each item carefully, then select the best response.

_____ **1.** Aside from EMS personnel, other rescuers at a crash scene include:
   A. firefighters.
   B. law enforcement.
   C. a rescue group.
   D. all of the above

_____ **2.** Of the four teams at a crash scene, the _____ team is responsible for investigating the crash or crime scene.
   A. firefighter
   B. law enforcement
   C. rescue group
   D. EMS personnel

CHAPTER 36

Gaining Access

_C_ **3.** Of the four teams at a crash scene, the _____ team is responsible for properly securing and stabilizing the vehicle.

    A. firefighter

    B. law enforcement

    C. rescue group

    D. EMS personnel

_A_ **4.** Of the four teams at a crash scene, the _____ team is responsible for washing down any spilled fuel.

    A. firefighter

    B. law enforcement

    C. rescue group

    D. EMS personnel

_D_ **5.** Before proceeding with an extrication you should:

    A. position your unit in a safe location.

    B. make sure the scene is properly marked.

    C. determine if any additional resources will be needed.

    D. all of the above

_A_ **6.** While you are gaining access to the patient and during extrication, you must make sure that the patient:

    A. remains safe.

    B. stays conscious.

    C. holds his/her head completely still.

    D. all of the above

_B_ **7.** When dealing with multiple patients, you should locate and rapidly _____ each patient.

    A. treat

    B. triage

    C. transport

    D. extricate

_____ **8.** When preparing for patient removal, you should determine:

    A. how urgently the patient must be extricated.

    B. where you should be positioned during extrication.

    C. how you will best move the patient from the vehicle.

    D. all of the above

_____ **9.** Once the patient has been extricated, additional assessment should be completed:

    A. once the patient has been placed on the stretcher.

    B. inside the ambulance inclement weather.

    C. en route if the patient's condition requires rapid transport.

    D. all of the above

_____ **10.** Even when a technical rescue group includes a paramedic or physician, generally nothing but essential _____ is provided until the rescuers can bring the patient to the nearest point where a safe, stable setting exists.

    A. bandaging

    B. triage

    C. simple care

    D. splinting

_____ **11.** When called to a person lost outdoors, your role involves:

    A. standing by at the search base until the lost person is found.

    B. preparing necessary equipment.

    C. obtaining any medical history from relatives on scene.

    D. all of the above

_____ **12.** Tactical situations involve all of the following, except:

    A. an armed hostage situation.

    B. a structure fire.

    C. the presence of a sniper.

    D. an exchange of shots.

Vocabulary emtb.com vocab explorer

Define the following terms using the space provided.

1. Entrapment:

_____

_____

_____

2. Technical rescue situation:

_____

_____

_____

3. Danger zone:

_____

_____

_____

4. Tactical situation:

_____

_____

_____

5. Technical rescue group:

_____

_____

_____

Fill-in

Read each item carefully, then complete the statement by filling in the missing word(s).

1. _____ is the final phase of extrication, and this usually results in the patient being placed on the ambulance stretcher.

2. During all phases of rescue, your primary concern is _____.

3. Good _____ among team members and clear leadership are essential to safe, efficient provision of proper emergency care.

4. You should not attempt to gain access to the patient or enter the vehicle until

you are sure that the vehicle is _____.

5. When gaining access, it is up to you to identify the _____, most efficient way to gain access.

6. Moving the patient in one fast, continuous step increases the risk of

_____ and confusion.

7. Search and rescue is performed by teams of firefighters wearing full turnout gear

and _____ _____ _____ _____
(SCBA), and carrying tools and fully charged hose lines.

7

True/False

If you believe the statement to be more true than false, write the letter "T" in the space provided. If you believe the statement to be more false than true, write the letter "F."

1. __F__ There should be no talking throughout the extrication process.

2. __F__ A team leader must be identified and agreed to before you arrive at the scene.

3. __T__ If you will be involved with extrication, you should wear leather gloves over your disposable gloves.

4. __F__ Once a physician arrives at an emergency scene and is properly identified, all care should be turned over to him or her.

5. __F__ The first step in simple access is to try to get to the patient as quickly as possible using tools or other forcible entry methods.

6. __T__ You should not try to access the patient until you are sure that the vehicle is stable and that hazards have been identified and rendered safe.

Short Answer

Complete this section with short written answers using the space provided.

1. Explain the four different basic functions that must be addressed at any crash scene.

_____

_____

_____

_____

2. To determine the exact location and position of the patient, you and your team should consider what questions?

_____

_____

_____

3. List the steps for assessing and caring for a patient who is entrapped once access has been gained.

_____

_____

_____

4. When examining the exposed area of the limb or other part of the patient that is trapped, explain what you are assessing for.

_____

_____

_____

5. Explain the proper technique for patient removal once they are disentangled.

_____

_____

_____

Word Fun

The following crossword puzzle is an activity provided to reinforce correct spelling and understanding of medical terminology associated with emergency care and the EMT-B. Use the clues in the column to complete the puzzle.

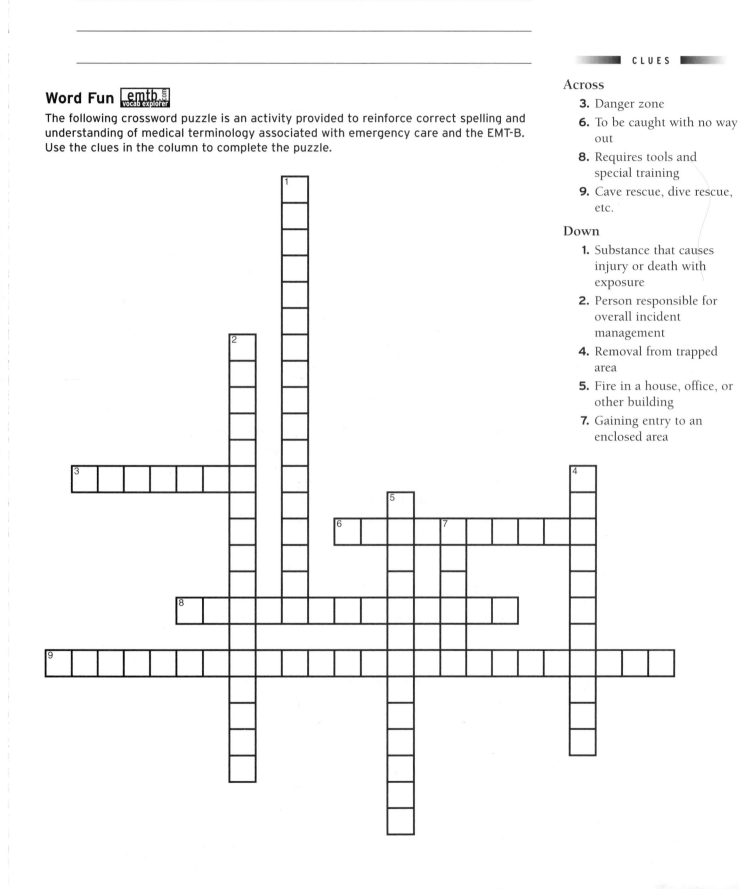

CLUES

Across

3. Danger zone

6. To be caught with no way out

8. Requires tools and special training

9. Cave rescue, dive rescue, etc.

Down

1. Substance that causes injury or death with exposure

2. Person responsible for overall incident management

4. Removal from trapped area

5. Fire in a house, office, or other building

7. Gaining entry to an enclosed area

7

Ambulance Calls

The following real case scenarios provide an opportunity to explore the concerns associated with patient management. Read each scenario, then answer each question in detail.

1. You are dispatched to the scene of a personal care home where an elderly resident has wandered off into nearby woods. The staff searched for half an hour before calling for assistance. You are in the staging area with the daughter and son-in-law. Staff members told you earlier that the patient has a history of Alzheimer's and some other details they could not remember.

   How would you best manage this situation?

   _____

   _____

   _____

   _____

   _____

   _____

   _____

   _____

   _____

   _____

2. You are called to the scene of a motor vehicle crash with a 24-year-old male who is pinned underneath the dashboard. Firefighters are setting up the equipment for extrication. The patient is alert but confused, and his breathing is slightly shallow at 24 breaths/min. He has weak radial pulses.

   How would you best manage this patient?

   _____

   _____

   _____

   _____

   _____

   _____

   _____

   _____

   _____

   _____

3. You are dispatched to a chemical spill where a train car derailed. The patient is the engineer who went back to look at the damage. He is lying beside the tracks and appears to be breathing from your vantage point at the staging area. HazMat team members are suiting up to go in and retrieve the patient. They will decontaminate him before bringing him to the staging area.

How will you best manage this situation and patient?

_____

_____

_____

_____

_____

_____

_____

_____

_____

_____

7

Workbook Activities

The following activities have been designed to help you. Your instructor may require you to complete some or all of these activities as a regular part of your EMT-B training program. You are encouraged to complete any activity that your instructor does not assign as a way to enhance your learning in the classroom.

Chapter Review

The following exercises provide an opportunity to refresh your knowledge of this chapter.

■ NOTES ■

Matching

Match each of the terms in the left column to the appropriate definition in the right column.

_J_ 1. Mass-casualty incident

_K_ 2. Incident command system

_M_ 3. Toxicity level

_D_ 4. Triage

_H_ 5. Chemical Transportation Emergency Center (CHEMTREC)

_E_ 6. Protection level

_B_ 7. Casualty collection area

_I_ 8. Disaster

_A_ 9. Hazardous materials incident

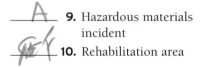 10. Rehabilitation area

A. incident in which a hazardous material is no longer properly contained and isolated

B. area where patients can receive further triage and medical care

C. individual who is in charge of and directs EMS personnel at the treatment area

D. the process of sorting patients based on the severity of injury and medical need, to establish treatment and transportation priorities

E. a measure of the amount and type of protective equipment that an individual needs to avoid injury during contact with a hazardous material

F. provides protection and treatment to firefighters and other personnel working at an emergency

G. area where ambulances and crews are organized

H. an agency that assists emergency personnel in identifying and handling hazardous materials transport incidents

I. widespread event that disrupts community resources and functions

J. an emergency situation involving more than one patient, and which can place such demand on equipment or personnel that the system is stretched to its limit or beyond

Special Operations

_G_ **11.** Transportation area

K. an organizational system to help control, direct, and coordinate emergency responders and resources; known more generally as an incident management system (IMS)

_C_ **12.** Treatment officer

L. a measure of the risk that a hazardous material poses to the health of an individual who comes in contact with it

Multiple Choice

Read each item carefully, then select the best response.

_D_ **1.** Functions normally centered at the command post include:
  A. information.
  B. safety.
  C. liaison with other agencies and groups who are responding.
  D. all of the above

_D_ **2.** In extended operations, the typical incident command structure may have multiple sectors including:
  A. operations.
  B. planning.
  C. logistics.
  D. all of the above

_D C_ **3.** With a major airplane crash, the leading agency is typically the:
  A. EMS department.
  B. law enforcement.
  C. fire department.
  D. HazMat team.

_A_ **4.** The _____ is a holding area for arriving ambulances and crews until they can be assigned a particular task.
  A. staging area
  B. treatment area
  C. transportation area
  D. rehabilitation area

5. The _____ provides protection and treatment to firefighters and other personnel working at the emergency scene.

   A. staging area

   B. treatment area

   C. transportation area

   D. rehabilitation area

6. The _____ is where ambulances and crews are organized to transport patients from the treatment area to local hospitals.

   A. staging area

   B. treatment area

   C. transportation area

   D. rehabilitation area

7. As patients are loaded into the ambulance, the transport officer logs:

   A. each patient's mass-casualty tag number.

   B. each patient's overall condition.

   C. the hospital to which they will be taken.

   D. all of the above

8. The _____ is where a more thorough assessment is made and on-scene treatment is begun while transport is being arranged.

   A. staging area

   B. treatment area

   C. transportation area

   D. rehabilitation area

9. Examples of mass-casualty incidents include:

   A. airplane crashes.

   B. earthquakes.

   C. railroad crashes.

   D. all of the above

10. To make decontaminating the ambulance easier after a HazMat incident:

   A. tape the cabinet doors shut.

   B. place any equipment that will not be used en route in the front of the truck.

   C. turn on the power vent ceiling fan and patient compartment air conditioning unit fan.

   D. all of the above

11. If you are treating and transporting a patient who has not been fully and properly decontaminated, you should do all of the following, except:

   A. wear two pairs of gloves.

   B. remove goggles.

   C. wear a protective coat.

   D. wear respiratory protection.

12. EMT-Bs providing care in the treatment area should assess and treat the patient:

   A. as contaminated.

   B. with respect.

   C. from the point where the previous caregiver left off.

   D. in the same way as a patient who has not previously been assessed or treated.

13. To avoid entrapment and communication of contaminants, only _____ are applied, until the "clean" patient has been moved to the treatment area.

    A. pressure dressings that are needed to control bleeding

    B. bandages

    C. splints

    D. cervical collars

14. When a material is hazardous because of its flammability or potential for explosion rather than its toxicity, you will need to be at an even greater distance and behind a windowless wall or other strong barrier that will shield you and others from:

    A. heat.

    B. the blast force.

    C. flying debris.

    D. all of the above

15. When toxic gas, fumes, or airborne droplets or particles are involved, the safe area is upwind and at least _____ from the site of any visible cloud or other discharge.

    A. 50'

    B. 100'

    C. 150'

    D. 200'

16. If you can see and read the placard or other warning sign, note _____, and, if included, the four-digit number that appears on it or on any orange panel near it.

    A. its color

    B. its wording

    C. any symbols that it contains

    D. all of the above

17. Once you have reached a safe place, try to rapidly assess the situation and provide as much information as possible when calling for the HazMat team, including:

    A. your specific location.

    B. the size and shape of the containers of the hazardous material.

    C. what you have observed and been told has occurred.

    D. all of the above

18. In the event of a leak or spill, a hazardous materials incident is often indicated by presence of:

    A. a visible cloud or strange-looking smoke resulting from the escaping substance.

    B. a leak or spill from a tank, etc., with or without HazMat placards or labels.

    C. an unusual, strong, noxious, acrid odor in the area.

    D. all of the above

19. Safety of _____ must be your most important concern.

    A. you and your team

    B. the other responders

    C. the public

    D. all of the above

7

B **20.** In some incidents, a large number of people are _____ and may be injured or killed before the presence of a hazardous materials incident is identified.

  A. transported

  B. exposed

  C. injected

  D. decontaminated

D **21.** Often, the presence of hazardous materials is easily recognized from warning signs, placards, or labels found:

  A. on buildings or areas where hazardous materials are produced, used, or stored.

  B. on trucks and railroad cars that carry any amount of hazardous material.

  C. on barrels or boxes that contain hazardous material.

  D. all of the above

For the remainder of the multiple-choice section, the following answers are to be applied:

  A. First priority (red)

  B. Second priority (yellow)

  C. Third priority (green)

  D. Fourth priority (black)

Classify the following emergencies according to triage priority:

22. Shock ___A___

23. Major or multiple bone or joint injuries _____

24. Cardiac arrest ___A___

25. Minor fractures ___A___

26. Decreased level of consciousness ___AC___

27. Obvious death ___D___

28. Airway and breathing difficulties ___A___

29. Burns without airway problems ___B___

30. Major open brain trauma ___D___

31. Minor soft-tissue injuries ___C___

Vocabulary

Define the following terms using the space provided.

1. Command post:

_____

_____

_____

2. Danger zone:

_____

_____

_____

3. Hazardous material:

_____

_____

_____

4. Incident commander:

_____

_____

_____

Fill-in

Read each item carefully, then complete the statement by filling in the missing word(s).

1. The _____ is more effective when used to organize large numbers of

   personnel at complex incidents such as hazardous materials spills and mass-

   casualty incidents.

2. The incident commander usually remains at a _____, the designated
   field command center.

3. The _____ is responsible for protecting all personnel and any victims
   of the incident.

4. When you arrive at the scene of a possible _____, you must first step
   back and assess the situation.

5. If patients are entrapped, _____ is required.

6. A _____ is a widespread event that disrupts functions and resources
   of a community and threatens lives and property.

7. _____ is the sorting of patients based on the severity of their
   conditions to establish priorities for care based on available resources.

8. Transporting a _____ patient merely increases the size of the event.

9. Most serious injuries and deaths from hazardous materials result from _____
   and _____ problems.

10. _____ is the process of removing or neutralizing and properly
    disposing of hazardous materials from equipment, patients, and rescue personnel.

11. When dealing with a HazMat situation, be sure to check the wind direction
    periodically, and _____ if a change in wind direction dictates.

12. Some substances are not hazardous; however, when mixed with another
    substance, they may become _____ or volatile.

7

13. In most cases, the package or tank must contain a certain amount of a hazardous material before a _____ is required.

True/False

If you believe the statement to be more true than false, write the letter "T" in the space provided. If you believe the statement to be more false than true, write the letter "F."

1. _____ When you are responding to a hazardous materials incident, you must first take time to accurately assess the scene.

2. _____ Moving patients from the contaminated area is your main responsibility in a hazardous materials situation.

3. _____ Toxicity level 1 is more dangerous than level 4.

4. _____ Protective clothing level A is the least level of protection.

5. _____ Patients with major or multiple bone or joint injuries should be assigned to the second priority triage category.

6. _____ Patients with severe burns should be assigned to the black triage category.

7. _____ A large number of hazardous gases and fluids are essentially odorless.

8. _____ Only the original patients who leave the hazard zone must pass through the decontamination area.

9. _____ Most hazardous materials have specific antidotes or treatments for exposure.

10. _____ The success of any incident command system depends on all personnel performing their assigned tasks and working within the system.

11. _____ When material is hazardous because of its flammability or potential for explosion rather than its toxicity, the damage or hazard zone is smaller.

Short Answer

Complete this section with short written answers using the space provided.

1. Define the five levels of toxicity as classified by the NFPA.

_____

_____

_____

_____

_____

2. Describe the four levels of protection and the type of protective gear required for each level.

_____

_____

_____

_____

3. List the eight major EMS-related positions within an incident command system.

_____

_____

_____

_____

_____

_____

_____

_____

4. List and define the four triage priorities.

_____

_____

_____

_____

5. For each of the following hazardous materials classifications, list the general category of hazard.

Class	Type
Class 1	
Class 2	
Class 3	
Class 4	
Class 5	
Class 6	
Class 7	
Class 8	
Class 9	

7

NOTES

6. Define and describe the process of decontamination and the decontamination area.

_____

_____

_____

_____

_____

CLUES

Across

3. Person responsible for overall incident management

5. Designated field command center

6. Zone where contaminants are removed

8. Any harmful substance

9. Determines amount of gear to be worn around a given hazard

Down

1. Zone for loading of patients into ambulances

2. Responsible for sorting of patients

4. Measures of health risk of a substance

6. Area of least safety, exposure to harm possible

7. Sorting by priority

Word Fun

The following crossword puzzle is an activity provided to reinforce correct spelling and understanding of medical terminology associated with emergency care and the EMT-B. Use the clues in the column to complete the puzzle.

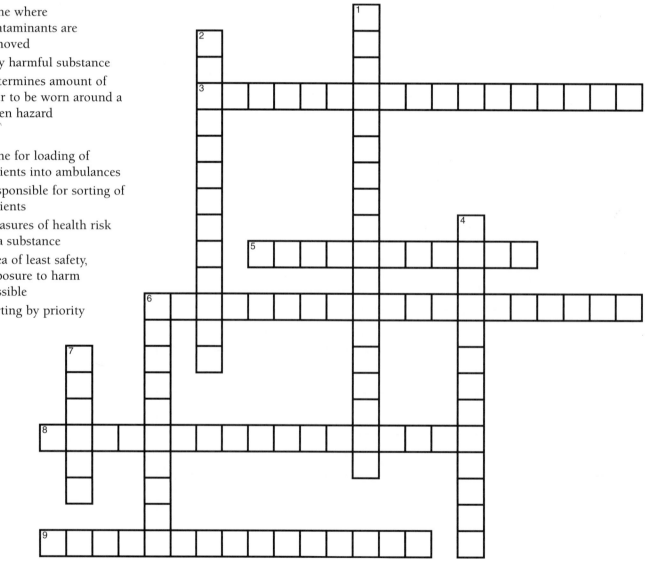

Ambulance Calls

The following real case scenarios will give you an opportunity to explore the concerns associated with patient management. Read each scenario, then answer each question in detail.

1. You are dispatched to a multi-vehicle crash where you encounter three patients: a 4-year-old male with bilateral femur fractures and absent radial pulse, a 27-year-old female with a laceration to the head and a humerus fracture, and a 42-year-old male who is apneic and pulseless with an open skull fracture.

   How should you triage these patients?

   _____

   _____

   _____

   _____

   _____

   _____

   _____

   _____

   _____

   _____

2. You respond to an overturned vehicle to find a tanker truck lying on its side with a white liquid pooling underneath. The driver is still restrained and is not moving. You see blood on the side of his head that faces you. There are no placards visible on the vehicle.

   How would you best manage this situation?

   _____

   _____

   _____

   _____

   _____

   _____

   _____

   _____

   _____

7

3. You are called to transport patients from a hazardous materials spill at a local chemical plant. The patients have gone through a partial decontamination, but are not "clean."

How would you best manage this situation?

_____

_____

_____

_____

_____

_____

_____

_____

_____

Notes

Workbook Activities

The following activities have been designed to help you. Your instructor may require you to complete some or all of these activities as a regular part of your EMT-B training program. You are encouraged to complete any activity that your instructor does not assign as a way to enhance your learning in the classroom.

Chapter Review

The following exercises provide an opportunity to refresh your knowledge of this chapter.

Matching

Match each of the terms in the left column to the appropriate definition in the right column.

_____ 1. Pharynx

_____ 2. Larynx

_____ 3. Epiglottis

_____ 4. Inhalation

_____ 5. Alveolar/capillary

_____ 6. Capillary/cellular

_____ 7. Exhalation

_____ 8. End tidal carbon dioxide detector

_____ 9. Cricoid

_____ 10. Stylet

_____ 11. Intubation

_____ 12. Laryngoscope

_____ 13. Vallecula

A. a plastic-coated wire that adds rigidity and shape to the endotracheal tube

B. passive phase of breathing

C. gas exchanged in body

D. an instrument that provides a direct view of the vocal cords

E. active phase of breathing

F. the space between the base of the tongue and the epiglottis

G. vocal cords

H. rigid, ring-shaped cartilage

I. the placement of a tube into the trachea to maintain the airway

J. gas exchanged in the lungs

K. throat

L. leaf-shaped structure that prevents food/liquids from entering the lower airway

M. a plastic disposable indicator that signals, by color change, that an ETT is in the proper place

Advanced Airway Management

Multiple Choice

Read each item carefully, then select the best response.

_____ 1. To maintain a patent airway, patients whose consciousness is altered may require:

A. an oropharyngeal airway.

B. a nasopharyngeal airway.

C. suctioning.

D. all of the above

_____ 2. The purpose of advanced airway management is to protect and improve _____ in patients by using a tube to create a direct channel to the trachea.

A. respiration

B. ventilation

C. oxygenation

D. patency

_____ 3. The upper airway includes all of the following, except:

A. nose.

B. mouth.

C. larynx.

D. pharynx.

_____ 4. The _____ is located at the glottic opening and prevents food and liquid from entering the lower airway during swallowing.

A. larynx

B. vocal cords

C. epiglottis

D. carina

_____ 5. Gas exchange occurs in the:

A. alveoli.

B. nares.

C. bronchioles.

D. trachea.

_____ 6. The mechanical process of breathing occurs through the nose of the diaphragm and:

   A. ribs.

   B. intercostal muscles.

   C. lung parenchyma.

   D. trachea.

_____ 7. The diaphragm and intercostal muscles _____ during inhalation, increasing the size of the chest cavity.

   A. expand

   B. contract

   C. dilate

   D. spasm

_____ 8. During inspiration:

   A. the diaphragm contracts and the ribs move up and out.

   B. the diaphragm relaxes and the ribs move up and out.

   C. the diaphragm contracts and the ribs move down and in.

   D. the diaphragm relaxes and the ribs move down and in.

_____ 9. During expiration:

   A. the diaphragm contracts and the ribs move up and out.

   B. the diaphragm relaxes and the ribs move up and out.

   C. the diaphragm contracts and the ribs move down and in.

   D. the diaphragm relaxes and the ribs move down and in.

_____ 10. After _____ minutes without oxygen, cells in the brain and nervous system may die.

   A. 2 to 3

   B. 3 to 5

   C. 4 to 6

   D. 5 to 8

_____ 11. The first step in airway management is:

   A. suctioning.

   B. c-spine control.

   C. applying oxygen.

   D. opening the airway.

_____ 12. Patients with gastric distention are prone to:

   A. gas.

   B. vomiting.

   C. sepsis.

   D. hypoxia.

_____ 13. A nasogastric tube can cause:

   A. nasal trauma.

   B. a basilar skull fracture.

   C. gastric distention.

   D. facial trauma.

_____ 14. To perform the Sellick maneuver, apply pressure to the:

   A. thyroid cartilage.

   B. cricoid cartilage.

   C. cricothyroid membrane.

   D. trachea.

_____ **15.** You should not immediately intubate a patient who is unresponsive or in cardiac arrest, but you must try:

A. to open the airway with the appropriate BLS maneuver.

B. to clear the airway.

C. to ventilate the patient with a BVM or oxygen-powered breathing device.

D. all of the above

_____ **16.** _____ is the most effective way to control a patient's airway and has many advantages over other airway management techniques.

A. An oropharyngeal adjunct

B. Orotracheal intubation

C. A nasopharyngeal adjunct

D. A jaw thrust

_____ **17.** Remember that _____ is the priority for patients in cardiac arrest from ventricular fibrillation.

A. airway

B. breathing

C. circulation

D. defibrillation

_____ **18.** The purpose of a _____ is to sweep the tongue out of the way and align the airway so that you can see the vocal cords and pass the ET tube through them.

A. laryngoscope

B. lighted stylet

C. Magill forceps

D. 10-mL syringe

_____ **19.** The curved laryngoscope blade is inserted just in front of the epiglottis, into the _____, allowing you to see the glottic opening and vocal cords.

A. uvula

B. vallecula

C. larynx

D. pharynx

_____ **20.** The proper sized tube for adult male patients ranges from _____ mm.

A. 6.5 to 8.0

B. 7.0 to 8.0

C. 7.5 to 8.5

D. 8.0 to 9.0

_____ **21.** The proper sized tube for adult female patients ranges from _____ mm.

A. 6.5 to 8.0

B. 7.0 to 8.0

C. 7.5 to 8.5

D. 8.0 to 9.0

_____ **22.** A good rule of thumb is to always have a _____ mm ETT on hand; this size tube will fit most male or female adult patients.

A. 6.5

B. 7.0

C. 7.5

D. 8.0

8

23. In children younger than 8 years, the circular narrowing of the trachea at the level of the _____ functions as a cuff.
    A. larynx
    B. cricoid cartilage
    C. pharynx
    D. thyroid cartilage

24. A plastic-coated wire called a _____ may be inserted into the ETT to add rigidity and shape to the tube.
    A. Murphy eye
    B. stylet
    C. pipe cleaner
    D. vallecula

25. You will use the _____ to test for air holes in the ETT before intubation.
    A. 10-mL syringe
    B. lighted stylet
    C. Murphy eye
    D. pilot balloon

26. Equipment used for airway and ventilation assistance include:
    A. oxygen.
    B. a suctioning unit.
    C. a ventilation device.
    D. all of the above

27. Do not let go of the ETT until:
    A. the distal cuff is inflated.
    B. placement is confirmed.
    C. it is secured.
    D. the stylet is removed.

28. Confirm placement of the ETT by listening with a stethoscope over both lungs and over the _____ as you ventilate the patient through the tube.
    A. stomach
    B. heart
    C. diaphragm
    D. ribs

29. The best way to confirm placement of an ETT is by:
    A. X-ray.
    B. auscultating over both lung fields and over the epigastrium.
    C. visualizing the cuff passing through the vocal cords.
    D. seeing the chest rise and fall.

30. The first step in nasotracheal intubation is to:
    A. check for a gag reflex.
    B. turn the stylet on.
    C. hyperventilate the patient.
    D. use BSI precautions.

_____ **31.** Complications of endotracheal intubation include:

    A. mechanical failure.

    B. intubating the esophagus.

    C. causing soft-tissue trauma.

    D. all of the above

_____ **32.** Benefits of endotracheal intubation include all of the following, except:

    A. it provides complete protection of the airway.

    B. patient intolerance.

    C. it prevents gastric distention and aspiration.

    D. it delivers better oxygen concentration than a BVM device.

_____ **33.** Always check for a gag reflex before intubation by using:

    A. a tongue blade.

    B. the laryngoscope blade.

    C. an oral airway.

    D. all of the above

_____ **34.** Benefits of using a multi-lumen airway include all of the following, except:

    A. no mask seal necessary.

    B. requires deeply comatose patient.

    C. may be inserted blindly.

    D. ease of proper placement.

_____ **35.** Contraindications of the ETC include:

    A. patients with a gag reflex.

    B. children younger than 16 years.

    C. patients who have ingested a caustic substance.

    D. all of the above

_____ **36.** To remove the ETC, you should _____ prior to removal.

    A. turn the patient on his or her side

    B. deflate both balloon cuffs

    C. have suction available

    D. all of the above

_____ **37.** Contraindications of the PtL include all of the following, except:

    A. patients with a gag reflex.

    B. children younger than 14 years.

    C. children older than 16 years.

    D. patients who have a known esophageal disease.

_____ **38.** Potential contraindications when using an LMA include:

    A. when positive pressure ventilation with high airway pressures occur, the mask may leak.

    B. active vomiting may dislodge the device.

    C. esophageal disease.

    D. all of the above

Labeling emt-b anatomy review

Label the following diagram with the correct terms.

1. The Upper and Lower Airways

A. _____

B. _____

C. _____

D. _____

E. _____

F. _____

G. _____

H. _____

I. _____

J. _____

K. _____

L. _____

M. _____

N. _____

O. _____

P. _____

Q. _____

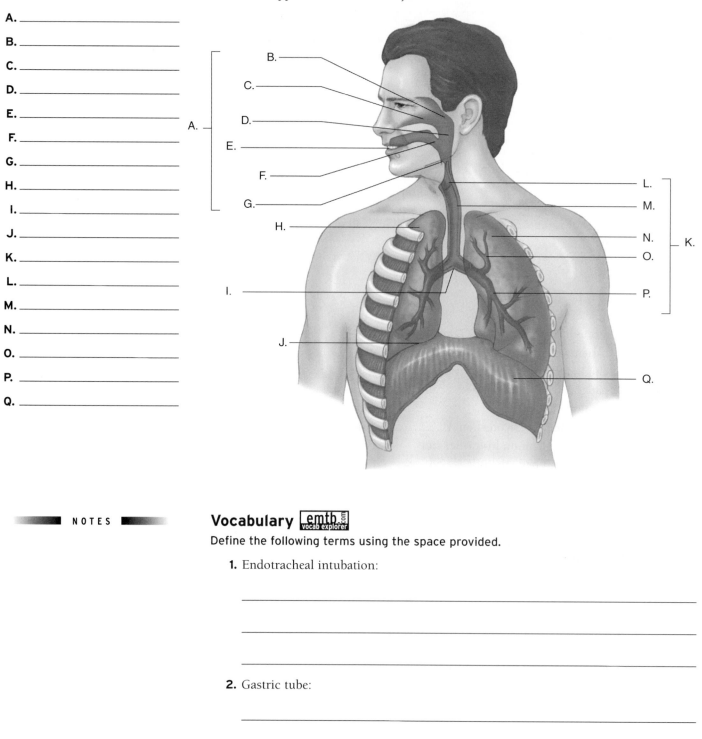

Vocabulary emtb. vocab explorer

Define the following terms using the space provided.

1. Endotracheal intubation:

_____

_____

_____

2. Gastric tube:

_____

_____

_____

3. Nasotracheal intubation:

_____

_____

_____

4. Orotracheal intubation:

_____

_____

_____

5. Sellick maneuver:

_____

_____

_____

Fill-in

Read each item carefully, then complete the statement by filling in the missing word.

1. During _____ , the diaphragm contracts.

2. The _____ branch off from the trachea.

3. During the _____ phase of breathing, the intercostal muscles relax.

4. _____ in the alveoli crosses over into the blood.

5. _____ in the blood crosses over into the alveoli.

6. Cells in the brain and nervous system may die after _____ minutes without oxygen.

7. The _____ are the smallest air passages leading to the alveoli.

8. Alveoli are surrounded by _____ , which bring deoxygenated blood to the lungs.

True/False

If you believe the statement to be more true than false, write the letter "T" in the space provided. If you believe the statement to be more false than true, write the letter "F."

1. _____ The light in the laryngoscope will not work unless the blade is attached correctly.

2. _____ Endotracheal intubation is usually most appropriate for unconscious patients.

3. _____ There are three different sizes of ETTs.

4. _____ The balloon cuff around the end of an ETT holds 25 mL of air.

8

5. _____   Uncuffed tubes are used in children younger than 8 years.

6. _____   When a wire stylet is used, it should stick out ½" beyond the tip of the ETT.

Short Answer

Complete this section with short written answers using the space provided.

1. Describe how to perform the Sellick maneuver.

_____

_____

_____

2. List three advantages of orotracheal intubation.

_____

_____

_____

3. List nine possible complications associated with endotracheal intubation.

_____

_____

_____

_____

_____

_____

_____

_____

4. List five contraindications for an Esophageal Tracheal Combitube.

_____

_____

_____

_____

5. List three benefits and three complications of multi-lumen airways.

_____

_____

_____

_____

_____

_____

Word Fun

The following crossword puzzle is an activity provided to reinforce correct spelling and understanding of medical terminology associated with emergency care and the EMT-B. Use the clues in the column to complete the puzzle.

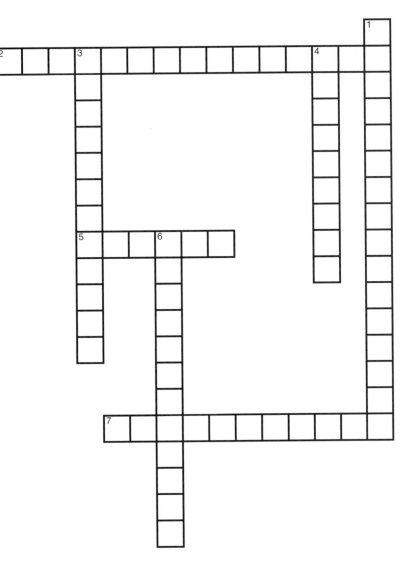

CLUES

Across

  2. Pressure on cricoid cartilage

  5. Small metal rod inserted into an ET tube

  7. Removes substances from stomach

Down

  1. Rigid, ring-shaped structure at larynx

  3. Vocal cord spasm

  4. Between base of tongue and epiglottis

  6. Used to gain direct view of patient's throat

Ambulance Calls

The following real case scenarios provide an opportunity to explore the concerns associated with patient management. Read each scenario, then answer each question in detail.

1. You are dispatched to a possible cardiac arrest. You are only a block from the call and arrive to find a 72-year-old male who is still pink and warm. No one is performing CPR.

   How would you best manage this patient?

   _____

   _____

   _____

   _____

   _____

   _____

   _____

   _____

   _____

2. You are called to a residence where an unresponsive 7-year-old has possibly ingested drain cleaner by accident. He is breathing shallowly at a rate of 8 breaths/min. He has no gag reflex.

   How would you best manage this patient?

   _____

   _____

   _____

   _____

   _____

   _____

   _____

   _____

   _____

3. You are dispatched to the scene of a motor vehicle crash where your patient, the unrestrained 18-year-old driver, is breathing at a rate of 4 breaths/min and has weak radial pulses. He is also trapped in the vehicle and the fire department has not yet arrived.

How would you best manage this patient?

_____

_____

_____

_____

_____

_____

_____

_____

_____

_____

Skill Drills emt-b video clips

Skill Drill 38-1: Performing the Sellick Maneuver
Test your knowledge of this skill drill by filling in the correct words in the photo captions.

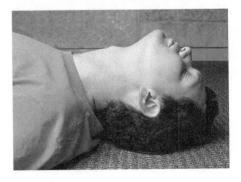

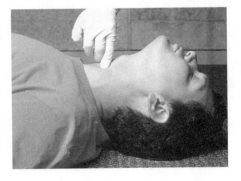

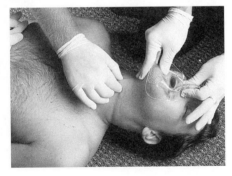

1. Visualize the _____ cartilage.

2. _____ to confirm its location.

3. Apply _____ pressure on the cricoid _____ with your thumb and index finger on either side of _____. Maintain pressure until _____.

8

Skill Drill 38-2: Performing Orotracheal Intubation

Test your knowledge of this skill drill by placing the photos below in the correct order. Number the first step with a "1," the second step with a "2," etc.

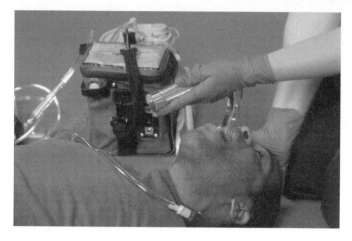

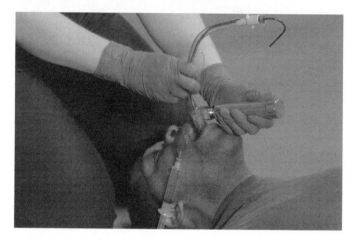

Confirm adequate hyperventilation and remove the oral airway.

If available, have another rescuer perform the Sellick manuever to improve visualization of the cords.

Use the head tilt–chin lift maneuver to position the non-trauma patient for insertion of the laryngoscope.

Insert the laryngoscope from the right side of mouth and move the tongue to the left. Lift the laryngoscope away from the posterior pharynx to visualize the vocal cords.

Insert the ETT from the right side. Remove laryngoscope and stylet. Hold the tube carefully until secured.

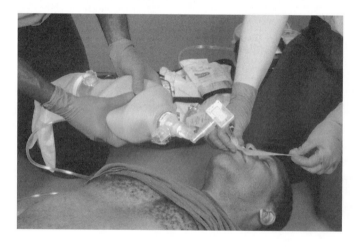

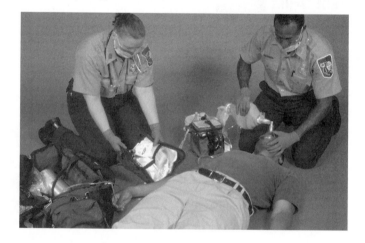

Secure the tube and continue to ventilate.

Note and record depth of insertion, and reconfirm position after each time you move the patient.

Open and clear the airway.

Place an oropharyngeal airway and hyperventilate with a BVM.

(continued)

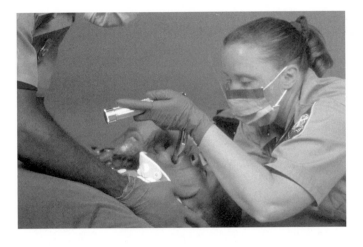

In a trauma patient, maintain the cervical spine in-line and neutral as your partner lies down or straddles the patient's head to visualize the vocal cords.

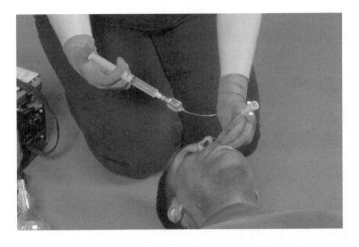

Inflate the balloon cuff and remove the syringe as your partner prepares to ventilate.

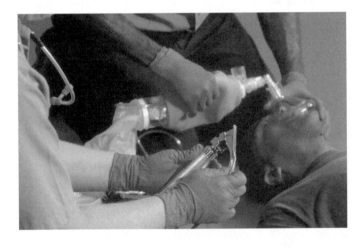

Assemble and test intubation equipment as your partner continues to ventilate.

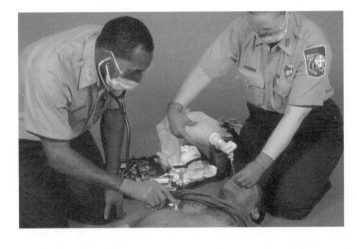

Begin ventilating and confirm placement of the ETT by listening over both lungs and the stomach. Also confirm placement with an end tidal carbon dioxide detector, if available.

8

Workbook Activities

The following activities have been designed to help you. Your instructor may require you to complete some or all of these activities as a regular part of your EMT-B training program. You are encouraged to complete any activity that your instructor does not assign as a way to enhance your learning in the classroom.

Chapter Review

The following exercises provide an opportunity to refresh your knowledge of this chapter.

Matching

Match each of the terms in the left column to the appropriate definition in the right column.

_____ **1.** Artificial ventilation

_____ **2.** Abdominal-thrust maneuver

_____ **3.** Basic life support (BLS)

_____ **4.** Advanced life support (ALS)

_____ **5.** Cardiopulmonary resuscitation (CPR)

A. steps used to establish artificial ventilation and circulation in a patient who is not breathing and has no pulse

B. noninvasive emergency lifesaving care used for patients in respiratory or cardiac arrest

C. procedures, such as cardiac monitoring, starting IV fluids, and using advanced airway adjuncts

D. a method of dislodging food or other material from the throat of a choking victim

E. opening the airway and restoring breathing by mouth-to-mask ventilation and by the use of mechanical devices

Multiple Choice

Read each item carefully, then select the best response.

_____ **1.** Basic life support is noninvasive emergency lifesaving care that is used to treat:

    A. airway obstruction.

    B. respiratory arrest.

    C. cardiac arrest.

    D. all of the above

_____ **2.** Exhaled gas from you to the patient contains _____ oxygen.

    A. 8%

    B. 12%

    C. 16%

    D. 21%

BLS Review

_____ **3.** BLS differs from advanced life support by involving advanced lifesaving procedures including all of the following, except:

A. cardiac monitoring.

B. mouth-to-mouth.

C. administration of IV fluids and medications.

D. use of advanced airway adjuncts.

_____ **4.** In some instances such as _____, early BLS measures may be all that a patient needs to be resuscitated.

A. choking

B. near drowning

C. lightning injuries

D. all of the above

_____ **5.** In addition to checking level of consciousness, it is also important to protect the _____ from further injury while assessing the patient and performing CPR.

A. spinal cord

B. ribs

C. internal organs

D. facial structures

_____ **6.** In most cases, cardiac arrest in children younger than 9 years results from:

A. choking.

B. aspiration.

C. congenital heart disease.

D. respiratory arrest.

_____ **7.** Causes of respiratory arrest in infants and children include:

A. aspiration of foreign bodies.

B. airway infections.

C. sudden infant death syndrome (SIDS).

D. all of the above

_____ **8.** Signs of irreversible or biological death include clinical death along with:

    A. rigor mortis.

    B. dependent lividity.

    C. decapitation.

    D. all of the above

_____ **9.** Once you begin CPR in the field, you must continue until:

    A. the fire department arrives.

    B. the funeral home arrives.

    C. a physician arrives who assumes responsibility.

    D. law enforcement arrives and assumes responsibility.

_____ **10.** Once the patient is properly positioned, you can easily assess:

    A. airway.

    B. breathing.

    C. disability.

    D. all of the above

_____ **11.** The chin lift has the added advantage of holding _____ in place, making obstruction by the lips less likely.

    A. loose dentures

    B. the tongue

    C. the mandible

    D. the maxilla

_____ **12.** To perform a _____, place your fingers behind the angles of the patient's lower jaw and then move the jaw forward.

    A. head tilt–chin lift maneuver

    B. jaw-thrust maneuver

    C. tongue-jaw lift maneuver

    D. all of the above

_____ **13.** Providing slow, deliberate inhalations over 2 seconds prevents:

    A. overexpansion of the lungs.

    B. rupture of the bronchial tree.

    C. gastric distention.

    D. rupture of the alveoli.

_____ **14.** A _____ is an opening that connects the trachea directly to the skin.

    A. tracheostomy

    B. stoma

    C. laryngectomy

    D. none of the above

_____ **15.** The _____ position helps to maintain a clear airway in a patient with a decreased level of consciousness who has not had traumatic injuries and is breathing on his or her own.

    A. recovery

    B. lithotomy

    C. Trendelenburg's

    D. Fowler's

_____ **16.** Excessive pressure applied to the carotid artery can:

    A. obstruct the carotid circulation.

    B. dislodge blood clots.

    C. produce marked reflex slowing of heart rate.

    D. all of the above

_____ **17.** The lower tip of the breastbone is the:

    A. xiphoid process.

    B. sternum.

    C. manubrium.

    D. intercostal space.

_____ **18.** Complications from chest compressions can include:

    A. fractured ribs.

    B. a lacerated liver.

    C. a fractured sternum.

    D. all of the above

_____ **19.** When checking for a pulse in an infant, you should palpate the
_____ artery.

    A. radial

    B. brachial

    C. carotid

    D. femoral

_____ **20.** The technique for chest compressions in infants and children differs
because of a number of anatomic differences, including:

    A. the position of the heart.

    B. the size of the chest.

    C. the fragile organs.

    D. all of the above

_____ **21.** The rate of compressions for an infant is at least _____ compressions
per minute.

    A. 70

    B. 80

    C. 90

    D. 100

_____ **22.** The rate of compression to ventilation for infants and children is
_____ for both one-rescuer and two-rescuer CPR.

    A. 1:5

    B. 5:1

    C. 15:2

    D. 2:15

_____ **23.** Sudden airway obstruction is usually easy to recognize in someone who is
eating or has just finished eating because they suddenly:

    A. are unable to speak or cough.

    B. turn cyanotic.

    C. make exaggerated efforts to breathe.

    D. all of the above

_____ **24.** You should suspect an airway obstruction in the unresponsive patient if:

    A. the standard maneuvers to open the airway and ventilate the lungs are
not effective.

    B. you feel resistance to blowing into the patient's lungs.

    C. pressure builds up in your mouth.

    D. all of the above

8

_____25. You should use _____ for women in advanced stages of pregnancy, patients who are very obese, and children younger than 1 year.

    A. the Heimlich maneuver

    B. chest thrusts

    C. the abdominal-thrust maneuver

    D. any of the above

_____26. For a patient with a partial airway obstruction, you should:

    A. perform the Heimlich maneuver.

    B. attempt a finger sweep to remove the foreign body.

    C. not interfere with the patient's attempt to expel the foreign body.

    D. all of the above

_____27. Transmission of infectious diseases such as _____ can occur while performing CPR.

    A. herpes simplex

    B. tuberculosis

    C. AIDS

    D. all of the above

Vocabulary emtb. vocab explorer

Define the following terms using the space provided.

1. Gastric distention:

_____

_____

_____

2. Head tilt–chin lift maneuver:

_____

_____

_____

3. Jaw-thrust maneuver:

_____

_____

_____

4. Recovery position:

_____

_____

_____

Fill-in

Read each item carefully, then complete the statement by filling in the missing word.

1. Permanent brain damage may occur if the brain is without oxygen for

_____ minutes.

2. CPR does not require any equipment; however, you should use a _____ device to perform rescue breathing.

3. Because of the urgent need to start CPR in a pulseless, nonbreathing patient, you must complete an initial assessment as soon as possible, evaluating the patient's

_____.

4. The most important element for successful CPR is immediate _____ of the airway.

5. Irreversible death is known as _____ death.

6. _____ is evidenced by the absence of a pulse and absence of breathing.

7. _____, such as living wills, may express the patient's wishes, but these documents are not binding for all health care providers.

8. For CPR to be effective, the patient must be lying supine on a _____ surface.

9. Without an open _____, rescue breathing will not be effective.

True/False

If you believe the statement to be more true than false, write the letter "T" in the space provided. If you believe the statement to be more false than true, write the letter "F."

1. _____ During the initial assessment, you need to quickly evaluate the patient's airway, breathing, circulation, and level of consciousness.

2. _____ All unconscious patients need all elements of BLS.

3. _____ A patient who is not fully conscious often needs some degree of BLS.

4. _____ The recovery position should be used to maintain an open airway in a patient with a head or spinal injury.

5. _____ You should always remove a patient's dentures before initiating artificial ventilation.

6. _____ You should not start CPR if the patient has obvious signs of irreversible death.

7. _____ After you apply pressure to depress the sternum, you must follow with an equal period of relaxation so that the chest returns to normal position.

8. _____ The ratio of compressions to ventilations for one-person CPR on an adult is 2:1.

9. _____ When performed correctly, external chest compressions provide 50% of the blood normally pumped by the heart.

10. _____ For infants, the preferred technique of artificial ventilation is mouth-to-nose-and-mouth ventilation with a mask or other barrier device.

11. _____ You need to use less ventilatory pressure to inflate a child's lungs because the airway is smaller than that of an adult.

8

Short Answer

Complete this section with short written answers using the space provided.

1. List the four obvious signs of death, in addition to absence of pulse and breathing, that are used as a general rule to not start CPR.

_____

_____

_____

_____

2. Complete the following table regarding pediatric BLS by listing the procedure parameters or guidelines for each age group as they relate to the action noted on the left.

Procedure	Infants (younger than age 1 year)	Children (age 1 to 8 years)
Airway		

Breathing
  Initial breaths
  Subsequent breaths		

Circulation
  Pulse check
  Compression area
  Compression width
  Compression depth
  Compression rate
Ratio of Compressions
to Ventilations
Foreign Body Obstruction		

3. List the four acceptable reasons for stopping CPR.

_____

_____

_____

_____

4. Describe how to perform a head tilt–chin lift maneuver.

_____

_____

_____

5. Describe how to perform a jaw-thrust maneuver.

_____

_____

_____

6. Describe the process of chest compressions during one-rescuer adult CPR.

_____

_____

_____

7. List and describe the method for "switching positions" during two-rescuer adult CPR.

_____

_____

_____

8. Describe the process of abdominal thrusts for a standing patient and a supine patient.

_____

_____

_____

9. Describe the process for chest thrusts on a standing and a supine patient.

_____

_____

_____

10. Describe the process for removing a foreign body airway obstruction in an infant.

_____

_____

_____

Word Fun

The following crossword puzzle is an activity provided to reinforce correct spelling and understanding of medical terminology associated with emergency care and the EMT-B. Use the clues in the column to complete the puzzle.

Across

3. Position after patient regains consciousness

4. Opening airway by moving lower bone forcibly forward

6. First level of care

Down

1. Preferred method to remove airway obstruction

2. Chest compression and artificial ventilation

5. High level of care, beyond basic

Ambulance Calls

The following real case scenarios provide an opportunity to explore the concerns associated with patient management. Read each scenario, then answer each question in detail.

1. You are dispatched to a person down. The dispatcher informs you that the caller said the patient is not breathing. Upon arrival, you find a 78-year-old female in bed, apneic and pulseless. In the process of moving the patient to place a CPR board underneath her, you note the discoloration of her back and hips known as dependent lividity.

How would you best manage this patient?

_____

_____

_____

_____

_____

_____

_____

_____

_____

_____

2. You are called to a business office downtown where CPR is in progress. Upon arrival, you immediately attach the pads of the AED to the patient while your partner checks the airway. He advises you that the patient is now breathing and palpation reveals a faint radial pulse.

How would you best manage this patient?

_____

_____

_____

_____

_____

_____

_____

_____

_____

_____

3. You arrive on the scene of a possible cardiac arrest to find a 48-year-old male with no related medical history, apneic and pulseless. Bystanders tell you he was breathing up until about 3 minutes ago.

How would you best manage this patient?

_____

_____

_____

_____

_____

_____

_____

_____

_____

8

Skill Drills emt-b

Skill Drill 39-1: Positioning the Patient

Test your knowledge of this skill drill by placing the photos below in the correct order. Number the first step with a "1," the second step with a "2," etc.

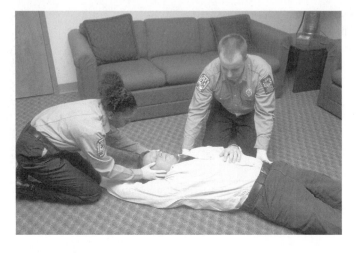

_____

Move the patient to a supine position with legs straight and arms at the sides.

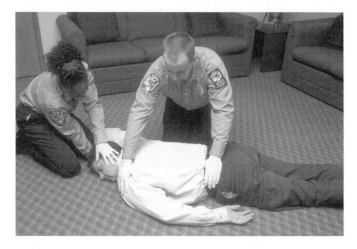

_____

Grasp the patient, stabilizing the cervical spine if needed.

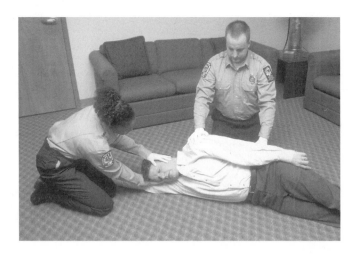

_____

Move the head and neck as a unit with the torso as your partner pulls on the distant shoulder and hip.

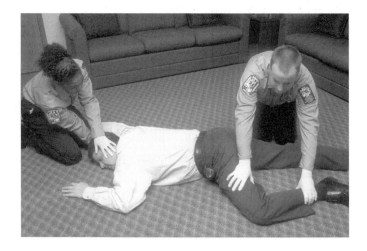

_____

Kneel beside the patient, leaving room to roll the patient toward you.

Skill Drill 39-2: Performing Chest Compressions
Test your knowledge of this skill drill by filling in the correct words in the photo captions.

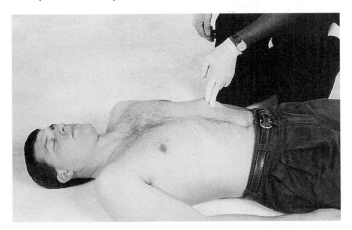

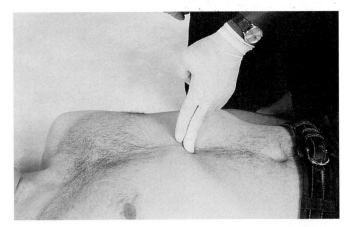

1. Slide your _____ and _____ fingers along the rib cage to the _____ in the _____ of the chest.

2. Push the middle finger high into the notch, and lay the index finger on the _____ _____ of the sternum.

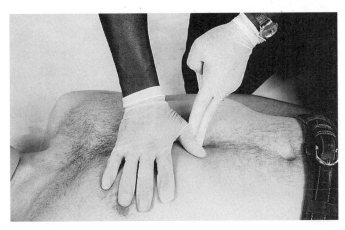

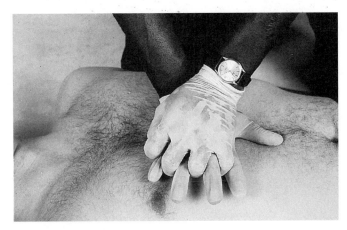

3. Place the _____ of the second hand on the lower half of the sternum, touching the _____ _____ of your first hand.

4. Remove your first hand from the notch, and place it over the _____ on the sternum.

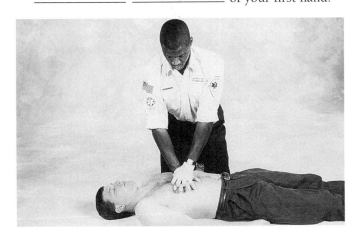

5. With your arms straight, _____ your elbows, and position your shoulders directly _____ your hands. Depress the sternum _____ inch(es) to _____ inch(es) using a rhythmic motion.

8

Skill Drill 39-3: Performing One-Rescuer Adult CPR

Test your knowledge of this skill drill by placing the photos below in the correct order.
Number the first step with a "1," the second step with a "2," etc.

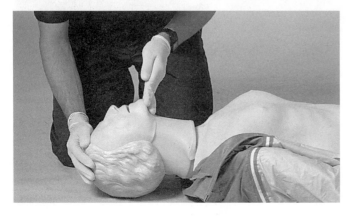

_____

Open the airway.

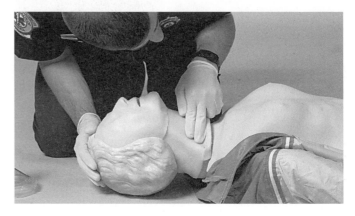

_____

Check for carotid pulse.

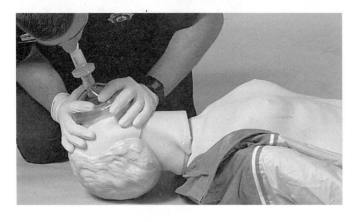

_____

If not breathing, give two breaths of 2 seconds each.

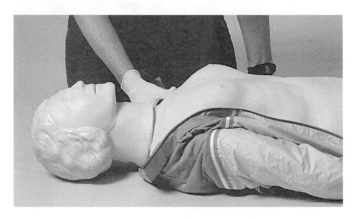

_____

Establish unresponsiveness and call for help.

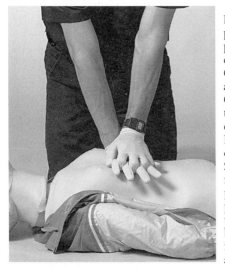

_____

If no pulse is found, place your hands in the proper position for chest compressions.
Give 15 compressions at about 100/min.
Open the airway and give two ventilations of 2 seconds each.
Perform four cycles of compressions.
Stop CPR and check for return of the carotid pulse. Depending on patient condition, continue CPR, continue rescue breathing only, or place in the recovery position and monitor.

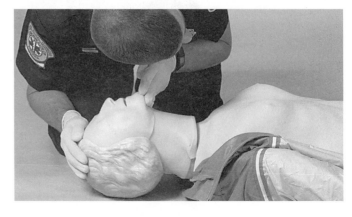

_____

Look, listen, and feel for breathing. If breathing, place in the recovery position and monitor.

Skill Drill 39-4: Performing Two-Rescuer Adult CPR
Test your knowledge of this skill drill by filling in the correct words in the photo captions.

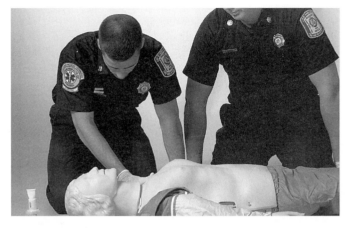

1. Establish _____ and take positions.

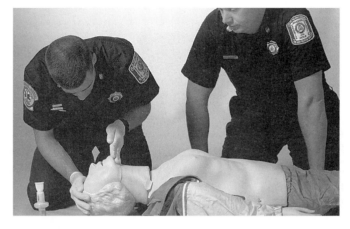

2. _____ the airway.

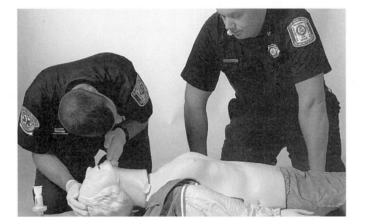

3. Look, listen, and feel for breathing. If breathing, place in the _____ position and _____.

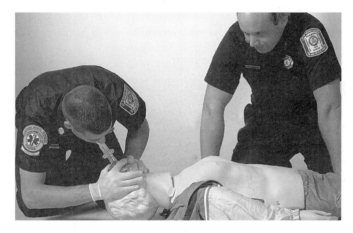

4. If not breathing, give _____ breaths of _____ seconds each.

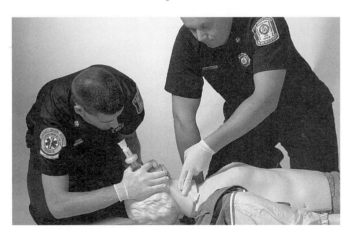

5. Check for _____ pulse.

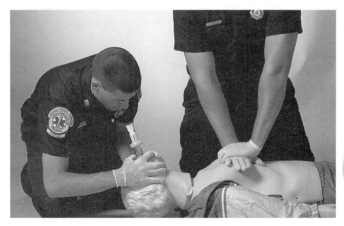

6. If no pulse, begin _____ _____ at about _____/min (_____ compressions to _____ ventilations). After _____ minute(s), check for _____ pulse. Check every few _____ thereafter. Depending on patient _____, continue CPR, continue rescue breathing only, or place in the recovery position and monitor.

8

Answer Key

Section 1 Preparing to be an EMT-B

Chapter 1: Introduction to Emergency Medical Care

Matching

1. H (page 8)
2. F (page 5)
3. M (page 4)
4. G (page 4)
5. A (page 4)
6. K (page 10)
7. B (page 11)

8. C (page 4)
9. L (page 15)
10. E (page 11)
11. I (page 10)
12. D (page 10)
13. J (page 6)

Multiple Choice

1. B – These are among the training or "scope of practice" of the EMT-Basic. Other levels of EMT (A, C, D) include advanced procedures in their practice. (page 6)

2. C – The EMT-B course requires 110 hours minimum training. (A), (B), and (D) are incorrect. (page 9)

3. A – Medical control may be on-line over radio or telephone, or off-line as protocols or standing orders, which describe the care authorized by the medical director. Dispatchers are not authorized to guide patient care by the EMS team. (page 10)

4. D – (A), (B), and (C) are among the circular system of CQI. The public is not a part of CQI. However, a complaint may lead to a CQI investigation. (page 11)

5. C – Quality control ensures that all staff members involved in caring for patients meet appropriate standards on each call. "Periodic" audits or reviews may be employed—these may be yearly, quarterly, or more frequently (A). EMT training in skills is a component of regulation and certification (B). Billing information (D) is an important aspect of documentation, but not a matter for the medical director in quality control. (page 11)

6. A – Ensuring your safety and that of your fellow EMTs is a primary responsibility. (B), (C), and (D) are among the concerns upon arrival at the scene during scene size-up. (page 14)

7. B – The emergency care of patients occurs in three phases: the first phase consists of assessment, packaging and transport, the second phase continues care in the emergency department (C), and in the third phase the patient receives the necessary definitive care. Public recognition (A) and accurate dispatching (D) are components of access to the EMS System. (page 14)

8. B – EMT-B training is divided into three main categories: life threats, non-life threats, and important non-medical issues as listed in the textbook. (pages 5 to 6)

9. C – EMT-B training is divided into three main categories: life threats, non-life threats and important non-medical issues as listed in the textbook. (pages 5 to 6)

10. B – The material and skills of the EMT-B training program was developed in the U.S. Department of Transportation 1994 EMT-Basic National Standard Curriculum. (A) The AAOS assists the development of EMS through research, quality standards, and publications. (C) The AHA conducts research, provides training standards, and certification in CPR. (D) The NAEMT represents EMS on the national, regional, and state levels. (page 5)

Vocabulary

1. Emergency medical service: A multidisciplinary system that represents the combined efforts of several professionals and agencies to provide prehospital emergency care to the sick and injured. (page 4)

2. First responder: The first trained individual, such as police officer, firefighter, or other rescuer to arrive at the scene to provide initial medical assistance.(page 8)

3. Primary service area: The designated area in which the EMS agency is responsible for the provision of prehospital emergency care and transportation to the hospital. (page 10)

4. Emergency medical technician: An EMS professional who is trained and licensed by the state to provide emergency medical care in the field. (page 4)

Fill-in

1. the U.S. DOT 1994 EMT-B national standard curriculum (page 5)

2. life-threatening, non-life threatening (page 5)

3. AED, medications (page 9)

4. 9-1-1 (page 10)

5. medical control (page 10)

True/False

1. F (page 4)

2. T (page 9)

3. F (page 11)

4. T (page 15)

5. T (page 15)

Short Answer

1. The EMT-B is one of the five levels of prehospital care. The EMT-B provides basic life support skills and may also provide some ALS skills, such as advanced airways and defibrillation. (page 9)

2. The Department of Transportation (DOT) has developed a series of guidelines, curriculum, funding sources, and assessment tools, all designed to develop and improve EMS in the United States. (page 8)

3. -Ensuring your own safety and the safety of your fellow EMT-Bs, the patient, and others at the scene

-Locating and safely driving to the scene

-Sizing up the scene and situation

-Rapidly assessing the patient's gross neurologic, respiratory, and circulatory status

-Providing any essential immediate intervention

-Performing a thorough, accurate patient assessment

-Obtaining an expanded SAMPLE history

-Reaching a clinical impression and providing prompt, efficient, prioritized patient care based on your assessment

-Communicating effectively with and advising the patient of any procedures you will perform

-Properly interacting and communicating with fire, rescue, and law enforcement responders at the scene

-Identifying patients who require rapid packaging and initiating transport without delay

-Identifying patients who do not need emergency care and will benefit from further detailed assessment and care before they are moved and transported

-Properly packaging the patients

-Safely lifting and moving the patient to the ambulance and loading the patient into it

-Providing safe appropriate transport to the hospital emergency department or other ordered facility

-Giving the necessary radio report to the medical control center or receiving hospital emergency department

-Providing any additional assessment or treatment while en route

-Monitoring the patient and checking vital signs while en route

-Documenting all findings and care on the run report

-Unloading the patient safely and, after giving a proper verbal report, transferring the patient's care to the emergency department staff

-Safeguarding the patient's rights (page 14)

4. On-line medical direction is provided through radio or telephone connections between the EMT-B and the medical control facility. Off-line medical direction is provided through written protocols, procedures, and standing orders. (page 11)

Word Fun

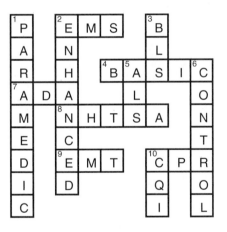

Ambulance Calls

1. Most services would require that you continue on the original call in order not to commit an act of negligence. The only exception would be if the route were blocked by the crash and another unit could reach the original call quicker from another direction. The crash should be reported to the dispatcher so that a unit could be dispatched and then proceed according to local protocols.

2. Call the dispatcher to send police if the environment appears hostile. Wait for them to arrive and then proceed to care for the patient. It would also be appropriate for the boy to come to you. Ask someone close by to bring the boy to the ambulance so that you can examine his injury. If in any doubt, wait for law enforcement.

3. The paramedics cannot transfer care of the patient to you in any area of the country. This would constitute abandonment by turning care over to someone with less training. You should tell the paramedics that you cannot accept transfer of this patient and notify your dispatcher to send an ALS unit for the transport. It is only acceptable to transport a patient with a BLS unit from one facility to another under a doctor's orders, not from the scene.

Chapter 2: The Well-Being of the EMT-B

Matching

1. A (page 58)
2. C (page 31)
3. F (page 38)
4. D (page 29)
5. B (page 39)
6. E (page 43)
7. N (page 38)
8. H (page 47)
9. M (page 47)
10. J (page 47)
11. K (page 38)
12. O (page 50)
13. I (page 47)
14. G (page 38)
15. L (page 47)

Multiple Choice

1. B (page 20)
2. C (page 21)
3. D (page 21)
4. A – This is a presumptive sign. (page 22)
5. D – Cases for the medical examiner also include those in which the patient is dead on arrival; death without previous medical care or when the physician is unable to state the cause of death; poisoning, known or suspected; and death resulting from accidents. (page 22)
6. C (page 22)
7. A (page 22)
8. D (page 22)
9. B (page 23)
10. A (page 23)
11. D (page 23)
12. C (page 24)
13. B (page 24)
14. D – Fear may also be expressed as withdrawal, tension, "butterflies" in the stomach, or nervousness. (page 24)
15. D (page 24)
16. B – Loss of contact with reality, regression, and diminished control of basic impulses and desires are other common characteristics of mental health problems. (page 25)
17. D (page 25)
18. D (page 26)
19. A – Other factors that influence how a patient reacts to the stress of an EMS incident include alcohol or substance abuse, history of chronic disease, mental disorders, reaction to medications, nutritional status, and feelings of guilt. (page 27)
20. B – This is a form of positive stress. (page 28)
21. D (page 28)
22. C – This is a sign, not a symptom. (page 28)
23. D – Prolonged or excessive stress has also been proven to be a strong contributor to alcoholism and depression. (page 29)
24. B (page 29)
25. D (page 29)
26. A (page 30)
27. B (page 30)
28. D – Stress management strategies also include changing or eliminating stressors; cutting back on overtime; stopping complaining and worrying about things you cannot change; adopting a more relaxed, philosophical outlook; expanding your social support system; sustaining friends and interests outside EMS; and minimizing the physical response to stress. (page 31)

29. D (page 31)

30. B (page 32)

31. C (page 33)

32. D (page 33)

33. D – Other components of the CISM system are on-scene peer support, one-on-one support, disaster support services, CISD, follow-up services, community outreach programs, and other health and welfare programs, such as wellness. (page 34)

34. D (page 35)

35. B – Drug and alcohol use in the workplace can lead to poor treatment decisions, not enhanced treatment decisions. (page 36)

36. C – Personal safety begins with mentally preparing yourself and wearing seat belts, etc. (page 37)

37. C (page 38)

38. B (page 38)

39. D – Modes of transmission also include oral contamination due to lack of or improper handwashing. (page 40)

40. A (page 42)

41. C (page 43)

42. B (page 43)

43. D – The tetanus-diphtheria booster is also recommended every ten years. (page 45)

44. B (page 47)

45. D (page 48)

46. B (page 51)

47. C (page 52)

48. D – Common hazards of fire also include oxygen deficiency and high ambient temperature. (page 53)

49. B (page 54)

50. D – Factors to take into consideration for potential violence also include the behavior triad of truancy, fighting, and uncontrollable temper, instability of family structure, tattoos, and functional disorder. (page 59)

Vocabulary

1. Critical incident stress management (CISM): A process that confronts the responses to critical incidents and diffuses them, directing the emergency services personnel toward physical and emotional equilibrium. (page 29)

2. Posttraumatic stress disorder (PTSD): A delayed stress reaction to a prior incident. This delayed reaction is the result of one or more unresolved issues concerning the incident that might have been alleviated with the use of critical incident stress management. (page 29)

3. Critical incident stress debriefing (CISD): A confidential group discussion of a severely stressful incident that usually occurs within 24 to 72 hours of the incident. (page 33)

Fill-in

1. well-being (page 20)

2. emotional stress (page 20)

3. heart disease (page 21)

4. physician (page 21)

5. warm (page 21)

6. medical examiner (page 22)

7. Fear (page 24)

8. depression (page 24)

9. minor (page 28)

10. high-stress (page 28)

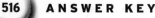

True/False

1. T   (page 20)
2. F   (page 22)
3. F   (page 22)
4. T   (page 22)
5. T   (page 22)
6. F   (page 39)
7. F   (page 28)
8. T   (pages 31 to 32)

Short Answer

1. An infection control practice that assumes all body fluids are potentially infectious. (page 39)
2. -Unresponsiveness to painful stimuli
   -Lack of a pulse or heartbeat
   -Absence of breath sounds
   -No deep tendon or corneal reflexes
   -Absence of eye movement; no systolic blood pressure
   -Profound cyanosis
   -Lowered or decreased body temperature. (page 21)
3. 1. Obvious mortal damage (decapitation)
   2. Dependent lividity
   3. Rigor mortis
   4. Putrefaction (page 22)
4. 1. Denial
   2. Anger/hostility
   3. Bargaining
   4. Depression
   5. Acceptance
   (pages 22 to 23)
5. -Irritability toward coworkers, family and friends
   -Inability to concentrate
   -Difficulty sleeping, increased sleeping, or nightmares
   -Anxiety; indecisiveness; guilt
   -Loss of appetite (gastrointestinal disturbances)
   -Loss of interest in sexual activities
   -Isolation
   -Loss of interest in work
   -Increased use of alcohol
   -Recreational drug use (page 31)
6. -Change or eliminate stressors
   -Change partners to avoid a negative or hostile personality
   -Change work hours
   -Cut back on overtime
   -Change your attitude about the stressor
   -Stop wasting your energy complaining or worrying about things you cannot change
   -Try to adopt a more relaxed, philosophical outlook
   -Expand your social support system apart from your coworkers
   -Sustain friends and interests outside emergency services
   -Minimize the physical response to stress by employing various techniques (page 31)

7. 1. Use soap and water.

2. Rub hands together for at least 10 to 15 seconds to work up a lather.

3. Rinse hands and dry with a paper towel.

4. Use paper towel to turn off faucet. (page 40)

8. (page 52)

Level	Hazard	Protection Needed
0	Little to no hazard	None
1	Slightly hazardous	SCBA only (level C suit)
2	Slightly hazardous	SCBA only (level C suit)
3	Extremely hazardous	Full protection, with no exposed skin (level A or B suit)
4	Minimal exposure causes death	Special HazMat gear (level A suit)

9. 1. Thin inner layer

2. Thermal middle layer

3. Outer layer

(pages 54 to 55)

10. 1. Past history

2. Posture

3. Vocal activity

4. Physical activity

(page 59)

Word Fun

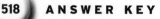

Ambulance Calls

1. -Say, "I'm sorry."

-Explain why CPR will be ineffective.

-Tell them, "If you want to cry, it's okay."

-Offer to call a relative or religious advisor.

-Notify your supervisor, dispatcher, or coroner as protocols dictate.

-Offer a hug and just listen.

2. -Your attention must focus on the mother—clearing her airway, assisting ventilations.

-Have a firefighter or bystander talk to the child and try to comfort him.

-Advise them NOT to remove him from his seat.

-Have the child brought to the ambulance in the seat and strapped into the Captain's chair—he should be immobilized in his seat.

-Continue your care of the mother and have help ride with you to care for the child.

3. -Continue to treat your patient appropriately, including c-spine stabilization and transport.

-Allow your cut to bleed as long as it is minimal—it will help to wash/clean it out.

-Clean your wound with an alcohol gel if available.

-Once patient care has been transferred at the receiving facility, immediately wash thoroughly with soap and water and report to your supervisor.

-Follow up with prompt medical attention.

Skill Drills

1. Skill Drill 2-1: Proper Glove Removal Technique (page 41)

1. Partially remove the first glove by pinching at the wrist. Be careful to touch only the outside of the glove.
2. Remove the second glove by pinching the exterior with the partially gloved hand.
3. Pull the second glove inside out toward the fingertips.
4. Grasp both gloves with your free hand touching only the clean, interior surfaces.

Chapter 3: Medical, Legal, and Ethical Issues

Matching

1. H   (page 71)

2. I   (page 69)

3. G   (page 72)

4. E   (page 71)

5. L   (page 68)

6. A   (page 72)

7. D   (page 69)

8. F   (page 68)

9. B   (page 69)

10. C   (page 70)

11. M   (page 70)

12. N   (page 70)

13. J   (page 68)

14. K   (page 66)

Multiple Choice

1. C – The scope of practice outlines the care the EMT-B is able to provide. (A) Duty to act is the responsibility to provide care. (B) Competency is the ability to make qualified decisions. (D) Certification is the process whereby an individual or agency is recognized for meeting a set of standards. (page 66)

2. A – The standard of care is a certain, definable way the EMT-B is required to act regardless of the activity involved. Generally, a standard of care is how a reasonably prudent individual with similar training would act with similar equipment and situation. (page 66)

3. D – Certification obliges the EMT to conform to predetermined standards. (page 68)

4. D – Negligence is proven when all four elements; duty, breach, cause and damages, are met, whether consent was expressed (A) or implied, care was terminated or not (B), or how the patient was transported (C). (pages 68 to 69)

5. A – Consent for emergency treatment of a mentally incompetent individual should be obtained from the person's legal guardian. In cases where this is not possible, many states have provisions to take the individual into protective custody. (B,C) expressed or informed consent require that the patient or guardian fully understand the condition and consequences of accepting or refusing treatment. Implied consent (D) is limited to true emergency situations where the individual is unconscious or delusional. (page 70)

6. C – If you leave patients alone, you risk being accused of negligence or abandonment. (page 71)

7. B – Good Samaritan laws protect individuals who provide care within their scope of practice and in good faith. Proper performance of CPR (A), improvising BLS materials (C), and supportive care given to a patient with an advance directive or DNR (D) are each appropriate to the scope of practice of the EMT-B. (page 72)

8. D – Confidential information includes history, assessment, treatments, diagnosis, mental or physical conditions (A, B, and C) and cannot be disclosed without authorization of the patient. The location of an emergency call is generally not considered to be confidential. (page 73)

9. D – Most experts agree that a complete and accurate record of an emergency incident is an important safeguard against legal complications. Not every call requires use of emergency lights and siren (A), or transport to an emergency department (C). Ambulance equipment and supplies should be checked (B) and restocked with every shift or after each run, but this is not necessarily related to legal issues. (page 74)

Vocabulary

1. Abandonment: Unilateral termination of care by the EMT-B without the patient's consent and without making provisions for transferring care to another medical professional with skills of the same level. (page 69)

2. Advance directive: Written documentation that specifies medical treatment for a competent patient should the patient become unable to make decisions. (page 72)

3. Assault: Unlawfully placing a patient in fear of bodily harm. (page 71)

4. Battery: Touching a patient or providing emergency care without consent. (page 71)

5. DNR order: Written documentation giving permission to medical personnel not to attempt resuscitation in the event of cardiac arrest. (page 72)

6. Certification: A process in which a person, an institution, or a program is evaluated and recognized as meeting certain predetermined standards to provide safe and ethical care. (page 68)

7. Duty to act: A medicolegal term relating to certain personnel who either by statute or by function have a responsibility to provide care. (page 68)

8. Expressed consent: A type of consent in which a patient gives express authorization for provision of care or transport. (page 69)

9. Good Samaritan laws: Statutory provisions enacted by many states to protect citizens from liability for errors and omissions in giving good faith emergency medical care, unless there is wanton, gross, or willful negligence. (page 72)

10. Implied consent: Type of consent in which a patient who is unable to give consent is given treatment under the legal assumption that he or she would want treatment. (page 69)

11. Negligence: Failure to provide the same care that a person with similar training would provide. (page 68)

12. Standard of care: Written, accepted levels of emergency care expected by reason of training and profession; written by legal or professional organizations so that patients are not exposed to unreasonable risk or harm. (page 66)

Fill-in

1. scope of practice (page 66)

2. standard of care (page 66)

3. duty to act (page 68)

4. negligence (page 68)

5. termination (page 69)

6. expressed, implied (page 69)

7. Assault, battery (page 71)

8. advance directive, DNR order (page 72)

9. refuse treatment (page 71)

10. special reporting (page 74)

True/False

1. T (page 69)
2. F (page 69)
3. T (pages 69 to 70)
4. T (page 70)
5. T (page 73)

Short Answer

1. If the minor is emancipated, married, or pregnant. (page 70)

2. You must continue to care for the patient until the patient is transferred to another medical professional of equal or higher skill level, or another medical facility. (page 69)

3. 1. Obtain refusing party's signature on an official medical release form that acknowledges refusal.

   2. Obtain a signature from a witness of the refusal.

   3. Keep the refusal form with the incident report.

   4. Note the refusal on the incident report.

   5. Keep a department copy of the records for future reference. (page 71)

4. 1. If it wasn't documented, it did not happen.

   2. Incomplete or disorderly records equate to incomplete or inexpert medical care. (page 74)

5. 1. Inform medical control.

   2. Treat the patient as you would any patient.

   3. Take any steps necessary to preserve life.

   4. If saving the patient is not possible, take steps to make sure the organ remains viable. (page 75)

Word Fun

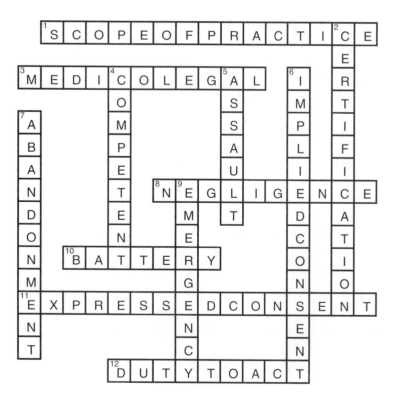

Ambulance Calls

1. You should not enter the scene until law enforcement arrives. The ambulance should "stage" several blocks from the residence and wait for police. Once the scene is safe, the EMTs will be summoned to enter the scene and provide care for the patient.

2. The child should be treated based on implied consent. A "loss of function" constitutes an emergency, and the child can be treated and transported without consent from the parent since not treating him could result in a loss of the extremity. The hospital will try to reach the child's mother once you arrive. You might also tell the other children to try and contact her and tell her to come to the hospital.

3. Assess the patient's mental status. If he is intoxicated, or has an altered mental status, he is treated under implied consent. If he is alert and oriented, you may attempt to talk him into being treated by explaining what you feel is necessary and what may happen if he does not receive care. If he has an altered mental status, orders from medical control may be obtained to restrain the patient with the help of law enforcement and transport him to the hospital.

Chapter 4: The Human Body

Matching

1. F	(page 80)	**20.** B	(page 92)
2. K	(page 108)	**21.** A	(page 93)
3. D	(page 80)	**22.** A	(page 93)
4. G	(page 81)	**23.** A	(page 96)
5. E	(page 80)	**24.** B	(page 97)
6. L	(page 108)	**25.** B	(page 97)
7. A	(page 81)	**26.** A	(page 96)
8. C	(page 81)	**27.** A	(page 96)
9. M	(page 108)	**28.** C	(page 97)
10. J	(page 81)	**29.** B	(page 97)
11. O	(page 108)	**30.** C	(page 97)
12. B	(page 81)	**31.** A	(page 96)
13. H	(page 80)	**32.** C	(page 97)
14. N	(page 108)	**33.** A	(page 86)
15. I	(page 80)	**34.** C	(page 110)
16. B	(page 91)	**35.** E	(page 112)
17. B	(page 92)	**36.** B	(page 112)
18. A	(page 87)	**37.** F	(page 110)
19. B	(page 92)	**38.** D	(page 112)

Multiple Choice

1. B – The parts that lie closer to the midline are called medial (inner) structures. (page 81)
2. C – Distal describes structures that are farther from the trunk or nearer to the free end of the extremity. (page 81)
3. B – Cricoid cartilage, a firm ridge of cartilage inferior to the thyroid cartilage, is somewhat more difficult to palpate. (page 85)
4. C – Smooth muscle carries out much of the automatic work of the body; therefore, it is also called involuntary muscle. (page 97)
5. D – The spinal column is the central supporting structure of the body and is composed of 33 bones, each called a vertebra. (page 86)
6. C – Much like the skull, each pelvic bone is formed by the fusion of three separate bones. These three bones are called the ilium, the ischium, and the pubis. (page 91)

7. B – Protecting the opening of the trachea is a thin, leaf-shaped valve called the epiglottis. (page 98)

8. B – The endocrine system is a complex message and control system that integrates many body functions. It releases substances called hormones, either by target organs or directly. (page 115)

9. D – The pulmonary artery begins at the right side of the heart and carries oxygen-poor blood to the lungs. (page 107)

10. C – The three major types of nerves are sensory nerves, motor nerves, and connecting nerves. Sensory nerves carry information from the body to the central nervous system. Motor nerves carry information from the central nervous system to the muscles of the body. Connecting nerves do just what their name implies: they connect the sensory and motor nerves. (page 112)

Labeling

1. Directional Terms (page 81)

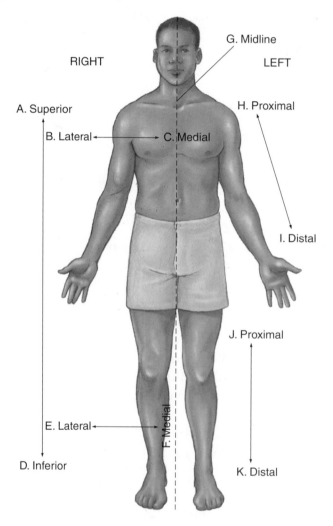

2. Anatomic Positions (page 83)

Prone

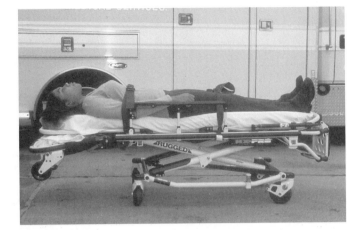

Supine

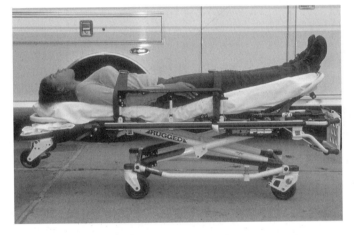

Shock position (modified Trendelenburg's position)

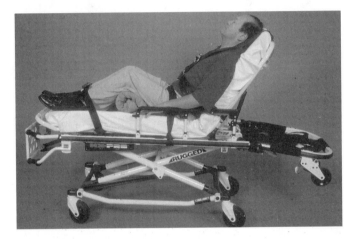

Fowler's position

3. Skeletal System (page 84)

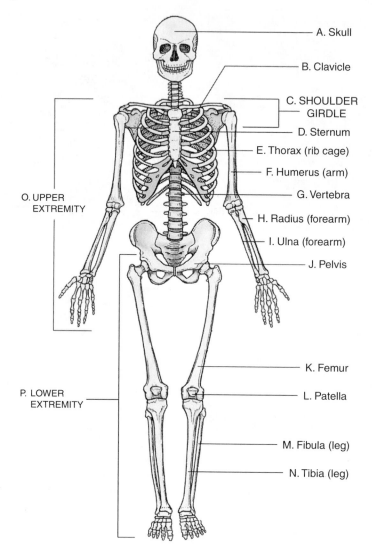

A. Skull
B. Clavicle
C. SHOULDER GIRDLE
D. Sternum
E. Thorax (rib cage)
F. Humerus (arm)
G. Vertebra
H. Radius (forearm)
I. Ulna (forearm)
J. Pelvis
O. UPPER EXTREMITY
P. LOWER EXTREMITY
K. Femur
L. Patella
M. Fibula (leg)
N. Tibia (leg)

4. The Skull (page 84)

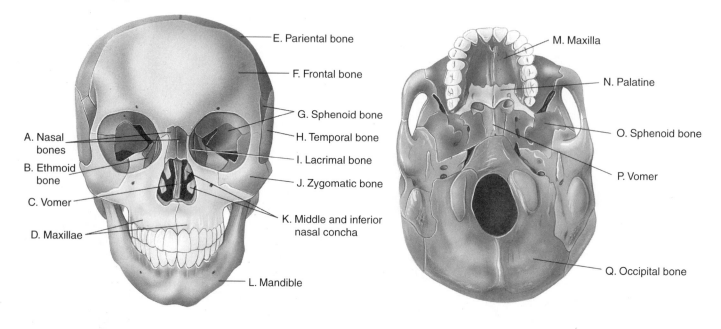

E. Pariental bone
F. Frontal bone
G. Sphenoid bone
H. Temporal bone
I. Lacrimal bone
J. Zygomatic bone
K. Middle and inferior nasal concha
A. Nasal bones
B. Ethmoid bone
C. Vomer
D. Maxillae
L. Mandible

M. Maxilla
N. Palatine
O. Sphenoid bone
P. Vomer
Q. Occipital bone

5. The Spinal Column (page 86)

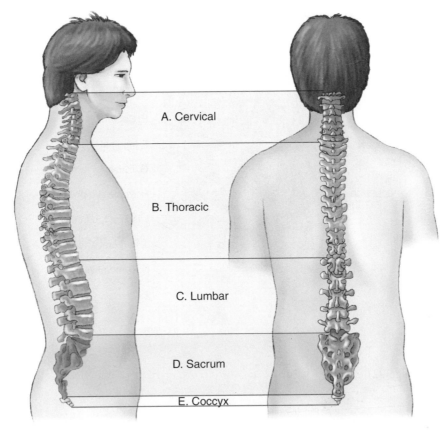

A. Cervical

B. Thoracic

C. Lumbar

D. Sacrum

E. Coccyx

6. The Thorax (page 87)

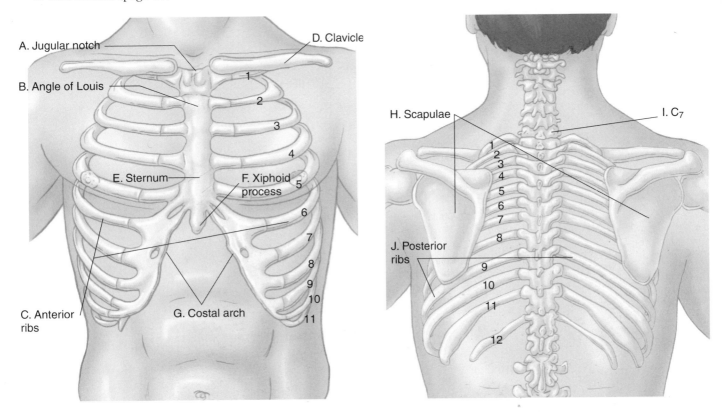

A. Jugular notch

D. Clavicle

B. Angle of Louis

E. Sternum

F. Xiphoid process

C. Anterior ribs

G. Costal arch

H. Scapulae

I. C₇

J. Posterior ribs

7. The Pelvis (page 91)

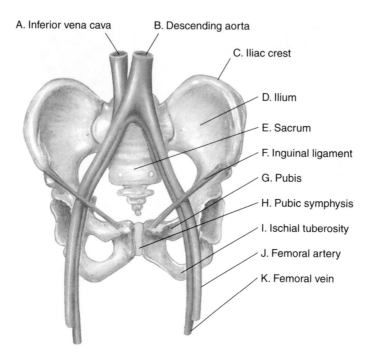

A. Inferior vena cava

B. Descending aorta

C. Iliac crest

D. Ilium

E. Sacrum

F. Inguinal ligament

G. Pubis

H. Pubic symphysis

I. Ischial tuberosity

J. Femoral artery

K. Femoral vein

8. The Lower Extremity (page 92)

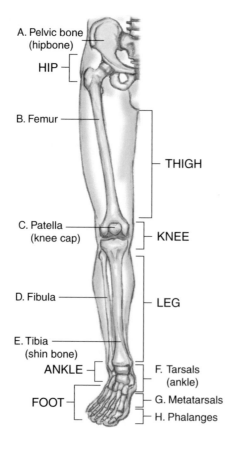

A. Pelvic bone (hipbone)

HIP

B. Femur

THIGH

C. Patella (knee cap)

KNEE

D. Fibula

LEG

E. Tibia (shin bone)

ANKLE

F. Tarsals (ankle)

FOOT

G. Metatarsals

H. Phalanges

9. The Shoulder Girdle (page 93)

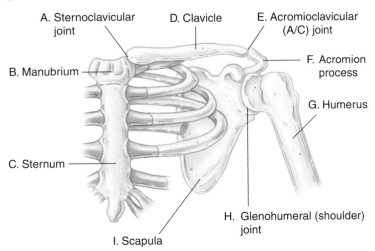

A. Sternoclavicular joint
B. Manubrium
C. Sternum
D. Clavicle
E. Acromioclavicular (A/C) joint
F. Acromion process
G. Humerus
H. Glenohumeral (shoulder) joint
I. Scapula

10. The Upper Extremity (page 94)

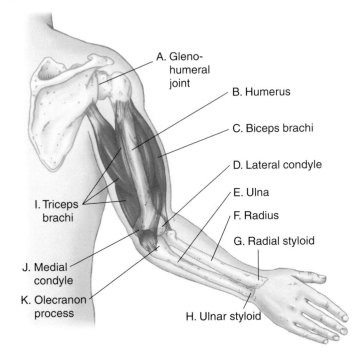

A. Gleno-humeral joint
B. Humerus
C. Biceps brachi
D. Lateral condyle
E. Ulna
F. Radius
G. Radial styloid
H. Ulnar styloid
I. Triceps brachi
J. Medial condyle
K. Olecranon process

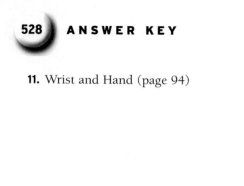

11. Wrist and Hand (page 94)

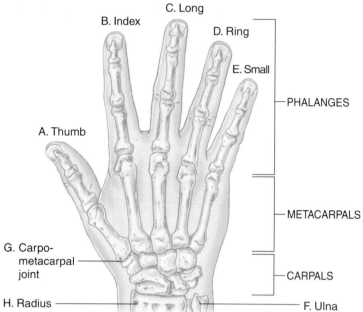

B. Index
C. Long
D. Ring
E. Small
A. Thumb
PHALANGES
METACARPALS
G. Carpo-
metacarpal
joint
CARPALS
H. Radius
F. Ulna

12. The Respiratory System (page 98)

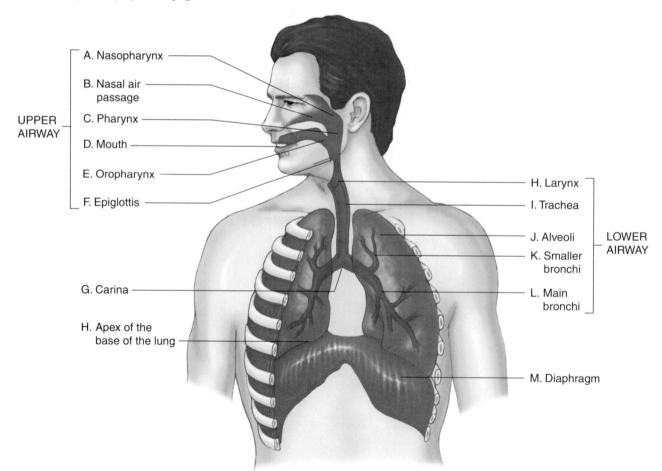

UPPER
AIRWAY

A. Nasopharynx
B. Nasal air
passage
C. Pharynx
D. Mouth
E. Oropharynx
F. Epiglottis

G. Carina

H. Apex of the
base of the lung

H. Larynx
I. Trachea
J. Alveoli
K. Smaller
bronchi
L. Main
bronchi

LOWER
AIRWAY

M. Diaphragm

13. The Circulatory System (page 104)

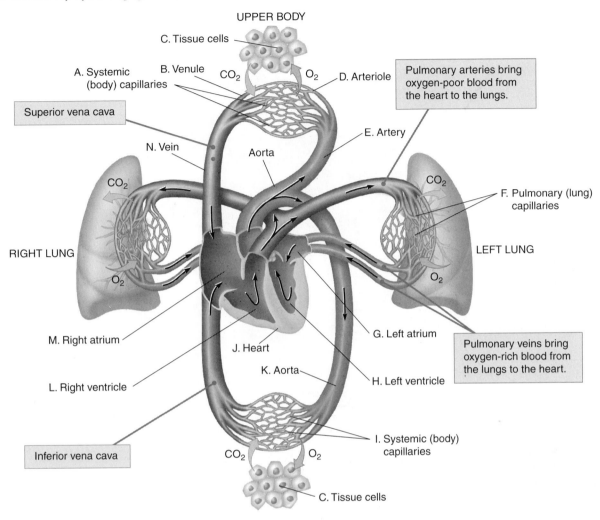

UPPER BODY

C. Tissue cells

A. Systemic (body) capillaries

B. Venule

CO_2

O_2

D. Arteriole

Pulmonary arteries bring oxygen-poor blood from the heart to the lungs.

Superior vena cava

E. Artery

N. Vein

Aorta

CO_2

CO_2

F. Pulmonary (lung) capillaries

RIGHT LUNG

LEFT LUNG

O_2

O_2

M. Right atrium

G. Left atrium

Pulmonary veins bring oxygen-rich blood from the lungs to the heart.

J. Heart

K. Aorta

H. Left ventricle

L. Right ventricle

I. Systemic (body) capillaries

Inferior vena cava

CO_2

O_2

C. Tissue cells

14. Electrical Conduction (page 106)

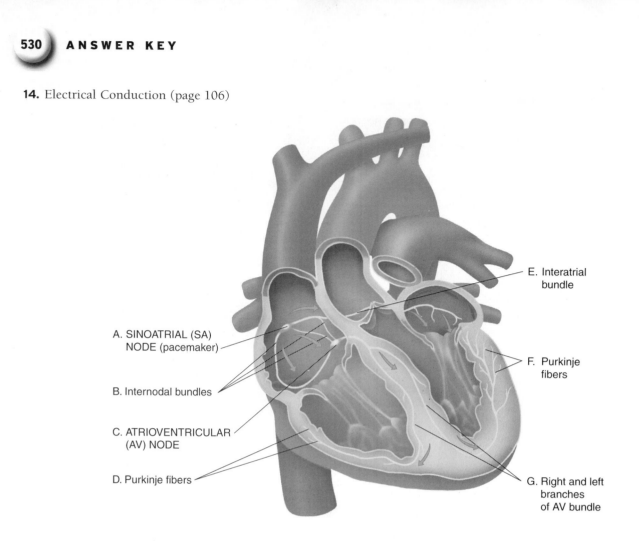

A. SINOATRIAL (SA)
NODE (pacemaker)

B. Internodal bundles

C. ATRIOVENTRICULAR
(AV) NODE

D. Purkinje fibers

E. Interatrial
bundle

F. Purkinje
fibers

G. Right and left
branches
of AV bundle

15. Central and Peripheral Pulses (page 109)

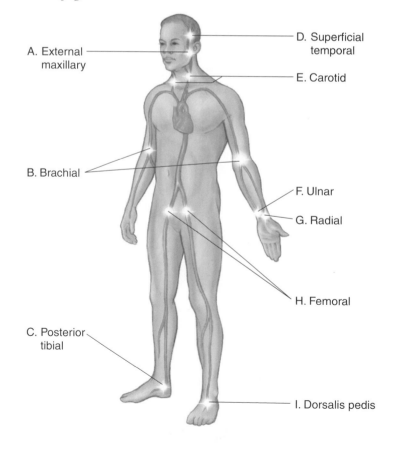

A. External
maxillary

B. Brachial

C. Posterior
tibial

D. Superficial
temporal

E. Carotid

F. Ulnar

G. Radial

H. Femoral

I. Dorsalis pedis

16. Brain (page 111)

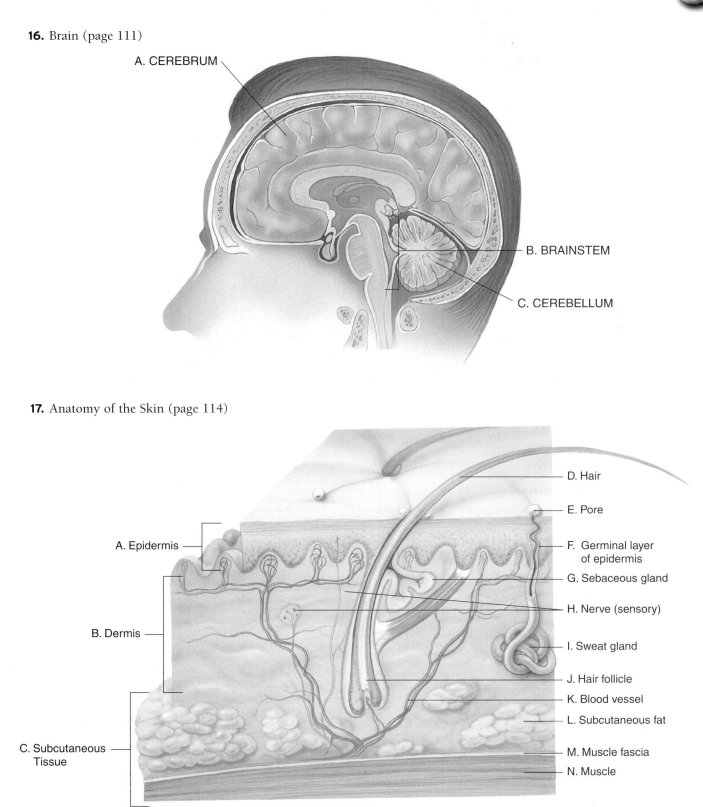

A. CEREBRUM

B. BRAINSTEM

C. CEREBELLUM

17. Anatomy of the Skin (page 114)

A. Epidermis

B. Dermis

C. Subcutaneous Tissue

D. Hair

E. Pore

F. Germinal layer of epidermis

G. Sebaceous gland

H. Nerve (sensory)

I. Sweat gland

J. Hair follicle

K. Blood vessel

L. Subcutaneous fat

M. Muscle fascia

N. Muscle

18. Male Reproductive System (page 120)

FRONT VIEW

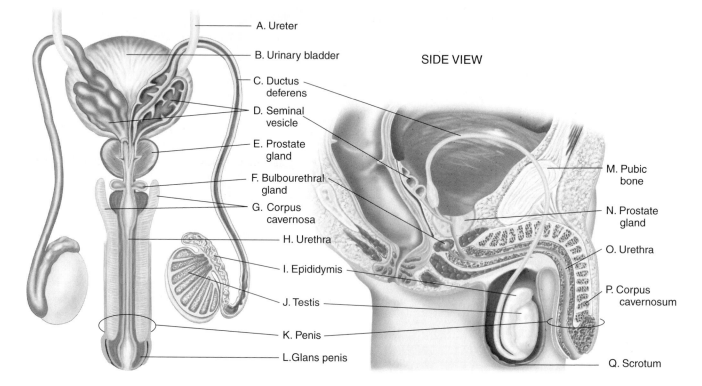

A. Ureter
B. Urinary bladder
C. Ductus deferens
D. Seminal vesicle
E. Prostate gland
F. Bulbourethral gland
G. Corpus cavernosa
H. Urethra
I. Epididymis
J. Testis
K. Penis
L. Glans penis

SIDE VIEW

M. Pubic bone
N. Prostate gland
O. Urethra
P. Corpus cavernosum
Q. Scrotum

19. Female Reproductive System (page 121)

FRONT VIEW

SIDE VIEW

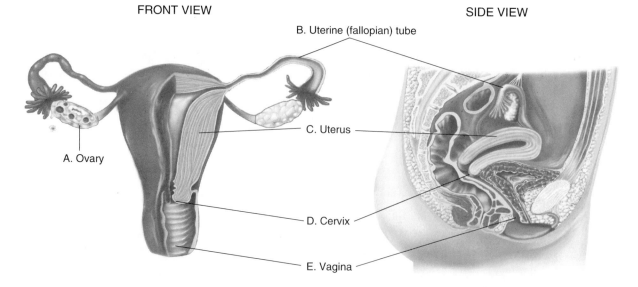

B. Uterine (fallopian) tube
A. Ovary
C. Uterus
D. Cervix
E. Vagina

Vocabulary

1. Perfusion: The circulation of blood within an organ or tissue in adequate amounts to meet the cells' current needs. (page 110)
2. Agonal respirations: Irregular, gasping respirations, sometimes heard in dying patients. (page 104)
3. Autonomic nervous system: The part of the nervous system that regulates many body functions that are not under voluntary control. (page 110)
4. Pleural space: Potential space between the parietal and visceral pleura. (page 100)
5. Trendelenburg's position: The body is supine with the head lower than the feet. (page 82)
6. Fowler's position: Patient is sitting up with the knees bent. (page 82)
7. Somatic nervous system: The part of the nervous system that regulates activities over which there is involuntary control. (page 110)
8. Endocrine system: A complex message and control system that integrates many body functions, including the release of hormones. (page 115)
9. Peripheral nervous system: The part of the nervous system that consists of 31 pairs of spinal nerves and 12 pairs of cranial nerves. These may be sensory or motor nerves. (page 112)
10. Epiglottis: The leaf-shaped valve that allows air to pass into the trachea, but prevents food or liquid from entering the airway. (page 98)
11. Metabolism: Chemical reactions that occur in the body. (page 113)
12. Brain stem: The area of the brain that lies deep within the cranium and is the best-protected part of the central nervous system. It controls vital body functions. (page 111)

Fill-in

1. 7 (page 86)
2. mandible (page 85)
3. 5 (page 99)
4. 12 (page 86)
5. 33 (page 86)
6. larynx (page 85)
7. talus (page 92)
8. floating ribs (page 88)

True/False

1. F (page 107)
2. T (page 92)
3. T (pages 92 to 94)
4. F (page 106)
5. F (page 88)

Short Answer

1. Plasma – is a sticky, yellow fluid that carries the blood cells and nutrients

   Red blood cells – give blood its red color and carry oxygen

   White blood cells – play a role in the body's immune defense mechanism against infection

   Platelets – essential in the formation of blood clots (pages 108 to 109)
2. Cervical spine – 7

   Thoracic spine – 12

   Lumbar spine – 5

   Sacrum – 5

   Coccyx – 4 (page 86)
3. RUQ – liver, gallbladder, portion of the colon

   LUQ – stomach, spleen, portion of the colon

   RLQ – large intestine, small intestine, appendix, ascending colon

   LLQ – large intestine, small intestine, descending and the sigmoid portions of the colon (page 90)

4. 1. superior and inferior
    vena cava
    2. right atrium
    3. right ventricle
    4. pulmonary artery
    5. lungs (pages 104 to 106)

6. pulmonary vein
7. left atrium
8. left ventricle
9. aorta

Word Fun

Ambulance Calls

1. Lacerated liver, gall bladder, small intestine, large intestine, pancreas, diaphragm, right lung if the pathway is up, right kidney depending on length of knife, you could also have involvement of the other four quadrants based on the direction of travel of the blade.

    The description would be a puncture wound or stab wound.

2. The patient has angulation and deformity, possibly with crepitus, to the left forearm, proximal to his left wrist or distal to his left elbow.

3. Deformity to left tibia/fibula with swelling. Possible fracture.

Chapter 5: Baseline Vital Signs and SAMPLE History

Matching

1. N (page 133)
2. E (page 137)
3. D (page 139)
4. L (page 134)
5. A (page 134)
6. H (page 143)
7. K (page 137)
8. F (page 135)

9. M (page 134)
10. O (page 135)
11. B (page 135)
12. G (page 139)
13. I (page 139)
14. C (page 135)
15. J (page 134)

Multiple Choice

1. D – When assessing the patient, look (A), listen (B), and feel (C), and think. (page 128)
2. C (page 129)
3. A – Dizziness (A) is a symptom. (page 129)
4. D (page 130)
5. B (page 130)
6. D – In addition, you should evaluate skin temperature and condition in adults. (page 130)
7. A – Inspiration is active (B). Oxygen is drawn in (C). (page 130)
8. D (pages 130 to 131)
9. D – The normal rate is 12 to 20 breaths/min. 15 to 30 breaths/min (B) is the rate for children. 25 to 50 breaths/min (C) is the rate for infants. (page 131)
10. D (page 132)
11. B (page 132)
12. C (page 132)
13. B – Dyspnea (B) is a symptom. (page 132)
14. C (page 133)
15. C (page 133)
16. A (page 133)
17. D (page 134)
18. B (page 134)
19. D (page 134)
20. D (page 134)
21. D (page 134)
22. D (page 135)
23. B (page 135)
24. D (page 135)
25. D (page 135)
26. B (page 136)
27. A (page 136)
28. D (page 136)
29. B (page 136)
30. A (page 137)
31. C (page 138)
32. A (page 139)
33. B (page 140)
34. C (page 141)
35. D (page 141)
36. D (page 142)
37. C (page 143)

Vocabulary

1. Glasgow Coma Scale: A method of assessing a patient's level of consciousness by scoring the patient's response to eye opening, motor response, and verbal response. (page 141)
2. AVPU scale: A method of assessing a patient's level of consciousness by determining whether the patient is awake or alert, responsive to verbal stimulus or pain, or is unresponsive. (page 140)
3. Chief complaint: The reason a patient called for help. Also, the patient's response to questions such as, "What's wrong?" or "What happened?" (page 129)
4. Stridor: Harsh, high-pitched, crowing inspiratory sound, such as the sound often heard in acute laryngeal (upper airway) obstruction. (page 132)

Fill-in

1. Tidal volume (page 133)
2. conjunctiva (page 134)
3. deductive (page 128)
4. symptom (page 129)
5. spontaneous respirations (page 130)
6. Vital signs (page 130)
7. quality (page 131)
8. labored breathing (page 132)
9. fluid (page 132)
10. perfusion (page 134)

True/False

1. T (page 133)
2. T (page 139)
3. F (page 139)
4. T (page 132)
5. T (page 134)
6. F (page 133)
7. F (page 134)
8. T (page 134)
9. T (page 135)
10. T (page 135)
11. F (page 135)
12. F (page 141)
13. F (page 131)

Short Answer

1. 1. Pulse
   2. Respirations
   3. Pupils
   4. Blood pressure
   5. Skin
   6. Level of consciousness
   7. Capillary refill (page 130)
2. 1. Flushed (red)
   2. Pale (white, ashen, or graying)
   3. Jaundice (yellow)
   4. Cyanotic (blue-gray) (page 135)
3. 1. Rate
   2. Rhythm
   3. Quality
   4. Depth (tidal volume) (pages 131 to 133)
4. Pressure exerted against the walls of the artery when the left ventricle contracts (page 137)
5. Pressure remaining against the walls of the artery when the left ventricle is at rest (page 137)
6. 1. Rate
   2. Strength
   3. Regularity (page 134)
7. 1. Color
   2. Temperature
   3. Moisture (pages 134 to 135)
8. Gently compress the fingertip until it blanches. Release the fingertip, and count until it returns to its normal pink color. (pages 135 to 136)
9. Pupils Equal And Round, Regular in size, react to Light (page 142)
10. A sign is a condition that can be seen, heard, felt, smelled, or measured. A symptom is something that the patient reports to you as a problem or feeling. (page 129)

Word Fun

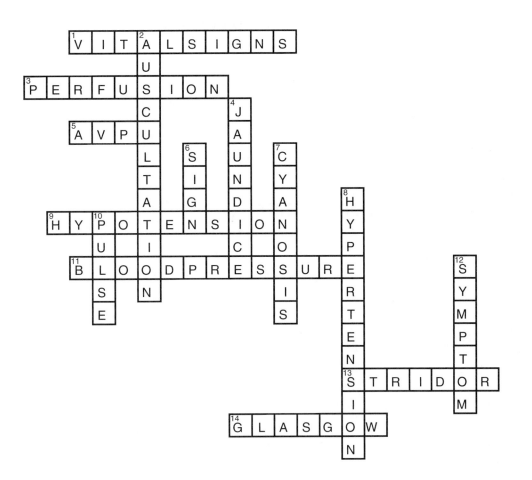

Ambulance Calls

1. -Place patient on the stretcher in the position of comfort.
   -Obtain a SAMPLE history.
   -Apply high-flow oxygen via nonrebreathing mask.
   -Obtain baseline vital signs.
   -Monitor patient and repeat vital signs every 15 minutes.
   -Normal transport

2. -Place patient on the stretcher in position of comfort.
   -Apply high-flow oxygen via nonrebreathing mask.
   -Obtain baseline vital signs.
   -Obtain SAMPLE history.
   -Monitor patient and repeat vital signs every 15 minutes.
   -Normal transport

3. -Open and assess the airway.
   -Apply high-flow oxygen via nonrebreathing mask or BVM.
   -Assess carotid pulse.
   -If no pulse, initiate chest compressions.
   -If patient has a pulse, treat for shock and rapid transport.
   -Place patient in Trendelenburg's position and cover to keep warm.
   -Continue assessment and obtain SAMPLE history en route.
   -Monitor vital signs and take a complete set every 5 minutes.

Skill Drills

1. Skill Drill 5-1: Obtaining a Blood Pressure by Auscultation or Palpation (page 138)

1. Apply the cuff snugly.
2. Palpate the brachial artery.
3. Place the stethoscope and grasp the ball-pump and turn-valve.
4. Close the valve and pump to 20 mm Hg above the point at which you stop hearing pulse sounds. Note the systolic and diastolic pressures as you let air escape slowly.
5. Open the valve and quickly release remaining air.
6. When using the palpatation method, you should place your fingertips on the radial artery so that you feel the radial pulse.

Chapter 6: Lifting and Moving Patients

Matching

1. C (page 168)
2. G (page 176)
3. H (page 178)
4. E (page 177)
5. A (page 177)
6. I (page 176)
7. F (page 166)
8. B (page 176)
9. D (page 155)

Multiple Choice

1. D (page 179)
2. D (page 161)
3. B (page 148)
4. D (page 149)
5. D (page 150)
6. D (page 152)
7. A (page 152)
8. B (page 154)
9. B (page 154)
10. D (page 156)
11. A – The command of execution is done during each phase of moving. (page 156)
12. D (page 157)
13. D – When carrying a patient in a stair chair, you should also bend at the knees. (page 157)
14. D (page 157)
15. C – When you can move no farther, stop and move back another 15" to 20." (page 159)
16. D (page 159)
17. B (page 160)
18. A (page 160)
19. A (page 161)
20. D (page 161)
21. D (page 161)
22. C (page 161)
23. D (page 163)
24. D – An urgent move may also be necessary if the patient needs immediate intervention that requires a supine position, if the patient's condition requires immediate transport to the hospital, or if there is an extreme weather condition. (page 164)

25. D (page 165)

26. C (page 170)

27. D – To move a patient from the ground or the floor onto the cot, you may also log roll the patient onto a blanket, centering the patient on the blanket and rolling up the excess material on each side. Lift the patient by the blanket, and carry him or her to the nearby cot. (page 170)

28. B (page 176)

29. B (page 176)

30. A (page 177)

31. D (page 177)

32. D (page 178)

33. D (page 178)

Vocabulary

1. Diamond carry: A carrying technique in which one EMT-B is located at the head end, one at the foot end, and one at each side of the patient; each of the two EMT-Bs at the sides uses one hand to support the stretcher so that all are able to face forward as they walk. (page 152)

2. Rapid extrication technique: A technique to move a patient from a sitting position inside a vehicle to supine on a backboard in less than 1 minute when conditions do not allow for standard immobilization. (page 163)

3. Power grip: A technique in which the litter or backboard is gripped by inserting each hand under the handle with the palm facing up and the thumb extended, fully supporting the underside of the handle on the curved palm with the fingers and thumb. (page 150)

4. Power lift: A lifting technique in which the EMT-B's back is held upright, with legs bent, and the patient is lifted when the EMT-B straightens the legs to raise the upper body and arms. (page 150)

5. Emergency move: A move in which the patient is dragged or pulled from a dangerous scene before initial assessment and care are provided. (page 161)

Fill-in

1. body mechanics (page 148)

2. upright (page 149)

3. power lift (page 150)

4. palm (page 151)

5. locked-in (page 154)

6. 250 (page 157)

7. sideways (page 159)

8. locked (page 160)

9. overhead (page 160)

10. less strain (page 161)

11. spine movement (page 164)

12. direct ground lift (page 166)

13. extremity lift (page 168)

14. fluid resistant (page 173)

15. scoop stretcher (page 178)

True/False

1. T   (page 178)

2. T   (page 176)

3. T   (page 150)

4. F   (page 157)

5. F   (page 170)

6. T   (page 161)

7. F   (page 178)

8. F   (page 155)

9. F   (page 163)

10. F   (page 165)

Short Answer

1. -Front cradle

  -Firefighter's drag

  -One-person walking assist

  -Firefighter's carry

  -Pack strap (page 163)

2. -The vehicle or scene is unsafe.

-The patient cannot be properly assessed before being removed from the car.

-The patient needs immediate intervention that requires a supine position.

-The patient's condition requires immediate transport to the hospital.

-The patient blocks the EMT-B's access to another seriously injured patient. (page 164)

3. -Make sure there is sufficient lifting power.

-Follow the manufacturer's directions for safe and proper use of the cot.

-Make sure that all cots and patients are fully secured before you move the ambulance. (page 174)

4. -Be sure that you know or can find out the weight to be lifted and the limitations of the team's abilities.

-Coordinate your movements with those of the other team members while constantly communicating with them.

-Do not twist your body as you are carrying the patient.

-Keep the weight that you are carrying as close to your body as possible while keeping your back in a locked-in position.

-Be sure to flex at the hips, not at the waist, and bend at the knees, while making sure that you do not hyperextend your back by leaning back from your waist. (page 152)

5. Always keep your back in a straight, upright position and lift without twisting. (page 149)

Word Fun

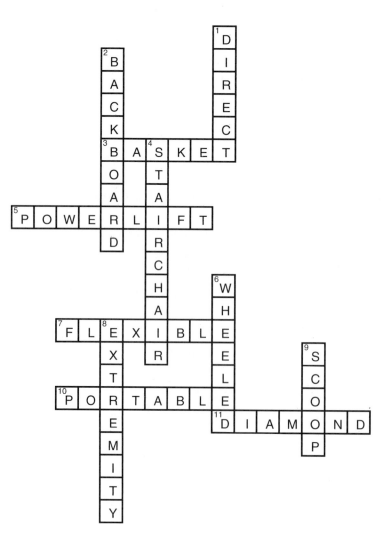

Ambulance Calls

1. -Immobilize the patient on a long spine board and apply high-flow oxygen.

   -Use four persons to carry the board back up the ledge.

   -Plan the route and brief your helpers before moving the patient.

   -Clarify whether you will move on "three" or count to three then move.

   -Coordinate the move until the patient is loaded into the ambulance.

2. -Move the patient's legs so they are clear of the pedals and are against the seat.

   -Rotate the patient so that his back is positioned facing the open car door.

   -Place your arms through the armpits and grasp either the patient's forearms or your own forearms.

   -Support the patient's head against your body.

   -While supporting the patient's weight, drag the patient from the seat.

   -Place him supine on the ground and secure the airway.

   -Assist ventilations with 100% oxygen until further help arrives.

Skill Drills

1. **Skill Drill 6-1: Performing the Power Lift (page 151)**

   1. Lock your back into an **upright**, inward curve. **Spread** and bend your legs. Grasp the cot, palms up and just in front of you. **Balance** and **center** the weight between your arms.
   2. Position your feet, **straddle** the object, and **distribute** the weight.
   3. **Straighten** your legs and lift, keeping your back locked in.

2. **Skill Drill 6-2: Performing the Diamond Carry (page 153)**

   1. Position yourselves facing the patient.
   2. After the patient has been lifted, the EMT-B at the foot turns to face forward.
   3. EMT-Bs at the side each turn the head-end hand palm down and release the other hand.
   4. EMT-Bs at the side turn toward the foot end.

3. **Skill Drill 6-3: Performing the One-Handed Carrying Technique (page 154)**

   1. **Face** each other and use both **hands**.
   2. Lift the backboard to **carrying height**.
   3. **Turn** in the direction you will walk and **switch** to using one hand.

4. **Skill Drill 6-4: Carrying a Backboard or Stretcher on Stairs (page 155)**

   1. **Strap** the patient securely.
   2. Carry a patient down stairs with the **foot** end first, **head** elevated.
   3. Carry the **head** end first going up stairs.

5. **Skill Drill 6-5: Using a Stair Chair (page 158)**

   1. Position and secure the patient on the chair, **arms** strapped down.
   2. Take your places at the **head** and **foot** of the chair.
   3. A third **rescuer** precedes and "backs up" the rescuer carrying the **foot**.
   4. **Lower** the chair to roll on landings, or for transfer to the cot.

6. Skill Drill 6-6: Performing Rapid Extrication Technique (pages 166 to 167)

1. First EMT-B provides in-line manual support of the head and cervical spine.

2. Second EMT-B gives commands, applies a cervical collar, and performs the initial assessment.

3. Second EMT-B supports the torso.
   Third EMT-B frees the patient's legs from the pedals and moves the legs together, without moving pelvis or spine.

4. Second and Third EMT-Bs rotate the patient as a unit in several short, coordinated moves.
   First EMT-B (relieved by Fourth EMT-B or bystander as needed) supports the head and neck during rotation (and later steps).

5. First (or Fourth) EMT-B places the backboard on the seat against patient's buttocks.
   Second and Third EMT-Bs lower the patient onto the long spine board.

6. Third EMT-B moves to an effective position for sliding the patient.
   Second and Third EMT-Bs slide the patient along the backboard in coordinated, 8" to 12" moves until the hips rest on the backboard.

7. Third EMT-B exits the vehicle, moves to the backboard opposite Second EMT-B, and they continue to slide the patient until patient is fully on the board.

8. First (or Fourth) EMT-B continues to stabilize the head and neck while Second and Third EMT-Bs carry the patient away from the vehicle.

7. Skill Drill 6-7: Extremity Lift (page 169)

1. Patient's hands are **crossed** over the chest.
   First EMT-B grasps patient's wrists or **forearms** and pulls patient to a **sitting** position.

2. When the patient is sitting, First EMT-B passes his or her arms through patient's **armpits** and grasps the patient's opposite (or his or her own) **forearms** or **wrists**.
   Second EMT-B kneels between the **legs**, facing the feet, and places his or her hands under the **knees**.

3. Both EMT-Bs rise to **crouching**.
   On **command**, both lift and begin to move.

8. Skill Drill 6-8: Scoop Stretcher (page 171)

1. Adjust stretcher **length**.
2. **Lift** patient slightly and **slide** stretcher into place, one side at a time.
3. **Lock** the stretcher ends together, avoiding **pinching**.
4. **Secure** the patient and **transfer** to the cot.

Section 2 Airway

Chapter 7: Airway

Matching

1. C	(page 188)	**7.** A (page 188)
2. I	(page 189)	**8.** F (page 188)
3. H	(page 190)	**9.** L (page 186)
4. G	(page 188)	**10.** D (page 186)
5. K	(page 191)	**11.** J (page 191)
6. E	(page 188)	**12.** B (page 193)

Multiple Choice

1. D – Normally the air we breathe contains 21% oxygen and 78% nitrogen (A). (B) is the amount of oxygen left after a rescue breath has been delivered. (C) is the amount of oxygen delivered by the rescuer during rescue breathing. (page 189)

2. B – Opening the airway to relieve obstruction can often be done using the head tilt–chin lift maneuver for patients who have not sustained trauma. Therefore, this procedure is not "always" used. (A), (C), and (D) are each true. (pages 194 to 196)

3. B – An adult who is breathing normally will have respirations of 12 to 20 breaths per minute. (page 192)

4. D – (A), (B), and (C) are among the conditions associated with hypoxia. (pages 191 to 192)

5. A – When the level of carbon dioxide becomes too high, the brain stem sends nerve impulses down the spinal cord that cause the diaphragm and intercostal muscles to contract, increasing respirations. (B), (C), and (D) do not affect respirations. (page 191)

6. A – (B), (C), and (D) are among the signs of inadequate breathing in an adult. Warm, dry skin (A) indicates adequate circulation and perfusion. (pages 192 to 193)

7. C – To select the proper size for an oropharyngeal airway, measure from the patient's earlobe to the corner of the mouth on the side of the face. (A), (B), and (D) are incorrect. (page 197)

8. D – It may be very difficult to maintain a proper seal between the mask and the face. It is not practical for the EMT-B to accurately measure tidal volumes (A). Positioning the patient's head (B) is easily accomplished with the head tilt–chin lift or modified jaw-thrust maneuver. All BVM devices should have the ability to perform under extreme environmental conditions (C). (page 211)

9. C – As you are assisting ventilations with a BVM device, you should evaluate how well the patient is breathing. You will know that artificial ventilation is not adequate if the patient's chest does not rise and fall with each ventilation. Inflation of the cheeks (A) may indicate obstruction of the air passage, requiring closer attention. Signs of spontaneous breathing (B) indicate how well the patient is breathing, but not how well the rescuer is assisting ventilations. Gurgling (D) indicates the need for immediate suctioning of the oral airway. (pages 212 to 213)

10. D – Do not suction an adult for more than 15 seconds. (A), (B), and (C) are incorrect. (page 200)

Labeling

1. Upper and Lower Airways (page 187)

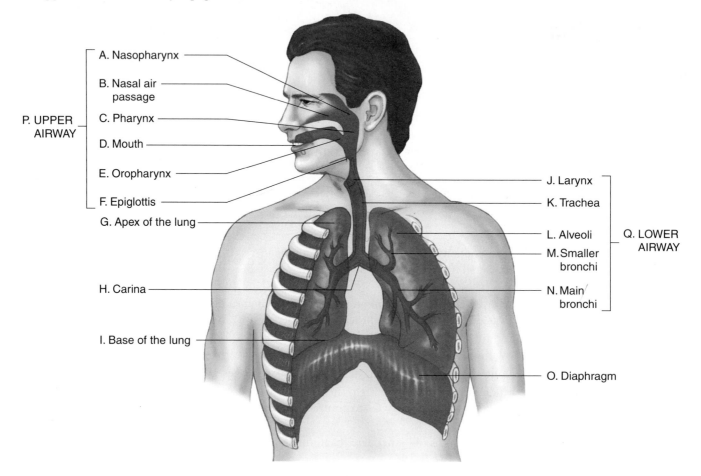

A. Nasopharynx

B. Nasal air passage

P. UPPER AIRWAY

C. Pharynx

D. Mouth

E. Oropharynx

F. Epiglottis

G. Apex of the lung

H. Carina

I. Base of the lung

J. Larynx

K. Trachea

L. Alveoli

M. Smaller bronchi

N. Main bronchi

Q. LOWER AIRWAY

O. Diaphragm

2. Thoracic Cage (page 187)

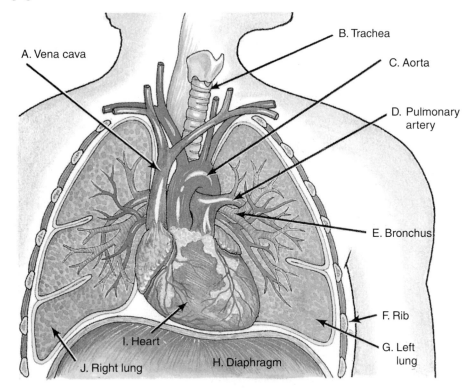

A. Vena cava

B. Trachea

C. Aorta

D. Pulmonary artery

E. Bronchus

F. Rib

G. Left lung

H. Diaphragm

I. Heart

J. Right lung

3. Cellular Exchange (page 190)

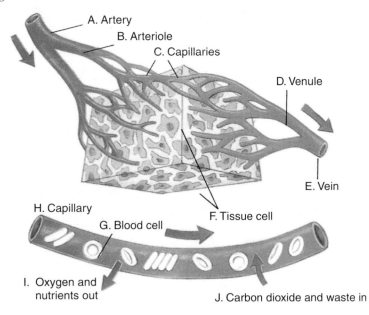

A. Artery

B. Arteriole

C. Capillaries

D. Venule

E. Vein

F. Tissue cell

G. Blood cell

H. Capillary

I. Oxygen and nutrients out

J. Carbon dioxide and waste in

4. Pulmonary Exchange (page 190)

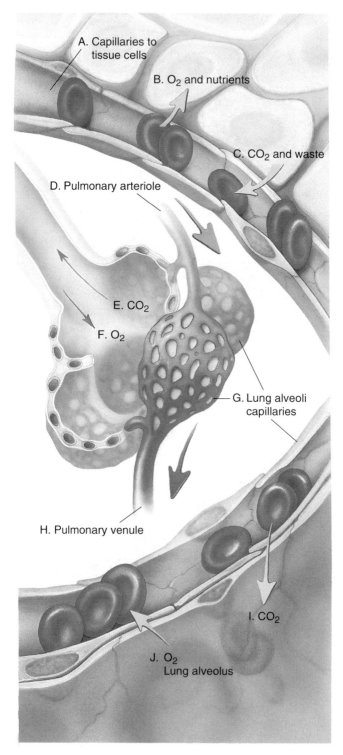

A. Capillaries to tissue cells

B. O_2 and nutrients

C. CO_2 and waste

D. Pulmonary arteriole

E. CO_2

F. O_2

G. Lung alveoli capillaries

H. Pulmonary venule

I. CO_2

J. O_2 Lung alveolus

Vocabulary

1. Gag reflex: Normal reflex mechanism that causes retching when the soft palate or the back of the throat is touched. (page 196)

2. Gastric distention: Air filling the stomach during artificial ventilation due to high volume and pressure. (page 213)

3. Stoma: Opening in the neck that connects the trachea directly to the environment. (page 214)

Fill-in

1. trachea (page 186)

2. higher (page 189)

3. 21, 78 (page 189)

4. carbon dioxide (page 191)

5. diaphragm, intercostal muscles (page 188)

6. low oxygen, high carbon dioxide (page 191)

7. hypoxia (page 191)

True/False

1. F   (page 196)

2. T   (page 207)

3. F   (page 197)

4. F   (pages 203, 206)

5. F   (page 204)

Short Answer

1. Nervousness, tachycardia, irritability, fear, apprehension. (Other signs: Mental status changes, use of accessory muscles for breathing, breathing difficulty, possible chest pain.) (page 191)

2. Adults: 12 to 20 breaths/min

Children: 15 to 30 breaths/min

Infants: 25 to 50 breaths/min (page 192)

3. Give slow, gentle breaths. (page 214)

4. 1. Kneel above the patient's head.

2. Extend the patient's neck unless you suspect a cervical spine injury.

3. Open the mouth and suction as needed. Insert an airway adjunct as needed.

4. Select a proper-sized mask.

5. Position the mask on the patient's face.

6. Use the C-clamp technique to hold the mask, then squeeze the bag every 5 seconds for adults, every 3 seconds for children and infants. (page 212)

5. 1. Respiratory rate of less than 8 breaths/min or greater than 24 breaths/min

2. Accessory muscle use

3. Skin pulling in around the ribs during inspiration

4. Pale, cyanotic, or cool (clammy) skin

5. Irregular pattern of inhalation and exhalation

6. Lung sounds that are decreased, unequal, or "wet"

7. Labored breathing

8. Shallow and/or uneven chest movement

9. Two- or three-word sentences spoken (page 193)

6. They are the secondary muscles of respiration. They are not used in normal breathing. They include:

1. Sternocleidomastoid (neck)

2. Pectoralis major (chest)

3. Abdominal muscles (page 192)

7. When the patient has severe trauma to the head or face. (page 198)

8. 1. Select the proper-size airway and apply a water-soluble lubricant.

2. Place the airway in the larger nostril with the curvature following the curve of the floor of the nose.

3. Advance the airway gently.

4. Continue until the flange rests against the skin. (pages 198 to 199)

9. Tonsil tips are best because they have a large diameter and do not collapse. In addition, they are curved, which allows easy, rapid placement. (page 200)

10. 15 seconds. (page 201)

Word Fun

Ambulance Calls

1. Maintain cervical spine stabilization.

Immediately open the airway with a modified jaw thrust.

Suction to remove obstruction.

Assess the airway for breathing (rate, rhythm, quality) and provide oxygen via nonrebreathing mask or BVM.

Continue initial assessment, rapid extrication, and rapid transport.

2. Assess breathing for rate, quality, and degree of distress.

Give 100% oxygen via nonrebreathing mask or BVM as appropriate.

Transport in position of comfort.

Rapid transport.

3. Open the airway and attempt to ventilate.

If air does not go in, reposition the airway and attempt to ventilate.

If air still does not go in, treat child for a FBAO.

Deliver 5 back blows and 5 chest thrusts.

Look in the mouth to see if there is a visible obstruction that can be removed.

If not, attempt to ventilate.

If air does not go in, continue sequence as above while transporting.

Rapid transport.

If obstruction is cleared, ventilate and assess circulation.

Continue to treat according to AHA guidelines.

Skill Drills

1. Skill Drill 7-1: Positioning an Unconscious Patient (page 194)

1. Support the head while your partner straightens the patient's legs.
2. Have your partner place his or her hand on the patient's far shoulder and hip.
3. Roll the patient as a unit with the person at the head calling the count to begin the move.
4. Open and assess the patient's airway and breathing status.

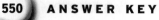

Section 3 Patient Assessment

Chapter 8: Patient Assessment

Matching

1. R (page 235)
2. D (page 243)
3. N (page 250)
4. O (page 250)
5. B (page 243)
6. H (page 250)
7. M (page 241)
8. A (page 250)
9. P (page 235)

10. L (page 260)
11. Q (page 260)
12. J (page 244)
13. G (page 243)
14. E (page 240)
15. F (page 258)
16. I (page 243)
17. K (page 240)
18. C (page 250)

Multiple Choice

1. C – Level of responsiveness is determined during the "initial assessment." (page 230)
2. D – Possible dangers you may observe in scene size-up also include leaking gasoline or diesel fuel, hostile bystanders/potential for violence, fire or smoke, possible hazardous or toxic materials, other dangers at crash or rescue scenes and crime scenes. (pages 230 to 232)
3. D (page 233)
4. B – Gunshot wounds (D) are penetrating trauma. (page 233)
5. C – Falls (D) may result in blunt or penetrating trauma, but are not classified as either. (page 233 to 234)
6. D (page 233)
7. A – The geographic location is irrelevant to the injury. (page 233)
8. C – As the speed of a crash increases, the forces that are exerted on the patient increase as well. (page 233)
9. D (page 233)
10. D (page 234)
11. B – The amount of force is directly related to the distance fallen, however, the other factors play a significant role in the extent of injury. (page 234)
12. D (page 235)
13. B – Rain is only a factor when it is raining in such a manner as to interfere with patient care. (page 235)
14. B – The pupils are assessed during the detailed exam. (page 236)
15. C – Mental status is the best indicator of cerebral perfusion. (page 240)
16. D – An altered mental status may also be caused by stroke, cardiac problems, or drug use. (page 241)
17. A – Tightness in the chest is a symptom. (page 241)
18. B – The tongue becomes an obstruction due to the relaxation of the muscles. (page 241)
19. D (page 241)
20. D (page 242)
21. B (page 243)
22. A – Use a gloved hand and a sterile dressing over the wound, and then bandage. (page 243)
23. C (page 243)
24. D (page 243)
25. D – When assessing for changes in skin color in deeply pigmented skin, also check the fingernail beds. (page 243)
26. D – Other conditions that may slow capillary refill but are not related to the body's circulation include frostbite and the patient's gender. (page 244)

27. B – Only apply an AED if the patient is unresponsive, apneic, and pulseless. (page 244)

28. C (page 244)

29. C – (A,B,D) are all treatments. (page 245)

30. D (page 246)

31. B (page 246)

32. C (page 248)

33. B – AVPU (A) evaluates level of responsiveness in the initial assessment, OPQRST (C) evaluates pain, and SAMPLE (D) gains the necessary history. (page 249)

34. C – Once the c-collar is in place it should not be removed and does not allow for palpation of the neck and cervical spine. (page 250)

35. A – (B,C) evaluate sensation, not motor function. (page 253)

36. C (page 255)

37. C – Evaluating whether the patient can move is irrelevant. When assessing a complaint of dizziness, also evaluate the rate and quality of respirations, monitor the level of consciousness and orientation, and check the head for signs of trauma. (page 256)

38. B – If the impact is minor and their ABCs are intact with no other complaint, they do not require a rapid trauma assessment and rapid transport. (page 257)

39. C (page 258)

40. B (page 260)

41. D – When assessing a chief complaint of chest pain, also evaluate blood pressure and look for trauma to the chest. (page 262)

42. D (page 263)

43. B (page 265)

44. D – When performing the detailed physical exam, depending on what you learn, you should also be prepared to perform spinal immobilization, and provide transport to an appropriate facility, or call for ALS backup. (page 266)

45. D (pages 268 and 271)

46. B – A diagnosis cannot be made in the field. (page 272)

47. D – When reevaluating interventions, also take a moment to ensure that the airway is still open. (page 274)

Vocabulary

1. Blunt trauma: A mechanism of injury in which force occurs over a broad area and the skin is not usually broken. (page 233)

2. Penetrating trauma: A mechanism of injury in which force occurs in a small point of contact between the skin and the object. The skin is broken and the potential for infection is high. (page 233)

3. Mechanism of injury: The way in which traumatic injuries occur; the forces that act on the body to cause damage. (page 233)

4. Capillary refill: A test that evaluates distal circulatory system function by squeezing (blanching) blood from an area such as a nail bed and watching the speed of its return after releasing the pressure. (page 244)

5. Golden Hour: The time from injury to definitive care, during which treatment of shock or traumatic injuries should occur because survival potential is the best. (page 246)

Fill-in

1. decisions (page 228)

2. body substance isolation (page 230)

3. victim (page 230)

4. safety (page 232)

5. properly (page 233)

6. three (page 234)

7. entrance wound, exit wound (page 234)

8. Triage (page 235)

9. general impression (page 236)

10. life-threatening (page 238)

11. patency (page 241)

12. reevaluate (page 242)

13. wrist (page 243)

14. initial assessment (page 245)

True/False

1. F (page 240)
2. T (page 266)
3. F (page 234)
4. T (page 244)
5. F (page 272)
6. T (page 238)
7. F (page 244)
8. T (page 243)
9. T (page 244)
10. T (page 249)
11. F (page 249)

Short Answer

1. The characteristics of the penetrating object, the amount of force or energy, and the part of the body affected. (page 233)

2. To identify and initiate treatment of immediate or potential life threats. (page 236)

3. Immediate assessment of the environment, the patient's presenting signs and symptoms, and the patient's chief complaint. (page 236)

4. A-Airway

    B-Breathing

    C-Circulation (page 240)

5. With focal pain, a patient is able to identify a single place or point of pain. With diffuse pain, patients are unable to point to a single location. Instead, they often move their finger around in a circle as they are asked to point to their pain. (page 260)

6. Orientation to person, place, time, and event. *Person* (name) evaluates long-term memory. *Place* and *time* evaluate intermediate-term memory. *Event* evaluates short-term memory. (page 240)

7. 1. Identify the patient's chief complaint.

    2. Understand the circumstances surrounding the chief complaint.

    3. Direct further physical examination. (page 246)

8. Deformities, Contusions, Abrasions, Punctures/penetrations, Burns, Tenderness, Lacerations, Swelling (page 249)

9. 1. Ejection from a vehicle

    2. Death in the passenger compartment

    3. Fall greater than 15 to 20 feet

    4. Vehicle rollover

    5. High-speed vehicle collision

    6. Vehicle-pedestrian collision

    7. Motorcycle crash

    8. Unresponsiveness or altered mental status following trauma

    9. Penetrating head, chest, or abdominal trauma (page 248)

Word Fun

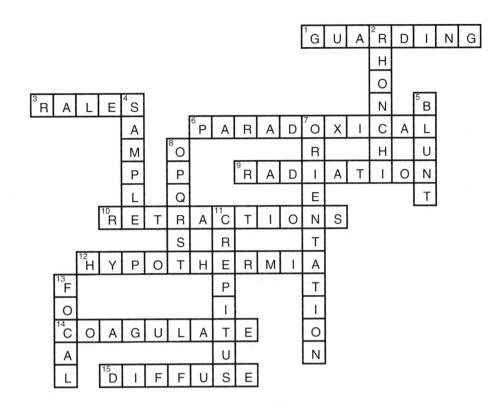

Ambulance Calls

1. -Maintain cervical spine control

   -Immediately manage the airway by suction and oxygen

   -Rapid survey and transport

   This patient is a load-and-go based on:

   Mechanism of injury

   Level of consciousness

   Airway compromise

   Damage to vehicle indicates possible occult injuries

2. -Assess level of responsiveness and the patient's orientation by asking her:

   The day of the week

   Her name

   Where she is

   If she knows why you are there

   -Apply high-flow oxygen via nonrebreathing mask

   -Complete a focused history and physical exam

   -Transport the patient in a position of comfort

   -Ask family members:

   What is her normal mental status?

   When did this start?

   Does she have any past history?

   Is she taking any new medications?

   Has she taken her medications?

   Has she eaten today?

   Anything else that might be pertinent based on your scene survey

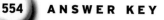

3. He is a rapid transport based on his altered mental status.

-En route, you should reassess all vital signs and interventions.

-Complete a detailed physical exam.

-Splint any fractures, bandage any minor wounds.

-Reassess vital signs at least every 5 minutes.

-Call in your radio report to the receiving facility.

-Reassess, reassess, reassess!

Skill Drills

1. Skill Drill 8-1: Performing a Rapid Trauma Assessment (pages 251 to 252)

1. Check ABCs, continue spinal immobilization, and assess mental status. Assess the head.
2. Assess the neck.
3. Apply a cervical collar.
4. Assess the chest, including breath sounds.
5. Assess the abdomen.
6. Assess the pelvis.
7. Assess the extremities.
8. Log roll the patient and assess the back.
9. Assess baseline vitals and SAMPLE history.

2. Skill Drill 8-2: Performing a Rapid Medical Assessment: Unresponsive Patient (page 264)

1. Assess the head.
2. Assess the neck.
3. Assess the chest.
4. Assess the abdomen.
5. Assess the pelvis.
6. Assess the extremities.
7. Assess the back.

3. Skill Drill 8-3: Performing the Detailed Physical Exam (pages 269 to 271)

1. Observe the face.
2. Inspect the eyelids and the area around the eyes.
3. Examine the eyes for redness, contact lenses. Check pupil function.
4. Look behind the ear for Battle's sign.
5. Check the ears for drainage or blood.
6. Observe and palpate the head.
7. Palpate the zygomas.
8. Palpate the maxillae.
9. Palpate the mandible
10. Assess the mouth.
11. Check for unusual breath odors.
12. Inspect the neck.
13. Palpate the neck, front and back.
14. Observe for jugular vein distention.
15. Inspect the chest and observe breathing motion.

16. Gently palpate the ribs.

17. Listen to anterior breath sounds (midaxillary, midclavicular).

18. Listen to posterior breath sounds (bases, apices).

19. Observe the abdomen and pelvis.

20. Gently palpate the abdomen.

21. Gently compress the pelvis from the sides.

22. Gently press the iliac crests.

23. Inspect the extremities; assess distal circulation and motor sensory function.

24. Log roll the patient and inspect the back.

Chapter 9: Communications and Documentation

Matching

1. M (page 283)

2. G (page 283)

3. J (page 283)

4. K (page 284)

5. H (page 284)

6. L (page 283)

7. I (page 283)

8. C (page 284)

9. A (page 283)

10. F (page 285)

11. E (page 285)

12. D (page 283)

13. B (page 293)

Multiple Choice

1. D (page 283)

2. A (page 283)

3. D (page 283)

4. B (page 284)

5. B (page 284)

6. D (page 285)

7. D (page 285)

8. C – This (C) is true of the duplex mode. (page 285)

9. D – Principle EMS-related responsibilities of the FCC also include licensing base stations and assigning appropriate radio call signs for those stations, and establishing licensing standards and operating specifications for radio equipment used by EMS providers. (page 286)

10. D – The dispatchers do not provide care—they provide instructions for care. Responsibilities of the dispatcher also include coordinating EMS response units with other public safety services until the incident is over. (pages 286 to 287)

11. D (page 287)

12. D – The determination of the level and type of response necessary is also based on the need for additional EMS units, fire suppression, rescue, a HazMat team, air medical support, or law enforcement. (page 287)

13. B – There is no way to know when the unit will arrive (B). The nature and severity of the injury, illness, or incident, special directions, or advisories, and the time at which the unit(s) are dispatched are also given. (page 287)

14. D (page 288)

15. A – The patient report also commonly includes your unit identification and level of services, the receiving hospital, the patient's chief complaint or your perception of the problem and its severity, a brief report of physical findings, and a brief summary of care given and any patient response. (page 288)

16. B (page 289)

17. D (page 298)

18. D (page 290)

19. D (page 289)

20. A (page 289)

21. A – If the patient wishes to be transported, you do not need permission (B). Dispatch coordinates the request of assistance from other agencies (C). You must have permission to restrain a patient, but not to immobilize (D). (page 289)

22. D (page 289)

23. A – It is very important to report changes, especially if the status is worse (B). You must check vital signs every 5 minutes in critical patients and every 15 minutes for stable patients (C). All treatments should be reassessed for the patient's response to care (D). (page 291)

24. C – As long as it is not on the road creating a hazard, there is no need to report an abandoned vehicle. (page 291)

25. D (page 290)

26. B (page 292)

27. D (page 292)

28. D – Components of the oral report also include a summary of the information you gave in your radio report, the patient's response to treatment, and any other information gathered that was not important enough to report sooner. (pages 292 to 293)

29. D (page 293)

30. C – Always face the person (A). Never shout (B). Never use baby talk (D). (page 294)

31. D (page 294)

32. C – Sign language only works for hearing-impaired patients, for obvious reasons (A). Stay in physical contact (B). If the patient can walk to the ambulance, place his or her hand on your arm, taking care not to rush (D). (page 296)

33. A – Use simple terms and phrases, not medical terms (B). Never shout (C). Positioning yourself so the patient can read your lips only works if the patient knows the language (D). (page 297)

34. B – This (B) is administrative information. (page 298)

35. A – (B), (C), and (D) are all patient information. (page 298)

36. D – Functions of the prehospital care report also include legal documentation, administrative functions, and evaluation and continuous quality improvement. (page 298)

37. D (page 299)

38. A – Include pertinent negative findings as well as pertinent positive findings (B). Never record conclusions, only findings (C). Avoid radio codes and use only standard abbreviations (D). (page 299)

39. D – You may also be required to file special reports for incidents involving certain infectious diseases. (page 300)

Vocabulary

1. Simplex: Single frequency radio; transmissions can occur in either direction but not simultaneously in both; when one party transmits, the other can only receive, and the party that is transmitting is unable to receive. (page 285)

2. Standing orders: Written documents, signed by the EMS system medical director, that outline specific directions, permissions, and sometimes prohibitions regarding patient care; also called protocols. (page 292)

3. Federal Communications Commission (FCC): The federal agency that has jurisdiction over interstate and international telephone and telegraph services and satellite communications, all of which may involve EMS activity. (page 286)

4. Duplex: The ability to transmit and receive simultaneously. (page 285)

Fill-in

1. patient care report (page 282)
2. transmitter, receiver (page 283)
3. dedicated line (page 283)
4. telemetry (page 284)
5. cell phones (page 284)
6. Pagers (page 287)
7. importance (page 287)
8. medical control (page 288)
9. slander (page 289)
10. medical control (page 289)
11. repeat (page 290)
12. standing orders (page 292)
13. eye contact (page 293)
14. honest (page 295)
15. interpreter (page 297)
16. minimum data set (page 298)
17. Competent (page 300)

True/False

1. T   (page 283)
2. T   (page 283)
3. T   (page 284)
4. F   (page 283)
5. T   (page 283)
6. T   (page 298)
7. F   (page 284)
8. F   (page 285)
9. T   (page 288)
10. F   (page 291)
11. F   (page 292)
12. T   (page 295)

Short Answer

1. -Allocating specific radio frequencies for use by EMS providers
   -Licensing base stations and assigning appropriate radio call signs for those stations
   -Establishing licensing standards and operating specifications for radio equipment used by EMS providers
   -Establishing limitations for transmitter power output
   -Monitoring radio operations (page 286)
2. -Monitor the channel before transmitting.
   -Plan your message.
   -Press the push-to-talk (PTT) button.
   -Hold the microphone 2 to 3 inches from your mouth and speak clearly.
   -Identify the person or unit you are calling, then identify your unit as the sender.
   -Acknowledge a transmission as soon as you can.
   -Use plain English.
   -Keep your message brief.
   -Avoid voicing negative emotions.
   -When transmitting a number with two or more digits, say the entire number first, then each digit separately.
   -Do not use profanity on the radio.
   -Use EMS frequencies only for EMS communications.
   -Reduce background noise.
   -Be sure other radios on the same frequency are turned down. (page 291)
3. -Continuity of care
   -Legal documentation
   -Education
   -Administrative
   -Research
   -Evaluation and continuous quality improvement (page 298)
4. -Traditional, written form with check boxes and a narrative section
   -Computerized, using electronic clipboard or similar device (page 299)

Word Fun

Ambulance Calls

1. -Go ahead and dispatch the closest ambulance for an emergency response.
   -Call for assistance from the fire department and local law enforcement.
   -Try to calm the caller down to obtain additional information.
   -If the caller is still of no help, ask her to get someone else to the phone.
   -Relay any additional information to the responding units.

2. -Obtain a general impression of the patient as you approach.
   -Squat down to be on the same level and smile.
   -Speak slowly and distinctly, and make sure you are positioned in front of the patient.
   -Use sign language, or point to the child's leg and shrug your shoulders with palms up to ask what's wrong.
   -Use a teacher as an "interpreter" to explain to the child what you will do.
   -Remember to maintain eye contact and smile to reduce anxiety.

3. -Explain who you are.
   -Keep in physical contact with the patient.
   -Explain each procedure you will perform.
   -Transport the dog along with the patient.

Section 4 Medical Emergencies

Chapter 10: General Pharmacology

Matching

1. I   (page 309)

2. J   (page 308)

3. F   (page 308)

4. G   (page 309)

5. D   (page 308)

6. H   (page 308)

7. B   (page 308)

8. C   (page 308)

9. E   (page 310)

10. A   (page 311)

Multiple Choice

1. D – However, a doctor's order must be obtained prior to assisting with any medication. (page 314)

2. C – Generic medications have the same active ingredients in the same concentrations as their brand name counterparts. (page 309)

3. A – Causing an increase in the strength and rate of contractions (B,C) will have an adverse affect on the heart as it increases the oxygen demand and worsens the pain. (page 308)

4. A – (page 308)

5. A – (page 309)

6. B – (page 309)

7. C – (page 310)

8. D – (page 310)

9. D – (page 311)

10. B – The only accepted use of a nasal cannula (A) in the prehospital setting is when the patient will not tolerate a nonrebreathing mask. A BVM (C) is used when the patient does not have a sufficient rate or tidal volume to support life. (page 313)

11. D – (page 315)

12. C – (page 315)

13. D – (page 315)

14. B – Epinephrine works to block the release of histamine from the MAST cells. (page 315)

15. A – (page 317)

16. D – Nitroglycerin is a potent vasodilator (B) that decreases blood pressure (A) by decreasing the venous return to the heart and thereby decreasing cardiac output. (page 317)

Vocabulary

1. Trade name: The brand name that a manufacturer gives to a drug. (page 308)

2. Generic name: The original chemical name of a drug. (page 309)

3. OTC: Over-the-counter: Medications that can be purchased without a prescription. (page 309)

4. Solution: A liquid mixture of one or more substances that cannot be separated by filtering or allowing the mixture to stand. (page 311)

5. Suspension: A mixture of ground particles distributed evenly throughout a liquid. (page 311)

6. Sublingual: Under the tongue. (page 309)

7. Metered-dose inhaler (MDI): A miniature spray canister, used to direct substances small enough to be inhaled through the mouth and into the lungs. An MDI delivers the same amount of medication each time it is used. (page 311)

Fill-in

1. Glucose (page 314)
2. Epinephrine (page 314)
3. sublingually (page 317)
4. intravenous injection (pages 309 to 310)
5. solutions (page 311)
6. medication (page 308)

True/False

1. F   (page 313)
2. F   (page 314)
3. F   (page 315)
4. T   (page 317)
5. F   (page 309)

6. T   (page 318)
7. T   (page 314)
8. F   (page 315)
9. F   (page 317)

Short Answer

1. Intravenous   Intramuscular   Transcutaneous

   Oral   Intraosseous   Inhalation

   Sublingual   Subcutaneous   Per rectum   (pages 309 to 310)

2. 1. Obtain an order from medical control.

   2. Verify the proper medication and prescription.

   3. Verify the form, dose, and route.

   4. Check the expiration date and condition of medication.

   5. Reassess vital signs, especially heart rate and blood pressure, at least every 5 minutes or as the patient's condition changes.

   6. Document. (pages 318 to 319)

3. Many drugs adsorb (stick to) activated charcoal, preventing the drugs from being absorbed by the body. It needs to be shaken because it is a suspension and should be given in a covered container with a straw. (page 314)

4. 1. Secreted by adrenal glands

   2. Decreases bronchial muscle tone

   3. Dilates lung passages

   4. Constricts blood vessels

   5. Increases heart rate and blood pressure (page 315)

5. By pressing it into the skin (page 316)

6. 1. Relaxes coronary arteries and veins

   2. Less blood is returned to the heart

   3. Decreases blood pressure

   4. Relaxes veins throughout the body

   5. May cause headaches (page 317)

7. To aim the spray properly (pages 316 to 317)

Word Fun

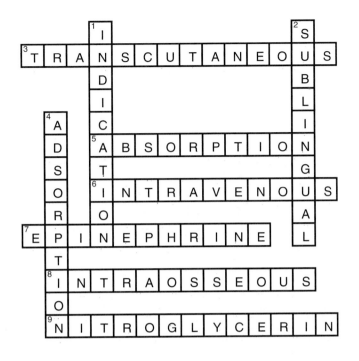

Ambulance Calls

1. Place the patient in the position of comfort.

Give 100% oxygen via nonrebreathing mask.

Contact medical control for permission to assist patient with 1 nitroglycerin tablet SL.

Monitor vital signs.

Rapid transport

2. Place patient in position of comfort.

Give 100% oxygen via nonrebreathing mask.

Contact medical control for permission to assist patient with inhaled medication.

Monitor vital signs.

If patient does not improve with MDI, contact medical control for permission to assist patient with her EpiPen.

Continue to frequently monitor her airway.

Rapid transport

3. Place patient in position of comfort.

Give 100% oxygen via nonrebreathing mask.

Check blood pressure!

Check expiration date on nitroglycerin.

Contact medical control for permission to assist patient with 1 nitroglycerin tablet SL.

Monitor vital signs.

Rapid transport

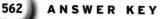

Chapter 11: Respiratory Emergencies

Matching

1. B (page 332)
2. D (page 327)
3. H (page 328)
4. J (page 330)
5. L (page 333)
6. C (page 331)
7. G (page 327)
8. E (page 328)
9. M (page 340)
10. A (page 330)
11. K (page 334)
12. F (page 332)
13. I (page 333)

Multiple Choice

1. D (page 324)
2. C (page 324)
3. D (page 326)
4. C – This will replace the carbon dioxide content in the CSF. (page 326)
5. B – Rapid and deep breathing helps to blow off excess carbon dioxide (page 326)
6. A – Pale or cyanotic skin (B), pursed lips and nasal flaring (C), and cool, damp skin (D) are signs of inadequate breathing. (page 326)
7. D (page 327)
8. B (page 327)
9. A (page 327)
10. B – Epiglottitis (A) and colds (C) create obstruction of the flow of air in the major passages. (page 327)
11. A (page 327)
12. D (page 329)
13. D (page 329)
14. C (page 330)
15. A (page 330)
16. D (page 330)
17. B – Blood pressure (A) of COPD patients is normal. The pulse (C) of COPD patients is rapid and occasionally irregular. (page 331)
18. C (page 331)
19. C (page 332)
20. B (page 332)
21. D (page 332)
22. C (page 333)
23. A (page 333)
24. D (page 334)
25. B (page 334)
26. B (page 335)
27. C (page 336)
28. D (page 336)
29. D (page 336)
30. C – Ventolin (A) and Metaprel (B) are trade names. (page 337)
31. D (page 338)
32. D (page 339)
33. C (page 341)

Labeling

The Upper Airway (page 325)

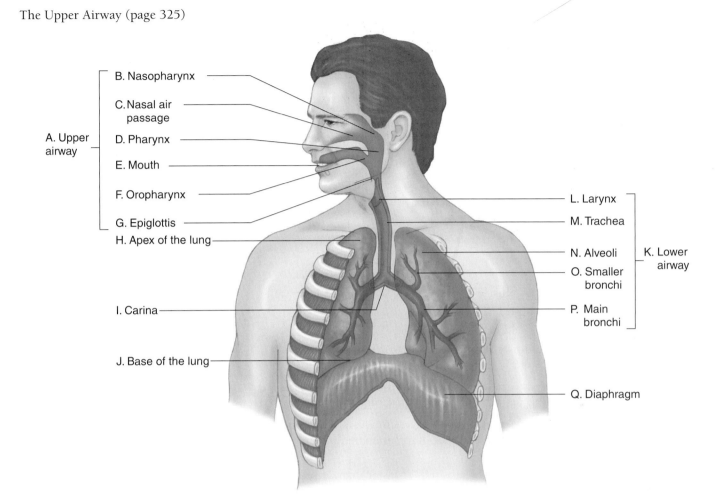

Vocabulary

1. Stridor: High-pitched, rough barking inspiratory sound often heard with upper airway obstruction. (page 329)
2. Croup: Inflammation and swelling of the lining of the larynx. (page 329)
3. Rales: Cracking, rattling breath sounds. (page 331)
4. Rhonchi: Coarse, gravelly, breath sounds. (page 331)
5. Diphtheria: An infectious disease in which a membrane lining the pharynx is formed that can severely obstruct passage of air into the larynx. (page 328)
6. Chronic Obstructive Pulmonary Disease (COPD): A slow process of dilation and disruption of the airways and alveoli, caused by chronic bronchial obstruction. (page 329)

Fill-in

1. carbon dioxide (page 326)
2. oxygen (page 327)
3. Oxygen, oxygen (page 324)
4. alveoli (page 324)
5. trachea (page 324)
6. 8, 24 (page 326)
7. carbon dioxide (page 324)

True/False

1. F   (page 330)
2. T   (page 331)
3. F   (page 332)
4. F   (page 332)
5. T   (page 334)

6. T   (page 334)
7. F   (pages 334 to 335)
8. T   (page 330)
9. F   (pages 330, 332)

Short Answer

1. 1. Normal rate and depth
   2. Regular pattern of inhalation and exhalation
   3. Good audible breath sounds on both sides of the chest
   4. Regular rise and fall on both sides of the chest
   5. Pink, warm, dry skin (page 326)

2. 1. Pulmonary vessels are obstructed from absorbing oxygen and releasing carbon dioxide by fluid, infection, or collapsed air spaces.
   2. Damaged alveoli
   3. Air passages obstructed by muscle spasm, mucus, weakened airway walls
   4. Blood flow to the lungs obstructed
   5. Pleural space is filled with air or excess fluid (page 326)

3. 1. Patient is unable to coordinate administration and inhalation
   2. Inhaler is not prescribed for patient.
   3. You did not obtain permission from medical control or local protocol.
   4. Patient has already met maximum prescribed dose before your arrival. (page 338)

4. An ongoing irritation of the respiratory tract; excess mucus production obstructs small airways and alveoli. Protective mechanisms are impaired. Repeated episodes of irritation and pneumonia can cause scarring and alveolar damage, leading to COPD. (page 330)

5. 1. Respiratory rate of slower than 8 breaths/min or faster than 24 breaths/min
   2. Muscle retractions above the clavicles between ribs, below rib cage, especially in children
   3. Pale or cyanotic skin
   4. Cool, damp (clammy) skin
   5. Shallow or irregular respirations
   6. Pursed lips
   7. Nasal flaring (pages 326, 327)

6. A condition characterized by a chronically high blood level of carbon dioxide in which the respiratory center no longer responds to high blood levels of carbon dioxide. In these patients, low blood oxygen causes the respiratory center to respond and stimulate respiration. If the arterial level of oxygen is then raised, as happens when the patient is given additional oxygen, there is no longer any stimulus to breathe; both the high carbon dioxide and low oxygen drives are lost. (page 327)

Word Fun

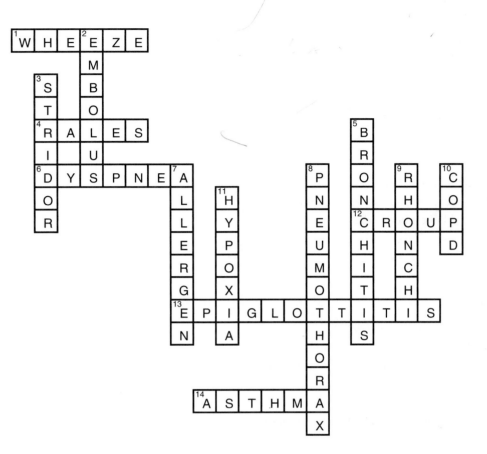

Ambulance Calls

1. -Transport in position of comfort
   -Maintain a clear airway
   -Rapid transport
   -High-flow oxygen and assist ventilations as needed
2. -Place patient in position of comfort.
   -Administer high-flow oxygen.
   -Try to calm the patient and talk her into slowing her respirations.
   -Monitor vital signs
   -Transport normal response
3. -Place patient in position of comfort.
   -High-flow oxygen via nonrebreathing mask
   -Monitor vital signs
   -Rapid transport

Skill Drills

Skill Drill 11-1: Assisting a Patient with a Metered-Dose Inhaler (page 339)

1. Ensure inhaler is at room temperature or **warmer**.
2. Remove oxygen mask.
   Hand inhaler to patient. Instruct about breathing and **lip seal**.
   Use a **spacer** if patient has one.

3. Instruct patient to press inhaler and inhale.

Instruct about **breath holding**.

4. Reapply **oxygen**.

After a few **breaths**, have patient repeat **dose** if order/protocol allows.

Chapter 12: Cardiovascular Emergencies

Matching

1. M (page 346)
2. D (page 347)
3. P (page 347)
4. H (page 347)
5. O (page 347)
6. L (page 348)
7. I (page 346)
8. N (page 346)
9. C (page 350)
10. E (page 352)
11. G (page 349)
12. B (page 355)
13. J (page 351)
14. K (page 352)
15. A (page 353)
16. F (page 353)

Multiple Choice

1. D – The number of deaths can also be reduced with increased numbers of lay people trained in CPR. (page 346)
2. D (page 346)
3. A (page 346)
4. B – Pulmonary veins transport blood from the lungs, where it has picked up oxygen. (page 346)
5. A (page 347)
6. B (page 347)
7. C (page 348)
8. C (page 348)
9. A – White blood cells (B) help fight infection. Platelets (C) help blood to clot. Veins (D) carry deoxygenated blood back to the heart. (page 348)
10. C (page 348)
11. C – Diastolic blood pressure is the resting phase of the ventricles. (pages 348 to 349)
12. A (page 350)
13. B (page 351)
14. B (page 351)
15. D (page 351)
16. B (page 351)
17. D – It is usually felt in the midchest, under the sternum, but may radiate. (page 351)
18. D (page 351)
19. C (page 352)
20. C (page 352)
21. C (page 352)
22. C – (A) and (B) refer to angina. (page 352)
23. D (page 352)
24. D (page 352)
25. A – Asystole (B) is absence of electrical activity. Ventricular stand still (C) is the same thing as asystole. Ventricular tachycardia (D) is a rapid heart rate greater than 100 beats/min. (page 352)

26. B (page 353)
27. A (page 353)
28. D (page 354)
29. B (page 355)
30. A (page 355)
31. C (page 355)
32. C (page 355)
33. D (page 355)
34. B – AVPU (A) is used to assess level of consciousness. SAMPLE (B) is used to assess history. CHART (D) is used for documentation. (page 356)
35. C (page 356)
36. D (page 356)
37. D (page 358)
38. C (page 359)
39. B (page 359)
40. D (page 360)
41. C (page 361)
42. C (page 362)
43. B (page 362)
44. C (page 362)
45. D (page 367)

Labeling

1. The Right and Left Sides of Heart (page 347)

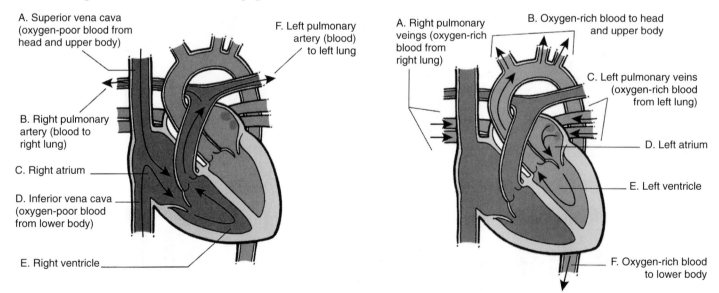

A. Superior vena cava (oxygen-poor blood from head and upper body)

F. Left pulmonary artery (blood) to left lung

B. Right pulmonary artery (blood to right lung)

C. Right atrium

D. Inferior vena cava (oxygen-poor blood from lower body)

E. Right ventricle

A. Right pulmonary veings (oxygen-rich blood from right lung)

B. Oxygen-rich blood to head and upper body

C. Left pulmonary veins (oxygen-rich blood from left lung)

D. Left atrium

E. Left ventricle

F. Oxygen-rich blood to lower body

2. Electrical Conduction (page 348)

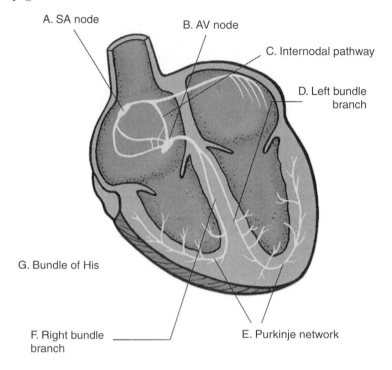

A. SA node

B. AV node

C. Internodal pathway

D. Left bundle branch

G. Bundle of His

F. Right bundle branch

E. Purkinje network

3. Pulse Points (page 350)

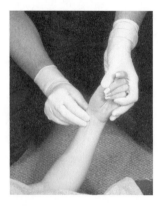

A. Carotid

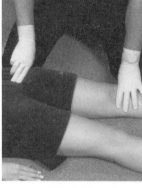

B. Femoral

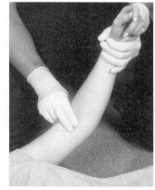

C. Brachial

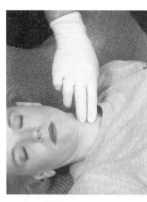

D. Radial

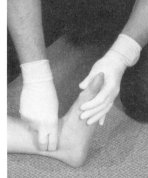

E. Posterior tibial

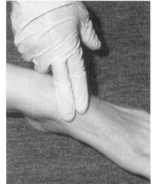

F. Dorsalis pedis

Vocabulary

1. Angina pectoris: Transient chest discomfort caused by partial or temporary blockage of blood flow to heart muscle. (page 351)
2. Ventricular fibrillation: Disorganized, ineffective twitching of the ventricles, resulting in no blood flow to the heart. (pages 352 to 353)
3. Cardiogenic shock: Body tissues do not get enough oxygen because the heart lacks enough power to force the proper volume of blood through the circulatory system. (page 353)
4. Acute myocardial infarction (AMI): Heart attack; death of myocardium following obstruction of blood flow to the heart muscle. (page 351)
5. Cardiac arrest: A state in which the heart fails to generate an effective and detectable blood flow. (page 352)
6. Syncope: Fainting spell or transient loss of consciousness. (page 355)
7. Congestive heart failure (CHF): A disorder in which the heart loses part of its ability to effectively pump blood, usually as a result of damage to the heart muscle and usually resulting in a backup of fluid into the lungs. (page 353)

Fill-in

1. septum (page 346)
2. aorta (page 346)
3. right (page 346)
4. AV (page 347)
5. dilation (page 347)
6. Red blood (page 348)
7. Diastolic (page 349)
8. four (page 346)
9. left (page 346)

True/False

1. F   (page 346)
2. F   (page 347)
3. T   (page 350)
4. F   (page 351)
5. T   (page 351)
6. F   (page 352)
7. F   (page 356)
8. F   (page 369)
9. T   (page 351)
10. F   (page 348)

Short Answer

1. Automated: Operator needs only to apply pads and turn on the machine. It performs all functions for analyzing and shocking. This type of defibrillator often has a computer voice synthesizer to advise the EMT which steps to take.

   Semi-automated: Operator applies pads, turns on the machine, and pushes button to shock. (page 361)
2. 1. Not having a charged battery

   2. Applying the AED to a patient who is moving.

   3. Applying the AED to a responsive patient with a rapid heart rate. (pages 361 to 362)
3. 1. If the patient regains a pulse.

   2. After six to nine shocks have been delivered.

   3. If the machine gives three consecutive "no shock" messages. (page 366)
4. 1. Place pads correctly.

   2. Make sure no one is touching the patient.

   3. Do not defibrillate a patient who is in pooled water.

   4. Dry the chest before defibrillating a wet patient.

   5. Do not defibrillate a patient who is touching metal that others are touching.

   6. Remove nitroglycerin patches and wipe the area with a dry towel before defibrillation. (page 367)
5. 1. Obtain an order from medical direction.

   2. Take the patient's blood pressure; continue with administration only if the systolic blood pressure is greater than 100 mm Hg.

   3. Check that you have the right medication, right patient, and right delivery route.

   4. Check the expiration date of the nitroglycerin.

5. Question the patient about the last dose he or she took and its effects.

6. Be prepared to have the patient lie down.

7. Give the medication sublingually.

8. Advise the patient to keep his or her mouth closed to allow the medication to dissolve.

9. Recheck blood pressure within 5 minutes.

10. Record each medication and the time of administration.

11. Perform continued assessment. (page 358)

6. 1. It may or may not be caused by exertion, but can occur at any time.

2. It does not resolve in a few minutes.

3. It may or may not be relieved by rest or nitroglycerin. (page 352)

7. 1. Sudden death

2. Cardiogenic shock

3. Congestive heart failure (page 352)

8. 1. Sudden onset of weakness, nausea, or sweating without an obvious cause

2. Chest pain/discomfort that does not change with each breath

3. Pain in lower jaw, arms, or neck

4. Sudden arrhythmia with syncope

5. Pulmonary edema

6. Sudden death

7. Increased and/or irregular pulse

8. Normal, increased, or decreased blood pressure

9. Normal or labored respirations

10. Pale or gray skin

11. Feelings of apprehension (page 355)

9. Remove clothing from the patient's chest area. Apply the pads to the chest: one just to the right of the sternum, just below the clavicle, the other on the left chest with the top of the pad 2 to 3 inches below the armpit. Ensure that the pads are attached to the patient cables (and that they are attached to the AED in some models). (page 363)

Word Fun

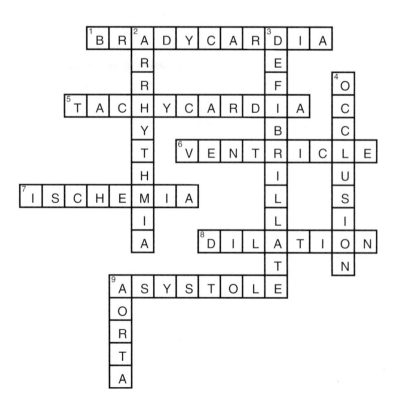

Ambulance Calls

1. -Place patient in the position of comfort.
   -High-flow oxygen via nonrebreathing mask
   -Monitor vital signs.
   -Normal transport

2. -Explain the need for seeking medical attention.
   -Place patient in position of comfort.
   -High-flow oxygen via nonrebreathing mask
   -Monitor vital signs.
   -Check a SAMPLE history and assist patient with nitroglycerin after contacting medical control if he currently has a prescription.
   -Rapid transport

3. -Sit the patient upright with legs down.
   -Administer high-flow oxygen and be prepared to ventilate with BVM if needed.
   -Monitor vital signs closely.
   -Take her medications along to the hospital.
   -Rapid transport

Skill Drills

1. ### Skill Drill 12-1: Caring for a Conscious Patient with Chest Pain (page 357)

   1. **Reassure** the patient as you perform the initial assessment.
      Apply **oxygen**.
      Place the patient in a **comfortable** position.
   2. Measure and record the **baseline vital signs**, obtain a **SAMPLE** history, and **monitor** the patient closely.
      Obtain a **focused** history and **physical exam**.
      Ask about the patient's discomfort using **OPQRST**.
   3. Check medication and **expiration date**.
   4. Help the patient administer **nitroglycerin**.
   5. Prepare to **transport** the patient.
      Report to **medical control**.

2. ### Skill Drill 12-2: AED and CPR (page 365)

   1. Check pulse.
      If pulse is present, check breathing.
   2. If pulseless, begin CPR.
      Prepare the AED pads.
      Turn on the AED; begin narrative if needed.
   3. Apply AED pads.
      Stop CPR.
   4. If breathing adequately, give oxygen and transport.
      If not, open airway, ventilate, and transport.
      If no pulse, perform CPR for 1 minute.
      Clear the patient and analyze again.
      If necessary, repeat one cycle of up to three shocks.
      Transport and call medical control.
      Continue to support breathing or perform CPR, as needed.

5. Verbally and visually clear the patient.

Push the Analyze button if there is one.

Wait for the AED to analyze rhythm.

If no shock advised, perform CPR for 1 min., if no pulse (go to step 18).

If shock advised, recheck that all are clear and push the Shock button.

Push the Analyze button, if needed, to analyze rhythm again.

Press Shock if advised (second shock).

Push the Analyze button, if needed, to analyze rhythm again.

Press Shock if advised (third shock).

6. Stop CPR if in progress.

Assess responsiveness.

Check breathing and pulse.

If unresponsive and not breathing adequately, give two slow ventilations.

Chapter 13: Neurologic Emergencies

Matching

1. D   (page 376)

2. G   (page 377)

3. E   (page 377)

4. B   (page 376)

5. I   (page 377)

6. F   (page 376)

7. J   (page 378)

8. C   (page 378)

9. M   (page 379)

10. H   (page 381)

11. L   (page 382)

12. K   (page 378)

13. A   (page 382)

Multiple Choice

1. D (page 376)

2. A – The cerebellum (B) controls muscle and body coordination. The cerebrum (C) controls emotion, thought, touch, movement, and sight. (page 376)

3. A (page 377)

4. C – (page 377)

5. C – A hemorrhagic stroke (A) is bleeding in the brain. Atherosclerosis (B) is usually the cause of the blockage. A cerebral embolism (D) may be the cause of the blockage. (page 378)

6. A – (B and C) may also result in a hemorrhagic stroke, but (D) leads to ischemic stroke. (page 378)

7. B (page 378)

8. B (page 378)

9. D (page 379)

10. A (page 379)

11. D – Epilepsy (A) is congenital in origin. A brain tumor (B) is a structural cause. A seizure due to a fever (C) is a febrile seizure. (page 379)

12. D (page 380)

13. D (page 380)

14. C – Unequal pupils may be seen in conjunction with a head injury, but they are not the cause of the altered mental status. (page 381)

15. B – A patient who has had a stroke may be alert and attempting to communicate normally, whereas a patient with hypoglycemia almost always has an altered or decreased level of consciousness. (page 381)

16. C (page 381)

17. D (page 382)

18. D (page 382)

19. B – Aphasia (A) is an inability to produce or understand speech. With expressive aphasia (C), the patient will be able to understand the question but cannot produce the right sounds in order to answer. With dysarthria (D), they will understand language and be able to speak, but their words may be slurred and hard to understand. (page 382)

20. D (page 383)

21. A (page 383)

22. B – The airway is assessed first with any patient. (page 383)

23. D (page 384)

24. A – Key physical tests for patients suspected of having a stroke include tests of speech, facial movement, and arm movement. (page 384)

25. C – A score of 11-13 indicates moderate to severe dysfunction. A score of 14-15 indicates mild dysfunction (B). A score of 10 or less indicates severe dysfunction (D). (page 385)

26. C (page 385)

27. B – The body is attempting to rid itself of excessive buildup of acid. (page 386)

28. C (page 387)

29. D (page 388)

Labeling

1. Brain (page 377)

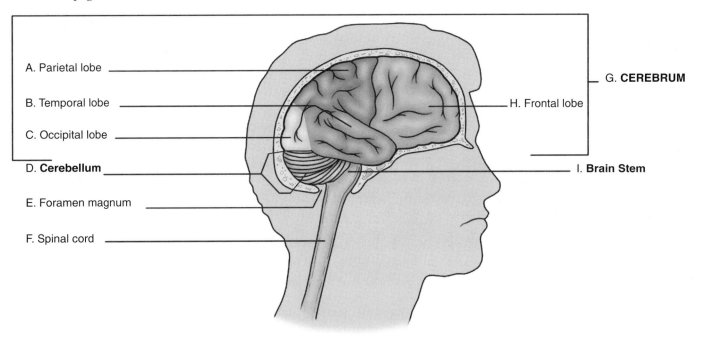

A. Parietal lobe

B. Temporal lobe

C. Occipital lobe

D. **Cerebellum**

E. Foramen magnum

F. Spinal cord

G. **CEREBRUM**

H. Frontal lobe

I. **Brain Stem**

2. Spinal Cord (page 377)

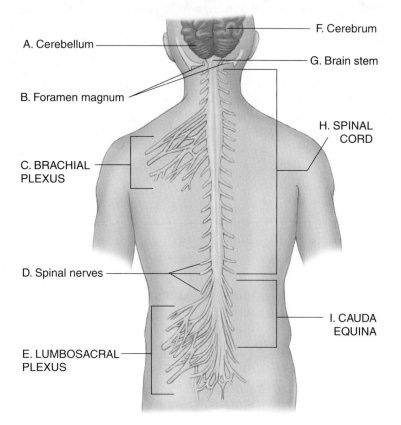

A. Cerebellum

F. Cerebrum

G. Brain stem

B. Foramen magnum

H. SPINAL CORD

C. BRACHIAL PLEXUS

D. Spinal nerves

I. CAUDA EQUINA

E. LUMBOSACRAL PLEXUS

Vocabulary

1. Cerebrovascular accident (CVA): An interruption of blood flow to the brain that results in the loss of brain function. (page 377)

2. Ischemic stroke: Occurs when blood flow to a particular part of the brain is cut off by blockage inside a blood vessel. (page 378)

3. Transient ischemic attack (TIA): Stroke symptoms that resolve spontaneously within 24 hours. (page 379)

4. Hemorrhagic stroke: Occurs as a result of bleeding inside the brain. (page 378)

5. Generalized seizure: Neurologic emergency characterized by unconsciousness and a generalized twitching of all the body's muscles that lasts several minutes or longer. (page 379)

6. Absence seizure: Seizure characterized by a brief lapse of attention during which the patient simply stares or does not respond to anyone. (page 379)

7. Atherosclerosis: A disorder in which cholesterol and calcium build up inside the walls of blood vessels, forming plaque, which eventually leads to partial or complete blockage of blood flow. (page 378)

8. Cerebral embolism: Obstruction of a cerebral artery caused by a clot that was formed elsewhere in the body and traveled to the brain. (page 378)

9. Febrile seizures: Convulsions that result from sudden high fevers, particularly in children. (page 379)

10. Thrombosis: Clotting of the cerebral arteries that may result in the interruption of cerebral blood flow and subsequent stroke. (page 378)

Fill-in

1. twelve (page 377)
2. cerebellum (page 376)
3. emotion and thought (page 376)
4. head (page 377)
5. three (page 376)
6. nerves (page 377)
7. opposite, same (page 376)
8. cerebrum (page 376)
9. Incontinence (page 380)
10. brain (page 376)
11. epidural (page 383)
12. hemiparesis (page 380 to 381)
13. altered mental status (page 382)

True/False

1. F (page 379)
2. F (page 379)
3. T (pages 379 to 380)
4. F (page 380)
5. T (page 383)
6. F (page 384)
7. F (page 385)

Short Answer

1. 1. Facial droop—Ask patient to show teeth or smile.
   2. Arm drift—Ask patient to close eyes and hold arms out with palms up.
   3. Speech—Ask patient to say, "The sky is blue in Cincinnati." (page 384)
2. Newer clot-busting therapies may be helpful in reversing damage in certain kinds of strokes, but treatment must be started within 3 hours after onset of the event. (page 388)
3. 1. Remove clothing.
   2. Spray/wipe with tepid water, particularly about the head/neck.
   3. Fan moistened areas. (page 388)
4. A period of time after a seizure, generally lasting from 5 to 30 minutes, that is characterized by some degree of altered mental status and labored respirations. (pages 378 to 379)
5. Infarcted cells are dead. Ischemic cells are still alive, although they are not functioning properly because of hypoxia. (page 378)
6. 1. Hypoglycemia
   2. Postictal state
   3. Subdural or epidural bleeding (page 383)

Word Fun

Ambulance Calls

1. High-flow oxygen.

Cool the patient by removing clothing and covering with wet towels.

Monitor vital signs.

Transport.

2. Maintain the airway—high-flow oxygen.

Suction if necessary or position lateral recumbent to clear secretions.

Check glucose level.

Rapid transport

3. High-flow oxygen.

Position of comfort—probably left side with the head elevated.

Protect the airway from secretions.

Monitor vital signs.

Rapid transport

Chapter 14: The Acute Abdomen

Matching

1. I (page 398)

2. D (page 394)

3. E (page 397)

4. M (page 397)

5. K (page 398)

6. A (page 394)

7. C (page 395)

8. J (page 395)

9. F (page 395)

10. B (page 394)

11. G (page 398)

12. N (page 397)

13. L (page 399)

14. O (page 394)

15. H (page 394)

Multiple Choice

1. C – Loss of body fluid into the abdominal cavity decreases the volume of circulating blood and may eventually cause hypovolemic shock. (page 395)

2. A – To gauge the degree of distention, simply look at the patient's abdomen. (page 395)

3. A – The mass or lump at times, will disappear back into the body cavity in which it belongs. If so, it is said to be reducible. If it cannot be pushed back within the body, it is said to be incarcerated. (page 399)

4. A – The spinal cord supplies sensory nerves to the skin and muscles; these nerves are called the somatic nervous system. The autonomic nervous system controls the abdominal organs and the blood vessels. (page 394)

5. D – Occasionally, an organ within the abdomen will be enlarged (swollen) and very fragile. The abdomen may be distended due to the swelling. Rough palpation could cause further damage and possibly rupture the organ. (page 397)

6. A – The aorta lies immediately behind the peritoneum on the spinal column. The patient may experience severe back pain due to the peritoneum being rapidly stripped away from the wall of the main abdominal cavity by the hemorrhage or the pressure of blood on the back itself. (page 398)

7. B – The kidneys, genitourinary structures, and large vessels (inferior vena cava, abdominal aorta) are found in the retroperitoneal space. The stomach, gall bladder, liver, pancreas, and uterus (answers A, C, D) are all found within the peritoneum. The adrenal glands (D) sit atop the kidneys in the retroperitoneal space. (page 397)

8. C – A hernia may result from a surgical wound that has failed to heal properly, a congenital defect, or a natural weakness in an area such as in the groin. (page 399)

9. B – The parietal peritoneum is supplied by the same nerves from the spinal cord that supply the skin of the abdomen; it can therefore perceive much the same sensations: pain, touch, pressure, heat, and cold. The visceral peritoneum is supplied by the autonomic nervous system and the nerves are far less able to localize sensation. (page 394)

10. D – All of these conditions (Answers A, B, C) may cause severe abdominal pain. (page 397)

11. C – The patient with peritonitis usually has abdominal pain, even when lying quietly. The patient may have difficulty breathing and may take rapid, shallow breaths because of the pain. (page 395)

12. C – Rebound tenderness and fever are signs and symptoms associated with inflammation of the peritoneum. The patient usually has abdominal pain, even when resting (A), and will present with hypotension and tachycardia (B) if associated with shock/fluid loss. (page 396)

13. C – Fever and distention (A, B) are common signs of peritonitis, but the degree of pain and tenderness is usually related directly to the severity of peritoneal inflammation. (page 395)

14. D – Pain associated with diverticulitis is usually felt in the left lower quadrant. (page 397)

15. C – The acute abdomen is associated with possible fluid loss and bleeding. You should anticipate the possible development of hypovolemic shock and be prepared to provide prompt treatment. (page 399)

Labeling

1. Solid Organs (page 396)

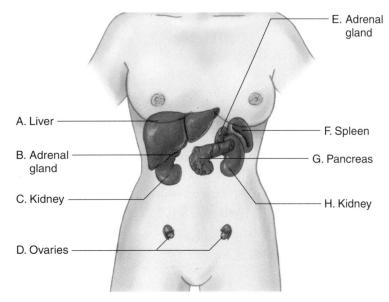

2. Hollow Organs (page 396)

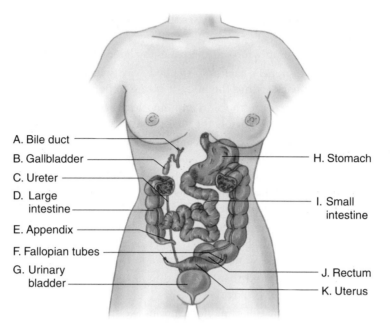

A. Bile duct
B. Gallbladder
C. Ureter
D. Large intestine
E. Appendix
F. Fallopian tubes
G. Urinary bladder
H. Stomach
I. Small intestine
J. Rectum
K. Uterus

3. Retroperitoneal Organs (page 397)

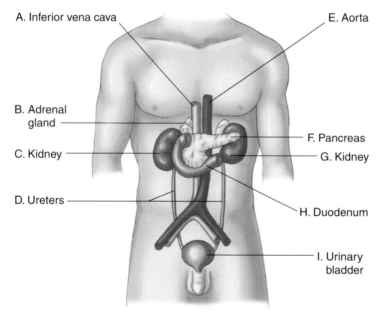

A. Inferior vena cava
B. Adrenal gland
C. Kidney
D. Ureters
E. Aorta
F. Pancreas
G. Kidney
H. Duodenum
I. Urinary bladder

Vocabulary

1. Acute abdomen: Sudden onset of abdominal pain indicating an irritation of the peritoneum. (page 394)

2. Diverticulitis: An inflammation of small pockets in the colon. (page 395)

3. Ectopic pregnancy: A fertilized egg implants outside the uterus. (page 398)

4. Cholecystitis: Inflammation of the gall bladder. (page 395)

5. Mittelschmerz: Common lower abdominal pain associated with the release of the egg from the ovary, characteristically occurring in the middle of the menstrual cycle. (page 398)

True/False

1. F   (page 394)
2. F   (page 398)
3. F   (page 397)
4. F   (page 394)
5. T   (page 394)

6. F   (page 395)
7. F   (page 397)
8. T   (page 398)
9. T   (page 398)

Short Answer

1. Occurs because of connections between the body's two nervous systems. The abdominal organs are supplied by autonomic nerves, which, when irritated, stimulate close-lying sensory (somatic) nerves. (page 394)

2. No. It is too complex and treatment is the same. (pages 394, 396)

3. Paralysis of muscular contractions in the bowel results in retained gas and feces. Nothing can pass through. (pages 394 to 395)

4. Bleeding and fluid shifts. (page 395)

5. Do not attempt to diagnose the cause.
- Clear and maintain the airway.
- Anticipate vomiting.
- Administer oxygen.
- Nothing by mouth.
- Document pertinent information.
- Anticipate shock.
- Keep comfortable.
- Monitor vital signs.

(page 399)

Word Fun

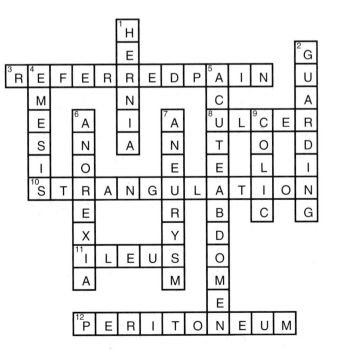

Ambulance Calls

1. Treat for hypovolemic shock.

   Place patient in the position of comfort.

   Apply high-flow oxygen.

   Keep patient warm.

   Rapid transport.

   Document information from the school nurse and try to obtain a SAMPLE history.

   Monitor vital signs and patient closely.

2. Possible appendicitis.

   Place patient in the position of comfort.

   Apply high-flow oxygen.

   Keep patient warm.

   Rapid transport.

   Obtain a SAMPLE history.

   Document OPQRST.

   Monitor patient closely.

3. Place patient in position of comfort.

   Apply high-flow oxygen.

   Normal transport.

   Obtain a SAMPLE history.

   Document OPQRST.

Chapter 15: Diabetic Emergencies

Matching

1. H   (page 404)
2. F   (page 406)
3. I   (page 406)
4. B   (page 405)
5. L   (page 406)
6. M   (page 404)
7. O   (page 407)
8. E   (page 406)
9. G   (page 405)
10. C   (page 406)
11. A   (page 408)
12. N   (page 404)
13. D   (page 408)
14. K   (page 405)
15. J   (page 404)

Multiple Choice

1. A – Patients with type I diabetes are more likely to have metabolic problems and organ damage, such as blindness, heart disease, kidney failure, and nerve disorders. (page 405)
2. A (page 405)
3. D – Diabetes mellitus is considered a metabolic disorder in which the body cannot metabolize glucose, usually because of the lack of insulin; the result is a wasting of glucose in the urine. (page 404)
4. C – The central problem in diabetes is the lack or ineffective action of insulin. (page 404)
5. B – Diabetes can have many severe complications, including blindness, cardiovascular disease, and kidney failure. (page 404)

6. A – When fat is used as an immediate energy source, chemicals called ketones and fatty acids are formed as waste products. As they accumulate in the blood and tissue, they can produce the dangerous form of acidosis seen in uncontrolled diabetes called diabetic ketoacidosis. (page 406)

7. C (page 406)

8. D – Insulin is a hormone that is normally produced by the pancreas that enables glucose to enter the cells. (page 404)

9. D – Diabetic coma occurs in the patient who is not under medical treatment, who takes insufficient insulin, who markedly overeats, or who is undergoing some sort of stress, such as infection, illness, overexertion, fatigue, or drinking alcohol. (pages 407 to 408)

10. B – With the exception of the brain, insulin is needed to allow glucose to enter individual body cells to fuel their functioning. (page 405)

11. B – A sweet or fruity (acetone) odor on the breath, caused by the unusual waste products in the blood (ketones). (page 408)

12. B (page 406)

13. D (page 405)

14. B – Diabetics must be willing to adjust their lives to the demands of the disease, especially their eating habits and activities. (page 404)

15. A – Glucose, or dextrose, is one of the basic sugars in the body. (page 404)

16. A (page 404)

17. A (page 406)

18. C – When fat is used as an immediate energy source, chemicals called ketones and fatty acids are formed as waste products and are hard for the body to excrete. (page 406)

19. B – Using a blood glucose self-monitoring unit is much simpler and more accurate than testing for glucose and acetone in the urine. Patients prick their finger and use a drop of blood. (page 406)

20. B – Insulin shock develops much more quickly than diabetic coma. In some instances, it can occur in a matter of minutes. (page 408)

21. D – Glucose is the major source of energy for the body, and all cells need it to function properly. A constant supply to the brain is as important as oxygen. (page 405)

22. C – Although brief seizures are not harmful, they may indicate a more dangerous and potentially life-threatening underlying condition. Kussmaul respirations and polydipsia (A, D) are signs of diabetic ketoacidosis. (page 411)

23. B – Normal is 80 to 120 mg/dL. (page 405)

24. A – Diabetic coma occurs in the patient who takes insufficient insulin. Too much insulin (B) results in insulin shock. (page 407)

25. B – If unable to perform a blood glucose test, you must always suspect hypoglycemia in any patient with altered mental status. (page 408)

26. C – The first step in caring for any patient is to perform an initial assessment to verify that the airway is open. All the others (A, B, D) are secondary to airway. (page 409)

27. D – You must use your knowledge of the signs and symptoms to decide whether the problem is diabetic coma or insulin shock when dealing with an unresponsive diabetic patient. However, this assessment should not prevent you from providing prompt treatment and transport (A, B, C). (page 409)

28. A – Confinement by police in a "drunk tank" because a person is thought to be intoxicated puts the hypoglycemic patient at risk of dying. Giving a patient who is hyperglycemic (B) or intoxicated (C) will not cause harm to the patient. However, lack of glucose to the brain can be fatal. (page 408)

29. A – The only contraindications to glucose are an inability to swallow or unconsciousness, since aspiration can occur. (page 410)

30. D – Maintain the airway and provide prompt transport to the hospital. (page 410)

31. D – All of the others (A, B, C) are signs of diabetic ketoacidosis. (page 408)

32. B – In insulin shock, the problem is hypoglycemia. The others (A, C, D) all equate to high glucose levels. (page 408)

33. C – Dehydration can be indicated by sunken eyes. Good skin turgor (A), would indicate sufficient hydration and elevated blood pressure (B), could indicate overhydration. (page 408)

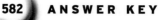

34. A – Diabetes may mask signs and symptoms of other problems. Any diabetic complaining of "not feeling well," with no mechanism of injury, should have their glucose level evaluated to rule out hypo- or hyperglycemia, followed by the appropriate medical assessment (B). As long as the mental status is intact, oral glucose (D) is not immediately indicated, nor is rapid transport to the closest facility (C). Transport, whether slow or rapid, should always be to the closest most appropriate facility. (page 408)

35. D (page 408)

36. A – Due to high energy levels and difficulty in keeping them on a strict schedule of medication and eating, management of children with diabetes poses a particularly troublesome task. (page 408)

37. C – Oral glucose (A) and a reduction in insulin (D) will cause an increase in glucose levels. Treatment includes a reversal of the condition under closely monitored conditions that may take many hours. (page 408)

38. C – If the patient has an altered mental status, a glucose test should be performed to rule out possible diabetic complications. (page 408)

39. B – Hypoglycemia (A) and insulin shock (C) are problems resulting from low glucose levels or high insulin levels. (page 408)

40. D (page 409)

41. D – You should NOT attempt to give anything by mouth to an unconscious patient. The risk of choking or aspirating liquid into the lungs outweighs the benefits of the small amount of glucose they would receive. (page 409)

42. D – Do not be afraid to give too much sugar. An entire candy bar or a full glass of sweetened juice is often needed. (page 410)

43. A – Medical protocols usually recommend sugar cubes, granulated sugar, honey, candy, oral glucose gel, etc. Do not give sugar-free drinks that are sweetened with saccharin or other synthetic sweetening compounds (B, C, D), as they will have little or no effect. (page 410)

Vocabulary

1. Diabetes mellitus: A metabolic disorder in which the body cannot metabolize glucose, usually because of the lack of insulin. (page 404)

2. Diabetes insipidus: A rare condition, also involves excessive urination, but here the missing hormone is one that regulates urinary fluid reabsorption. (page 404)

3. Juvenile diabetes: Type I diabetes, so named because it is the type that strikes children. (page 405)

4. Ketones: Chemicals, along with fatty acids, formed as waste products when the body uses fat as an immediate energy source. They are hard for the body to excrete and an accumulation in the blood and tissue can lead to acidosis. (page 406)

5. Glucometer: An elaborate machine, or unit, that automatically analyzes test strips and provides a digital readout of the blood glucose level. (page 406)

Fill-in

1. diabetes mellitus (page 404)

2. autoimmune (page 405)

3. ineffective (page 405)

4. diabetic coma (page 407)

5. sugar, insulin (page 410)

True/False

1. T   (page 406)

2. F   (page 411)

3. T   (page 405)

4. T   (page 408)

5. T   (page 407)

6. T   (page 404)

7. T   (page 405)

8. F   (page 405)

9. F   (page 404)

10. F   (page 405)

11. T   (page 405)

12. T   (page 405)

Short Answer

1. Insulin is a hormone that enables glucose to enter body cells. (page 404)

2. 1. Glucose

2. Insta-Glucose (page 409)

3. A patient who is unconscious or not able to swallow should not be given oral glucose. (page 409 to 410)

4. 1. Diabinase

2. Orinase

3. Micronase

4. Glucotrol (page 405)

5. 1. Ketoacidosis

2. Dehydration

3. Hyperglycemia (page 407)

6. Kussmaul respirations; dehydration; fruity odor on breath; rapid, weak pulse; normal or slightly low blood pressure; varying degrees of unresponsiveness. (page 408)

7. Insulin shock; it develops rapidly as opposed to diabetic coma, which takes longer to develop. (pages 407 to 408)

8. It will immediately benefit the patient in insulin shock and is unlikely to worsen the condition of the patient in a diabetic coma. (page 410)

Word Fun

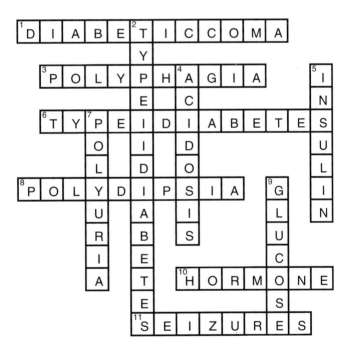

Ambulance Calls

1. Turn the patient on her side immediately or use suction to clear the airway.

Insert an oral or nasal airway and apply high-flow oxygen.

Attempt to obtain a blood glucose level.

Transport patient rapidly, since you should never give anything by mouth to an unresponsive patient.

Monitor the patient closely.

2. Consider cervical spine immobilization.

Apply high-flow oxygen.

Try to obtain a SAMPLE history, but do not delay treatment or transport.

Attempt to obtain a blood glucose level.

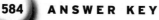

With permission from medical control, administer oral glucose.

Monitor the patient.

3. Apply high-flow oxygen and maintain the airway.

Obtain a SAMPLE history from the staff while loading the patient.

Transport in a semi-Fowler's position.

Support vital signs and monitor the patient en route to the hospital.

Skill Drills

1. Skill Drill 15-1: Administering Glucose (page 411)

1. Make sure that the tube of glucose is intact and has not **expired**.

2. Squeeze the entire tube of oral glucose onto the **bottom third** of a **bite stick** or tongue depressor.

3. Open the patient's **mouth**. Place the tongue depressor on the **mucous membranes** between the cheek and the gum with the **gel side** next to the cheek.

Comparison Table (page 407)

	Diabetic Coma	Insulin Shock
History Food intake Insulin dosage Onset Skin Infection	Excessive Insufficient Gradual Warm and dry Common	Insufficient Excessive Rapid, within minutes Pale and moist Uncommon
Gastrointestinal Tract Thirst Hunger Vomiting	Intense Absent Common	Absent Intense Uncommon
Respiratory System Breathing Odor of breath	Rapid, deep (Kussmaul respirations) Sweet, fruity	Normal or rapid Normal
Cardiovascular System Blood pressure Pulse	Normal to low Normal or rapid and full	Low Rapid, weak
Nervous System Consciousness	Restless merging to coma	Irritability, confusion, seizure, or coma
Urine Sugar Acetone	Present Present	Absent Absent
Treatment Response	Gradual, within 6 to 12 hours following medication and fluid	Immediately after administration of glucose

Chapter 16: Allergic Reactions and Envenomations

Matching

1. E   (page 416)
2. B   (page 416)
3. I   (page 416)
4. J   (page 421)
5. F   (page 416)

6. C   (page 419)
7. H   (page 416)
8. D   (page 418)
9. A   (page 423)
10. G   (page 418)

Multiple Choice

1. C – (A) Epinephrine can be given repeatedly. (B) The patient cannot be given too much oxygen. (page 422)

2. D – Side effects may also include increased blood pressure, pallor, dizziness, headache, vomiting, and anxiety. (page 421)

3. D – Black widows prefer dry, dim places and are found in every state except Alaska. (page 423)

4. A – (B) The tip is placed against the lateral thigh. (C) Needles are never to be recapped. (page 420)

5. A – Coral snakes live primarily in Florida and in the desert Southwest. (page 426)

6. B – Cytotoxins cause local tissue damage, whereas neurotoxins affect the nervous system. (page 423)

7. B – Ticks are only a fraction of an inch long and can easily be mistaken for a freckle due to their brown color, small size, and round shape. (page 427)

8. A – Rocky Mountain spotted fever and Lyme disease are both spread through the tick's saliva, which is injected into the skin when the tick attaches itself. (page 427)

9. D – Additional signs, which may or may not occur, include weakness, sweating, fainting, and shock. (pages 425 to 426)

10. C – (A) Suffocating it with gasoline or (B) trying to burn it with a lighted match will only succeed in burning the patient. (C) Pulling it straight out will usually remove the whole tick. (page 428)

11. D – Almost any substance can trigger the body's immune system and cause an allergic reaction: animal bites, food, latex gloves, or even semen can be an allergen. (page 416)

12. B – The wasp's stinger is unbarbed, meaning that it can inflict multiple stings. (page 417)

13. A – Severe cases can rapidly result in death. (page 416)

14. D – The patient may also experience sudden pain, redness, and itching. (page 418)

15. C – The patient becomes tachycardic due to the body's attempt to circulate the decreased oxygen supply to the brain and other vital organs. (page 419)

16. B – More than two thirds of patients who die of anaphylaxis do so within the first half hour, so speed on your part is essential. (pages 418 to 419)

17. D – It is also important to determine what the effects of the exposure have been and how they have progressed, what interventions have been completed, and if the patient has prescribed medication for allergic reactions. (page 419)

18. D – Symptoms may also include fainting. (page 429)

19. D – Maintaining the ABCs is essential for all patients. (page 423)

20. D (page 426)

21. C – 0.15 mg (B) is the pediatric dose. (page 420)

22. D – Epinephrine constricts the blood vessels, which raises blood pressure. Epinephrine also raises the pulse rate, inhibits allergic reactions, and dilates the bronchioles. (page 420)

23. A – In some cases, a bite on the abdomen causes muscle spasms so severe that the patient may be thought to have an acute abdomen, possibly peritonitis. (page 423)

24. B – Epinephrine (C) should only be administered in extreme reaction cases, and the patient with no signs or symptoms should be placed in the position of comfort (A). (page 422)

25. C – Localized swelling and ecchymosis (A, B) are local signs as opposed to systemic. (page 426)

26. B – Systemic signs of envenomation by a coral snake include (A, C, D) respiratory problems, bizarre behavior, and paralysis of the nervous system. (page 426)

27. C – Because the stinger of the honeybee remains in the wound, it can continue to inject venom for up to 20 minutes after the bee has flown away. (page 418)

28. A – Generally, you should not use tweezers or forceps, as squeezing may cause the stinger to inject still more venom into the wound. (page 418)

29. D – Any patient experiencing an allergic reaction should receive a complete assessment after maintaining the ABCs. Be sure to assess mental status as well. (page 419)

30. C – By going through the digestive tract it takes longer to get into the system. The person may also be unaware of the exposure or inciting agent. (page 417)

31. C – Certain allergens may cause swelling of the airway resulting in partial or complete obstruction. (page 416)

32. A – Wheezing occurs because excessive fluid and mucus are secreted into the bronchial passages, and muscles around these passages tighten in reaction to the allergen. (page 419)

Vocabulary

1. Anaphylaxis: An extreme, possibly life-threatening systemic allergic reaction that may include shock and respiratory failure. (page 416)

2. Histamine: A substance released by the immune system in allergic reactions that is responsible for many of the symptoms of anaphylaxis. (page 416)

3. Epinephrine: A substance produced by the body (adrenaline), and a drug produced by pharmaceutical companies that increases pulse rate and blood pressure; the drug of choice for an anaphylactic reaction. (page 420)

4. Envenomation: When an insect bites and injects the bite with its venom. (page 417)

5. Rabies: An acute, fatal viral infection of the central nervous system that can affect all warm-blooded animals. (page 428)

6. Nematocysts: The stinging cells of the coelenterates. (page 429)

Fill-in

1. bronchial passages (page 419)
2. urticaria (page 416)
3. barbed (page 418)
4. systemic (page 417)
5. hypoperfusion (page 419)
6. supine (page 419)
7. bronchioles (page 420)
8. imminent death (page 416)
9. rabid (page 428)

True/False

1. F   (page 420)
2. F   (page 425)
3. T   (page 425)
4. F   (page 428)
5. F   (page 429)
6. T   (page 416)
7. T   (page 416)
8. T   (page 419)
9. F   (page 419)
10. F   (page 429)

Short Answer

1. Increased blood pressure, tachycardia, pallor, dizziness, chest pain, headache, nausea, vomiting (page 421)

2. 1. Insect bites/stings
   2. Medications
   3. Plants
   4. Food
   5. Chemicals (page 417)

3. 1. Obtain order from medical control.

   2. Follow BSI techniques.

   3. Make sure medication was prescribed for that patient.

   4. Check for discoloration or expiration of medications.

   5. Remove cap.

   6. Wipe thigh with alcohol if possible.

   7. Place tip against lateral midthigh.

   8. Push firmly until activation.

   9. Hold in place until medication is injected.

   10. Remove and dispose.

   11. Record the time and dose.

   12. Reassess and record patient's vital signs. (page 420)

4. Respiratory: Sneezing or itchy, runny nose; chest or throat tightness; dry cough; hoarseness; rapid, noisy, or labored respirations; wheezing and/or stridor

Circulatory: Decreased blood pressure; increased pulse (initially); pale skin and dizziness; loss of consciousness and coma (page 419)

5. 1. Have the patient lie flat and stay quiet.

   2. Wash the bite area with soapy water.

   3. Splint the extremity.

   4. Mark the skin with a pen to monitor advancing swelling. (page 426)

6. Black widow: Bite has a systemic effect (venom is neurotoxic).

Brown recluse: Bite destroys tissue locally (venom is cytotoxic). (page 423)

7. Because dog and human mouths contain virulent bacteria. (page 428)

8. 1. Limit further discharge of nematocysts.

   2. Keep the patient calm.

   3. Reduce motion of the extremity.

   4. Apply alcohol (isopropyl or rubbing, or any kind available).

   5. Remove remaining tentacles by carefully scraping.

   6. If necessary, immerse injury in hot water for 30 minutes.

   7. Transport. (page 429)

Word Fun

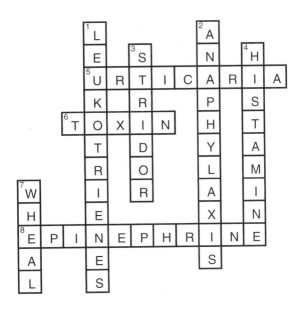

Ambulance Calls

1. - Cover the bites with dry, sterile dressings
   - Splint the extremity
   - Prompt transport
   - Reassure the patient
   - Monitor vital signs

2. - Provide BLS—high-flow oxygen
   - Treat for anaphylaxis
   - Rapid transport

3. - Check the EpiPen for clarity, expiration date, etc.
   - Obtain a physician's order to administer the EpiPen to the patient
   - Administer the EpiPen and promptly dispose of auto-injector
   - Apply high-flow oxygen
   - Rapid transport
   - Monitor the patient and assess vital signs frequently

Skill Drills

1. Skill Drill 16-1: Using an Auto-Injector (page 421)

1. Remove the auto-injector's **safety cap** and quickly wipe the thigh with **antiseptic**.
2. Place the tip of the auto-injector against the **lateral** thigh.
3. Push the **injector** firmly against the **thigh** and hold it in place until all the medication is injected.

2. Skill Drill 16-2: Using an AnaKit (page 422)

1. Prepare the injection site with antiseptic and remove the needle cover.
2. Hold the syringe upright and carefully use the plunger to remove air.
3. Turn the plunger one-quarter turn.
4. Quickly insert the needle into the muscle.
5. Hold the syringe steady and push the plunger until it stops.
6. Have the patient take the Chlo-Amine tablets provided in the kit
7. If available, apply a cold pack to the sting site.

Chapter 17: Substance Abuse and Poisoning

Matching

1. G   (page 434)
2. I   (page 434)
3. H   (page 435)
4. E   (page 439)
5. K   (page 444)
6. J   (page 434)

7. L   (page 440)
8. F   (page 442)
9. B   (page 434)
10. A   (page 441)
11. D   (page 447)
12. C   (page 444)

Multiple Choice

1. B (page 437)
2. B – The presence of such injuries at the mouth strongly suggests the swallowing of a poison, such as lye. (page 434)

3. D – Assess ABCs (A) in all patients. Take the plant with you (B) for identification of the poison, and always provide prompt transport (C) for suspected or known poisonings. (page 448)

4. C – You cannot treat the patient if you become a victim. Maintaining the airway (A) and applying oxygen (B) come after scene safety. (page 445)

5. D – In addition to these, clues to help determine the nature of the poison include an overturned bottle, an overturned or damaged plant, the remains of any food or drink found nearby, and any containers such as pill bottles. (page 435)

6. C – You can administer oxygen for inhaled poisons (B), and give activated charcoal for ingested poisons (A). You can also flush the skin/eyes for absorbed poisons (D), but it is very difficult to remove or dilute injected poisons (C). (page 436)

7. C – It may also cause nausea and vomiting. (page 439)

8. B – Alcohol dulls the sense of awareness (A) and decreases reaction time (C). (page 440)

9. C – Heroin (A) and morphine (B) are natural, not synthetic. (page 440)

10. D – Maintain the ABCs of all patients and provide rapid transport (C) for any respiratory problem. (page 442)

11. A – Anticholinergics are antagonists to the parasympathetic (A) division of the autonomic nervous system. (page 444)

12. C – In the smoked (C) form, crack reaches the capillary network of the lungs and can be absorbed into the body in seconds. (page 443)

13. D (page 444)

14. D (page 435)

15. B – (A) and (C) are symptoms of poisoning by botulism. (page 447)

16. A (page 442)

17. D (page 443)

18. D – Almost any substance can be abused. (page 439)

19. D – Patients who are unable to swallow (C), or have a decreased LOC (B), may aspirate the charcoal into their lungs. (page 439)

20. D – These would be unusual responses to the drug and require assessment in an emergency department. (page 443)

21. C – Adrenalin causes an increase in the heart's rate, automaticity, contractility, and conductivity, thereby significantly increasing the workload and oxygen demand. (page 442)

22. B – (A and C) are signs seen with depressant use. (page 442)

23. D – The patient's signs and symptoms are a direct result of the hypoxia. (page 437)

24. C – Chlorine is very irritating and can cause airway obstruction as well as pulmonary edema. (page 437)

25. B – (A) is not a sign or symptom, but part of the interview process. (C) Dyspnea is a systemic rather than localized problem. (page 438)

26. C (page 438)

27. D – Do not use water to flush the skin as it will ignite these substances and cause severe burns. (page 439)

28. A (page 438)

29. D – Many injuries and illnesses cause the patient to have an altered mental status. Remember to do a thorough assessment on each patient including checking glucose levels (C), mechanism of injury (A), and obtaining a complete history of the illness (B) when possible. (page 440)

30. D – Also confusion, disorientation, delusions, and hallucinations. (page 440)

31. A – Move the patient into fresh air (A) immediately. Rescuers should wear SCBAs (B) if indicated by the situation. (page 437)

32. D – They may also present with burning eyes, sore throat, hoarseness, headache, respiratory distress, dizziness, confusion, stridor, seizures, or an altered mental status. (page 437)

33. D (page 436)

34. A – Accounts for approximately 80%. (page 436)

35. D (page 437)

36. A – Venom (B) is injected and dieffenbachia (C) is ingested. (page 437)

37. A (page 437)

38. C – Most poisons do not have a specific antidote (A). Syrup of ipecac (D) is only used for specific instances. Oxygen (B) is never contraindicated, but it will not solve the problem this patient is experiencing. (page 436)

Vocabulary

1. Sedative-hypnotic: A drug class that produces CNS depression and an altered level of consciousness with effects similar to those of alcohol. (page 441)
2. Anticholinergic: Drug that blocks the parasympathetic nerves, such as atropine, Benadryl, jimsonweed, and certain cyclic antidepressants. (page 444)
3. Delirium tremens: Syndrome seen with alcohol withdrawal, characterized by restlessness, fever, sweating, disorientation, agitation, and convulsions. (page 440)
4. Hallucinogen: An agent that produces false perceptions in any one of the five senses. (page 443)
5. Addiction: An overwhelming desire or need to continue using a drug or agent. (page 439)
6. Substance abuse: The knowing misuse of any substance to produce a desired effect. (page 434)
7. Hypnotic: A sleep-inducing effect or agent. (page 440)

Fill-in

1. ABCs (page 437)
2. alcohol (page 440)
3. adsorbing (page 439)
4. 5 to 10 minutes, 15 to 20 minutes (page 438)
5. respiratory depression (page 440)
6. hypoglycemia (page 440)
7. recognize (page 434)
8. 1 gram, kilogram (page 439)
9. outward (page 438)
10. ingestion (page 436)
11. delirium tremens (DTs) (page 440)
12. ignite (page 439)
13. addiction (page 439)
14. Hypovolemia (page 440)

True/False

1. T   (page 439)
2. F   (pages 436, 439)
3. F   (pages 436 to 437)
4. F   (page 437)
5. F   (page 439)
6. T   (page 441)
7. T   (page 444)
8. F   (page 440)
9. T   (page 440)
10. T   (page 443)

Short Answer

1. Activated charcoal adsorbs (binds to) the toxin and keeps it from being absorbed in the gastrointestinal tract. (page 437)
2. 1. Ingestion
   2. Inhalation
   3. Injection
   4. Absorption (page 436)
3. Hypertension, tachycardia, dilated pupils, and agitation/seizures (page 443)
4. 1. The organism itself causes the disease.
   2. The organism produces toxins that cause disease. (page 446)
5. Symptoms of acetaminophen overdose do not appear until the damage is irreversible, up to a week later. Finding evidence at the scene can save a patient's life. (page 445)
6. They describe patient presentation in cholinergic poisoning (organophosphate insecticides, wild mushrooms).
   DUMBELS: Defecation, urination, miosis, bronchorrhea, emesis, lacrimation, salivation
   SLUDGE: Salivation, lacrimation, urination, defecation, gastrointestinal irritation, eye constriction/emesis (page 445)
7. 1. Opioid analgesics (page 440)
   2. Sedative-hypnotics (page 441)
   3. Inhalants (page 442)
   4. Sympathomimetics (page 442)
   5. Hallucinogens (page 443)

6. Anticholinergic agents (page 444)

7. Cholinergic agents (page 444)

8. 1. What substance did you take?

2. When did you take it or become exposed to it?

3. How much did you ingest?

4. What actions have been taken?

5. How much do you weigh? (page 434)

9. Because they ignite when they come into contact with water. (page 439)

Word Fun

Ambulance Calls

1. Maintain c-spine control.

Assess ABCs and provide support.

Control bleeding if necessary.

Check glucose level.

Monitor vital signs and transport patient promptly due to alteration in mental status.

2. Assess the ABCs.

Keep the patient as calm as possible.

Take the plant, or at least part of it, with you for identification.

Monitor the vital signs and provide support en route.

Rapid transport

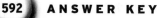
3. Maintain the airway with an adjunct and high-flow oxygen via BVM or nonrebreathing mask with 100% oxygen.

Monitor vital signs and provide supportive measures.

Rapid transport.

Take the pill bottle along to the emergency department.

Be alert for possible vomiting.

Monitor patient closely and be prepared for the possible need for CPR.

Chapter 18: Environmental Emergencies

Matching

1. J   (page 454)
2. H   (page 469)
3. K   (page 454)
4. F   (page 460)
5. A   (page 467)
6. N   (page 455)
7. I   (page 454)

8. C   (page 464)
9. B   (page 461)
10. E   (page 454)
11. G   (page 454)
12. M   (page 460)
13. L   (page 461)
14. D   (page 464)

Multiple Choice

1. D – With convection (A), heat is transferred directly to circulating air. Conduction (B) is the direct transfer of heat from the body to a colder object by touch. Radiation (C) is the loss of body heat directly to colder objects in the environment. (page 454)

2. A (page 454)

3. D (page 455)

4. C – Venous blood is naturally low in oxygen (A). Frostbite (B) causes a white, waxy appearance. (page 455)

5. C – Confusion is seen in moderate hypothermia. (pages 455 to 456)

6. D (page 455)

7. A – It is necessary to assess the trunk of the body to get a true feel for the extent of the cold emergency. The cooler the core temperature, the more serious the emergency. (page 456)

8. C – The golden rule of hypothermia states, "They are not dead until they are warm and dead." Patients may survive even severe hypothermia if proper emergency measures are carried out. (page 456)

9. D (page 457)

10. B – The patient could be bradycardic and initiating chest compressions could cause cardiac arrhythmias. (page 457)

11. D (page 458)

12. B (page 458)

13. C – (A), (B), and (D) are all localized signs and symptoms. (page 455)

14. A – (B), (C), and (D) are conditions that result from hyperthermia. (page 460)

15. D (page 460)

16. D (page 461)

17. A (page 461)

18. D (page 462)

19. A – The patient should also be transported if the level of consciousness decreases; if the temperature remains elevated; or if the person is very young, elderly, or has any underlying medical condition, such as diabetes, cardiovascular disease, or another worrisome condition. (page 462)

20. A (page 462)

21. D (page 462)

22. B (page 464)

23. D (page 465)

24. A – Take care to stabilize and protect the patient's spine since associated cervical spine injuries are possible. (page 465)

25. C (page 465)

26. D (page 465)

27. D (page 467)

28. B – (A) Heat is transferred from the body to the water, resulting in hypothermia. (C) Hypothermia lowers the metabolic rate in an effort to preserve body heat. (page 467)

29. B (page 469)

30. A – The skin, joints, and vision are areas with signs and symptoms of air embolism. (page 469)

31. D (page 469)

32. A – The brain and spinal cord require a constant supply of oxygen. (page 469)

33. B (page 471)

Vocabulary

1. Hyperbaric chamber: A chamber pressurized to more than atmospheric pressure for the treatment of diving injuries (page 470)

2. Decompression sickness: Condition seen in divers in which gas, usually nitrogen, forms bubbles that obstruct blood vessels. (page 470)

3. Heat exhaustion: The result of the body losing so much water and so many electrolytes through very heavy sweating that hypovolemia occurs. (page 461)

4. Frostbite: The most serious local cold injury. Because the tissues are actually frozen, the freezing permanently damages the cells. (page 458)

5. Near drowning: Survival, at least temporarily, after suffocation in water. (page 464)

6. Pneumomediastinum: Air entering the mediastinum and resulting in pain and severe dyspnea. (page 469)

Fill-in

1. moderate to severe (page 457)

2. ascent (page 469)

3. rewarming (page 460)

4. self-protection (page 460)

5. Shivering (page 455)

6. diving reflex (page 467)

7. mild (page 457)

True/False

1. T (page 460)

2. T (pages 456)

3. F (page 456)

4. T (page 460)

5. T (page 461)

6. T (page 462)

7. F (page 471)

8. T (page 467)

9. T (page 462)

Short Answer

1. 1. Increase heat production: shiver, jump, walk around, etc.

2. Move to another area where heat loss decreases: out of wind, into sun, etc.

3. Wear insulated clothing: layer with wool, down, synthetics, etc. (page 455)

2. 1. Move the patient out of the hot environment and into the ambulance.

2. Set air conditioning to maximum cooling.

3. Remove patient's clothing.

4. Administer high-flow oxygen.

5. Apply cool packs to patient's neck, groin, and armpits.

6. Cover patient with wet towels, or spray with cool water and fan.

7. Keep fanning.

8. Transport immediately.

9. Notify the hospital. (pages 463 to 464)

3. An air embolism is a bubble of air in the blood vessels caused by breath-holding during rapid ascent. The resulting high pressure in the lungs causes alveolar rupture. (page 469)

4. Treatment of air embolism and decompression sickness (page 470)

5. 1. Remove the patient from the cold.

2. Handle injured part gently and protect from further injury.

3. Administer oxygen.

4. Remove wet or restricting clothing. (page 459)

6. 1. Do not break blisters.

2. Do not rub or massage area.

3. Do not apply heat or rewarm unless instructed by medical control.

4. Do not allow patient to stand or walk on a frostbitten foot. (page 460)

7. Blotching; froth at nose and mouth; severe pain in muscles, joints, abdomen; dyspnea and/or chest pain; dizziness; nausea; vomiting; dysphasia; difficulty with vision; paralysis and/or coma; and irregular pulse with possible cardiac arrest. (pages 469 to 470)

Word Fun

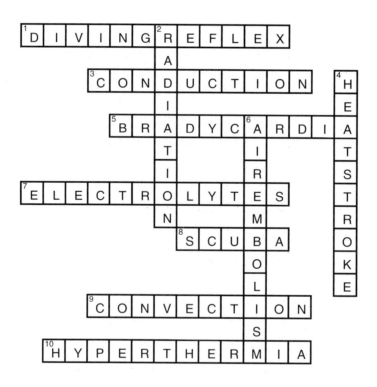

Ambulance Calls

1. Provide BLS.

Administer oxygen.

Transport the patient in the left lateral recumbent position with the head down.

Transport to a facility with hyperbaric chamber access.

2. Wrap patient in blankets.

Apply heat packs to the groin, axillary, and neck regions.

Turn the heat up in the ambulance.

Administer warmed/humidified oxygen.

Monitor vital signs.

3. Remove the patient from the hot environment, preferably into an air-conditioned office.

Rest—have the patient lie down with knees pulled up to relieve pressure on the abdomen.

Replace fluid loss with water, diluted Gatorade, etc.

Transport if patient still complains of cramps after this treatment.

Monitor vital signs.

Skill Drills

1. Skill Drill 18-1: Treating for Heat Exhaustion (page 463)

1. Remove **extra clothing**
2. Move the patient to a **cooler environment.**
   Give **oxygen.**
   Place the patient in a **supine** position, elevate the legs, and **fan** the patient.
3. If the patient is **fully alert**, give water by mouth.
4. If nausea develops, **transport** on the side.

2. Skill Drill 18-2: Stabilizing a Suspected Spinal Injury in the Water (page 468)

1. Turn the patient to a supine position by rotating the entire upper half of the body as a single unit.
2. As soon as the patient is turned, begin artificial ventilation using the mouth-to-mouth method or a pocket mask.
3. Float a buoyant backboard under the patient.
4. Secure the patient to the backboard.
5. Remove the patient from the water.
6. Cover the patient with a blanket and apply oxygen if breathing. Begin CPR if breathing and pulse are absent.

Chapter 19: Behavioral Emergencies

Matching

1. C   (page 479)
2. F   (page 482)
3. B   (page 479)
4. D   (page 479)
5. E   (page 480)
6. A   (page 478)

Multiple Choice

1. D (page 478)
2. B – A and C are both indications that the patient may indeed have a behavioral problem. (page 478)
3. D (page 478)
4. B (page 479)

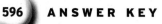

5. A (page 479)

6. B (page 479)

7. A (page 479)

8. D (page 479)

9. C – A, B, and D would have no influence on the mental status because they are all normal measures. (page 479)

10. D - All are correct because anything that creates a temporary or permanent dysfunction of the brain may cause OBS. (page 479)

11. A – (B) and (C) are both diseases or dysfunction of the brain. Schizophrenia (A) cannot be identified as a problem with the brain itself. (page 480)

12. B – (A) All findings should be documented objectively; (C) Quote the patient's own words using quotation marks. (page 479)

13. D – (A) Scene safety is top priority on any call; (B) It may take longer to assess, listen to, and prepare the patient for transport; (C) Help the patient to prepare by dressing and gathering appropriate belongings to take to the hospital. (page 480)

14. B – (A), (C), and (D) are all important aspects of the assessment, but scene safety (B) is first on any call. (page 481)

15. D (page 482)

16. C – (A) Substance abuse and (B) divorce may be considered risk factors. (page 483)

17. A – Suicidal patients may have no qualms about taking others with them who try to interfere with their plans. (page 483)

18. D – All of these as well as heat- and cold-related illnesses, poisoning or overdose, TIAs, and infection may cause altered behavior. (page 484)

19. D – Ordinarily, a restraint of a person must be ordered by a physician, a court order, or a law enforcement officer. (page 485)

20. C – Legal actions may involve charges of assault, battery, false imprisonment, and violation of civil rights. (page 485)

21. B – The airway (B) must be assessed frequently due to the possibility of obstruction from vomit or inability to maintain their own airway due to mental status or positioning. (page 486)

Vocabulary

1. Mental disorder: An illness with psychological or behavioral symptoms that may result in an impairment in functioning, caused by a social, psychological, genetic, physical, chemical, or biological disturbance. (page 479)

2. Activities of daily living (ADL): The basic activities a person usually accomplishes during the normal day. (page 479)

3. Altered mental status: A change in the way a person thinks or behaves. (page 480)

4. Implied consent: When a patient is not mentally competent to grant consent for emergency medical care, the law assumes that there is implied consent. Consent is implied because of the necessity for immediate emergency treatment. (page 484)

Fill-in

1. behavior (page 478)

2. behavioral crisis (page 479)

3. depression (page 479)

4. Organic brain syndrome (page 479)

5. coping mechanisms (pages 481 to 482)

6. suicide (page 482)

True/False

1. T   (page 478)
2. F   (page 479)
3. F   (page 479)
4. F   (page 485)
5. F   (page 480)
6. F   (page 482)

7. F   (page 483)
8. T   (page 484)
9. F   (page 478)
10. F   (page 479)
11. T   (page 485)

Short Answer

1. A behavioral crisis is a temporary change in behavior that interferes with ADL or that is unacceptable to the patient or others. A mental health problem is this kind of behavioral change recurring on a regular basis. (page 479)

2. 1. Improper functioning of the central nervous system.
   2. Drugs or alcohol
   3. Psychogenic circumstances (page 482)

3. 1. The degree of force necessary to keep the patient from injuring self or others
   2. Patient's gender, size, strength, and mental status
   3. The type of abnormal behavior the patient is exhibiting (pages 485 to 486)

4. 1. Be prepared to spend extra time.
   2. Have a definite plan of action.
   3. Identify yourself calmly.
   4. Be direct.
   5. Assess the scene.
   6. Stay with the patient.
   7. Encourage purposeful movement.
   8. Express interest in the patient's story.
   9. Do not get too close to the patient.
   10. Avoid fighting with the patient.
   11. Be honest and reassuring.
   12. Do not judge. (page 480)

5. 1. Depression at any age
   2. Previous suicide attempt
   3. Current expression of wanting to commit suicide or sense of hopelessness
   4. Family history of suicide
   5. Age older than 40 years, particularly for single, widowed, divorced, alcoholic, or depressed individuals
   6. Recent loss of spouse, significant other, family member, or support system
   7. Chronic debilitating illness or recent diagnosis of serious illness
   8. Holidays
   9. Financial setback, loss of job, police arrest, imprisonment, or some sort of social embarrassment
   10. Substance abuse, particularly with increasing usage
   11. Children of an alcoholic parent
   12. Severe mental illness
   13. Anniversary of death of loved one, job loss, marriage, etc.
   14. Unusual gathering or new acquisition of things that can cause death, such as purchase of a gun, a large volume of pills, or increased use of alcohol (page 483)

Word Fun

Ambulance Calls

1. - Call for police back up.
- Calmly speak to the patient without being judgmental.
- Stay out of his "personal space"—leave yourself an out.
- Try to obtain the patient's consent for treatment and transport.

2. - Apply high-flow oxygen via nonrebreathing mask for possible hypoxia.
- Attempt to obtain a blood glucose level to rule out hypoglycemia.
- Obtain vital signs.
- Monitor airway and vital signs en route.

3. - Be understanding and listen.
- Explain to the patient that she needs medical care.
- Monitor vital signs and reassure patient en route.

Chapter 20: Obstetrics and Gynecological Emergencies

Matching

1. D   (page 492)
2. C   (page 492)
3. L   (page 492)
4. B   (page 493)
5. N   (page 492)
6. H   (page 492)
7. M   (page 492)
8. E   (page 492)
9. F   (page 492)
10. O   (page 505)
11. K   (page 505)
12. I   (page 493)
13. A   (page 499)
14. G   (page 504)
15. J   (page 494)

Multiple Choice

1. D (pages 499 to 500)
2. B (page 499)
3. B – Aggressive suctioning of the baby's mouth and oropharynx before delivery of the baby may prevent meconium aspirations and respiratory distress. (page 499)
4. C – By suctioning the nose first (B), the baby may be stimulated to breathe and aspirate fluid into its lungs. (page 500)
5. D (page 500)
6. B (page 501)
7. A (page 502)
8. C (page 502)
9. D (page 503)
10. D – 90 compressions to 30 ventilations. (page 504)
11. A (page 505)
12. D (page 506)
13. B (page 507)
14. A – After scene safety, airway is the first priority for every patient. (page 509)
15. D (page 493)
16. B (page 493)
17. D (page 493)
18. A (page 493)
19. C (page 493)
20. C (page 493)
21. B – (A), (C), and (D) are signs of pre-eclampsia. (page 494)
22. B (page 494)
23. B (page 495)
24. D (page 495)
25. C – Never leave the mother once the decision has been made to deliver at the scene (A). The only acceptable reason for inserting fingers into the vagina is a breech delivery (to provide an airway) and prolapsed cord (B). (page 496)
26. A (page 494)
27. D (page 494)
28. D (page 505)
29. C (page 495)
30. B (page 495)
31. A (page 495)
32. C (page 507)

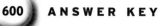

Tables

Apgar Scoring System

Area of Activity	Score		
	2	1	0
Appearance	Entire infant is pink.	Body is pink, but hands and feet remain blue.	Entire infant is blue and pale.
Pulse	More than 100 beats/min	Fewer than 100 beats/min	Absent pulse.
Grimace or Irritability	Infant cries and tries to move foot away from finger snapped against its sole.	Infant gives a weak cry in response to stimulus.	Infant does not cry or react to stimulus.
Activity or Muscle Tone	Infant resists attempts to straighten out hips and knees.	Infant makes weak attempts to resist straightening.	Infant is completely limp, with no muscle tone.
Respiration	Rapid respirations	Slow respirations	Absent respirations

Labeling

1. Anatomic Structures of the Pregnant Woman (page 492)

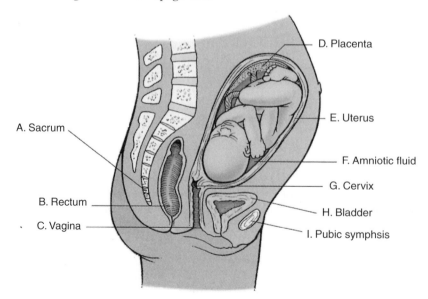

Vocabulary

1. Primigravida: A woman who is experiencing her first pregnancy. (page 493)
2. Multigravida: A woman who has had previous pregnancies. (page 493)
3. Ectopic pregnancy: A pregnancy that develops outside the uterus, typically in a fallopian tube. (page 494)
4. Crowning: The appearance of the top of the infant's head at the vaginal opening during labor. (page 493)
5. APGAR score: A scoring system for assessing the status of a newborn that assigns a number value to each of five areas of assessment. (page 502)
6. Bloody show: A plug of pink-tinged mucus that is discharged when the cervix begins to dilate. (page 492)

Fill-in

1. placenta (page 492)
2. arteries, vein (page 492)
3. 500 to 1000 mL (page 493)
4. 36, 40 (page 493)
5. trimesters (page 493)
6. body fluids (page 493)
7. ectopic pregnancy (page 494)
8. resuscitate (page 495)
9. fontanels (page 499)

True/False

1. T   (page 492)
2. F   (page 493)
3. F   (page 493)
4. F   (page 493)
5. F   (page 495)
6. T   (page 505)
7. T   (page 500)
8. F   (page 501)
9. F   (page 501)
10. T   (page 505)
11. T   (page 507)

Short Answer

1. Early: spontaneous abortion (miscarriage) or ectopic pregnancy
   Later: Placenta previa or placenta abruptio (page 494)
2. On the left side, to prevent supine hypotensive syndrome (low blood pressure occurring from the weight of the fetus compressing the inferior vena cava) (page 494)
3. 1. Uterine contractions
   2. Bloody show
   3. Rupture of amniotic sac (page 493)
4. 1. When delivery can be expected in a few minutes.
   2. When some natural disaster or catastrophe makes it impossible to reach the hospital.
   3. When no transportation is available. (page 495)
5. Exert gentle pressure on the head as it emerges to prevent rapid expulsion with a strong contraction. (page 499)
6. The brain is covered by only skin and membrane at the fontanels. (page 499)
7. Exerting gentle pressure horizontally across the perineum with a sterile gauze pad may reduce the risk of perineal tearing. (page 499)
8. 1. During a breech delivery to protect the infant's airway
   2. When the umbilical cord is prolapsed (pages 505, 506)
9. 1. Prematurity
   2. Low birth weight
   3. Severe respiratory depression (page 507)

Word Fun

Ambulance Calls

1. -Place the mother on her left side.
   -Maintain the airway, suctioning as needed.
   -Provide high-flow oxygen.
   -Monitor vital signs.
   -Protect the mother from harm if seizing starts again.
   -Rapid transport

2. -Place patient on her left side.
   -Give high-flow oxygen. The baby needs extra oxygen if the placenta is abruptio, or there are other serious conditions.
   -Place a sterile pad or sanitary napkin over the vagina.
   -Take any clots to the hospital.
   -Transport promptly.

3. -Position the mother for delivery and apply high-flow oxygen.
   -As crowning occurs, use a clamp to puncture the sac, away from the baby's face.
   -Push the ruptured sac away from the infant's face as the head is delivered.
   -Clear the baby's mouth and nose immediately.
   -Continue with the delivery as normal.

Skill Drills

1. Skill Drill 20-1: Delivering the Baby (page 498)

1. Support the **bony** parts of the head with your hands as it emerges.
   Suction fluid from the **mouth**, then **nostrils**.
   If the cord is wrapped around the neck, gently pull it over the head and off.
2. As the **upper shoulder** appears, guide the head **down** slightly, if needed to deliver the **shoulder**.
3. Support the head and upper body as the **lower shoulder** delivers, guiding the head **up** if needed.
4. Handle the slippery delivered infant firmly but gently, keeping the neck in **neutral** position to **maintain** the airway.
5. Place clamps **2"** to **4"** apart and **cut** between them.
6. Allow the **placenta** to deliver itself. Do not pull on the **cord** to speed delivery.

2. Skill Drill 20-2: Giving Chest Compressions to an Infant (page 504)

1. Find the proper position: just below the **nipple line**, middle of **lower third** of the sternum.
2. Wrap your hands around the body, with your **thumbs** resting at that position.
3. Press your thumb gently against the sternum, compressing ½" to ¾" deep.

Chapter 21: Kinematics of Trauma

Matching

1. H (page 524)
2. G (page 520)
3. F (page 516)
4. E (page 517)

5. C (page 517)
6. A (page 517)
7. D (page 524)
8. B (page 516)

Multiple Choice

1. B – Thermal energy causes burns. (page 516)
2. C (page 516)
3. C – Energy can be neither created nor destroyed. (page 516)
4. B (page 517)
5. A (page 517)
6. D – Rear-end and rotational are types of motor vehicle crashes as well. (page 517)
7. C (pages 517 to 518)
8. C (page 517)
9. D (page 518)
10. A (page 518)
11. C (page 518)
12. B (page 518)
13. D – Significant mechanisms of injury also include severe deformities of the frontal part of the vehicle, with or without intrusion to the passenger compartment. (page 520)
14. A – Other passengers have likely experienced the same amount of force that caused the passenger's death. (page 520)
15. D (page 520)
16. B – You should still suspect that other serious injuries to the extremities and to internal organs have occurred. (page 520)
17. C (page 521)
18. C (page 522)
19. B – The cervical spine has little tolerance for lateral bending. Approximately 25% of all severe injuries to the aorta that occur in motor vehicle crashes are a result of lateral collisions. (page 522)
20. C – Rollover crashes are particularly dangerous for both restrained and, to a greater degree, unrestrained passengers because these crashes provide multiple opportunities for second and third collisions. (page 522)
21. B (page 523)
22. D (page 523)
23. D (page 524)
24. B – This is one reason that exit wounds are often many times larger than entrance wounds. (page 524)
25. B (page 525)

Vocabulary

1. Newton's First Law: Objects at rest tend to stay at rest, and objects in motion tend to stay in motion unless acted upon by some force. (page 525)
2. Newton's Second Law: Force equals mass times acceleration (F = MA). (page 525)
3. Newton's Third Law: For every action, there is an equal and opposite reaction. (page 525)

Fill-in

1. Injuries (page 516)
2. injury patterns, injury events (page 516)
3. Traumatic injury (page 516)
4. $KE = \frac{1}{2} MV^2$ (page 516)
5. Polaroids (page 520)
6. deceleration (page 520)
7. submarining (page 520)

True/False

1. T   (page 516)
2. F   (page 516)
3. F   (page 516)
4. T   (page 522)
5. T   (page 522)
6. T   (page 523)
7. T   (page 516)

Short Answer

1. Potential energy is the product of mass (weight), force of gravity, and height, and is mostly associated with the energy of falling objects. (page 517)
2. 1. Collision of the car against another car or other object
   2. Collision of the passenger against the interior of the car
   3. Collision of the passenger's internal organs against the solid structures of the body (pages 517 to 518)
3. 1. The height of the fall
   2. The surface struck
   3. The part of the body that hits first, followed by the path of energy displacement (page 523)
4. A bullet, because of its speed, creates pressure waves that emanate from its path, causing distant damage. (page 524)
5. The size (mass) and speed (velocity) of the projectile affect the potential damage. If the mass is doubled, the potential energy is doubled. If the velocity is doubled, the potential energy is quadrupled. (page 524)

Word Fun

Ambulance Calls

1. -Maintain cervical spine control.

-Apply high-flow oxygen.

-Keep a high index of suspicion for life-threatening injuries.

-Monitor vital signs.

-Rapid transport

2. -Maintain cervical spine control.

-Apply high-flow oxygen.

-Fully immobilize the patient and splint the leg.

-Monitor vital signs.

-Normal transport

3. -Apply high-flow oxygen.

-Stabilize object in place with bulky dressings.

-Monitor vital signs.

-Transport in a supine position.

-Rapid transport due to abdominal penetration.

Chapter 22: Bleeding

Matching

1. J   (page 530)
2. E   (page 530)
3. K   (page 530)
4. F   (page 531)
5. C   (page 530)
6. H   (page 530)
7. B   (page 535)

8. I   (page 542)
9. M   (page 540)
10. A   (page 542)
11. D   (page 535)
12. L   (page 534)
13. G   (page 534)

Multiple Choice

1. D (page 530)
2. D (page 530)
3. C (page 530)
4. C (page 530)
5. B – Blood returns to the heart from the lungs via the pulmonary veins. (page 530)
6. A – Arteries carry blood away from the heart. The pulmonary arteries carry blood away from the heart to the lungs to be oxygenated. (page 530)
7. D – The left ventricle is the thickest chamber of the heart because it must pump blood throughout the body. (pages 530 to 531)
8. B (page 531)
9. D – The muscles at the arterial ends of the capillaries also dilate and constrict in response to heat and cold. (page 532)
10. C (page 532)
11. B (page 532)
12. C – The lungs (A) and kidneys (B) require a constant blood supply. (page 533)
13. C (page 533)
14. A – Shock is also called hypoperfusion. (page 533)
15. D (page 533)
16. B (page 533)
17. B (page 534)
18. D – They may also present with an altered mental status, cool/clammy skin, and cyanosis. (page 534)
19. A (page 534)
20. B (page 534)
21. C (page 534)
22. C (page 535)
23. D (page 535)
24. B (page 535)
25. C (page 535)
26. A (page 536)
27. A (page 536)
28. C (page 537)
29. D – Contraindications also include acute heart failure, groin injuries, and major head injuries. (pages 537 to 538)
30. C (page 538)
31. D – Never cover a tourniquet (B). Wide padding may actually help protect the tissues and help with arterial compression (C). (page 540)
32. D – Also, facial injuries, hypertension, and digital trauma. (page 540)

33. B (page 541)

34. B (page 542)

35. D – Nontraumatic internal bleeding in the abdomen can also result from irritable bowel syndrome. (page 542)

36. D – (A), (B), and (C) are signs, not symptoms. (page 542)

37. D – Also includes hematuria, hematemesis, nonmenstrual vaginal bleeding, and pain, tenderness, bruising, guarding, or swelling. (pages 542 to 543)

38. C – Blood pressure changes (D) are a late sign. (page 543)

Labeling

1. The left and right sides of the heart (page 531)

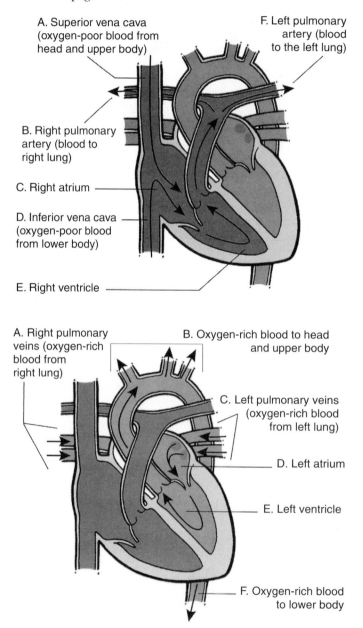

2. Perfusion (page 533)

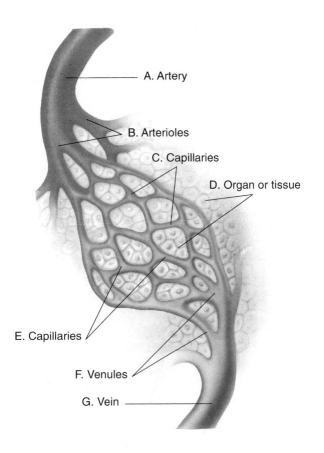

A. Artery

B. Arterioles

C. Capillaries

D. Organ or tissue

E. Capillaries

F. Venules

G. Vein

3. Arterial pressure points (page 537)

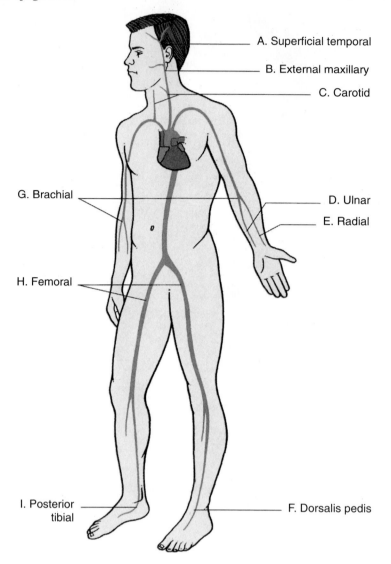

A. Superficial temporal

B. External maxillary

C. Carotid

G. Brachial

D. Ulnar

E. Radial

H. Femoral

I. Posterior tibial

F. Dorsalis pedis

Vocabulary

1. Perfusion: Circulation of blood within an organ or tissue in adequate amounts to meet the cells' current needs for oxygen, nutrients, and waste removal. (page 532)

2. Shock: A condition in which the circulatory system fails to provide sufficient circulation for every body part to perform its function. Low blood volume results in inadequate perfusion, also known as hypovolemic shock. (page 533)

3. Melena: Black, foul smelling, tarry stool that contains digested blood. (page 543)

4. Hematemesis: Vomited blood (page 543)

5. Hemoptysis: Bright red blood coughed up by a patient. (page 543)

Fill-in

1. perfusion (page 532)

2. involuntary (page 530)

3. lungs (page 530)

4. hypoperfusion (page 533)

5. inferior (page 530)

6. white cells, red cells, platelets, and plasma (page 532)

7. oxygenated (page 530)

8. heart, brain, lungs, and kidneys (page 533)

9. 4 to 6 (page 533)

True/False

1. F (pages 534, 535) **5.** F (page 540)

2. F (page 534) **6.** T (page 541)

3. F (page 535) **7.** T (page 537)

4. T (page 535) **8.** F (page 536)

Short Answer

1. It redirects blood away from nonessential organs to the heart, brain, lungs, and kidneys. (page 533)

2. Artery: Bright red, spurting

Vein: Dark color with steady flow

Capillary: Darker color, oozes (pages 534, 535)

3. 1. Direct pressure and elevation

2. Pressure dressings

3. Pressure points (for upper and lower extremities)

4. Splints

5. Air splints

6. PASG

7. Tourniquets (page 535)

4. 1. Change in mental status (restlessness, anxiety, combativeness)

2. Weakness, fainting, or dizziness on standing (early sign) or at rest (later sign)

3. Tachycardia

4. Thirst

5. Nausea and vomiting

6. Cold, moist (clammy) skin

7. Shallow, rapid breathing

8. Dull eyes

9. Slightly dilated pupils, slow to respond to light

10. Capillary refill in infants and children of more than 2 seconds

11. Weak, rapid (thready) pulse

12. Decreasing blood pressure

13. Altered level of consciousness (page 543)

5. 1. BSI

2. Maintain the airway.

3. Administer high-flow oxygen.

4. Control all obvious external bleeding.

5. Apply splints to extremity, if a limb is involved.

6. Monitor and record the patient's vital signs.

7. Give the patient nothing by mouth.

8. Elevate the legs 6 to 12 inches in nontrauma patients.

9. Keep the patient warm.

10. Provide immediate transport. (page 543)

Word Fun

Ambulance Calls

1. Control bleeding with direct pressure, elevation, pressure point, and a tourniquet as a LAST resort.
   Apply high-flow oxygen.
   Place patient in the position of comfort.
   Monitor vital signs.
   Rapid transport
2. High-flow oxygen.
   Maintain c-spine immobilizations and place patient in the Trendelenburg's position.
   Cover to keep warm.
   Consider the mechanism of injury.
   Monitor vital signs.
   Rapid transport
3. Control bleeding with direct pressure, elevation, pressure point, or tourniquet as a LAST resort.
   Apply high-flow oxygen and treat for shock as signs and symptoms present.
   Monitor vital signs.
   Normal transport unless patient shows signs and symptoms of shock.

Skill Drills

1. **Skill Drill 22-1: Controlling External Bleeding (page 536)**

1. Apply **direct pressure** over the wound.
   Elevate the injury above the **level** of the **heart** if no **fracture** is suspected.
2. Apply a **pressure dressing.**
3. Apply pressure at the appropriate **pressure point** while continuing to hold **direct pressure** and **elevation.**

2. **Skill Drill 22-2: Applying a Pneumatic Antishock Garment (PASG) (page 539)**

1. Apply the garment so that the top is below the lowest rib.

2. Enclose both legs and the abdomen.

3. Open the stopcocks.

4. Inflate with the foot pump, and close the stopcocks either when the patient's systolic blood pressure reaches 100 mm Hg or the Velcro crackles.

5. Check the patient's blood pressure again. Monitor vital signs.

Chapter 23: Shock

Matching

1. B   (page 548)

2. G   (page 548)

3. H   (page 548)

4. C   (page 548)

5. E   (page 548)

6. A   (page 552)

7. F   (page 551)

8. I   (page 552)

9. D   (page 553)

Multiple Choice

1. D (page 548)

2. B (page 548)

3. D (page 548)

4. C – It is accomplished by vessel constriction or dilation, together with sphincter constriction or dilation. (page 548)

5. D – Adequate waste removal is primarily through the lungs. (page 549)

6. D – C is peripheral vasoconstriction. (page 549)

7. D (page 549)

8. C – Sepsis (A) and metabolic (B) shock may result in fluid loss or poor pump function. Hypovolemia (D) is due to fluid loss. (page 550)

9. A (page 550)

10. B (page 550)

11. C (page 550)

12. A (pages 550 to 551)

13. C (page 551)

14. D (page 551)

15. A (page 551)

16. B (page 551)

17. A (page 552)

18. A (page 552)

19. D (page 552)

20. B (page 553)

21. A – With allergic reactions, you should suspect shock only in cases of anaphylaxis. (page 553)

22. B (page 555)

23. C – Because the first 60 minutes after the injury are critical, rapid evaluation, stabilization, and transport are important. (page 555)

24. B (page 557)

25. D – With too little circulating blood, additional oxygen may be lifesaving. (page 558)

26. B (page 559)

Vocabulary

1. Edema: The presence of abnormally large amounts of fluid between cells in body tissues, causing swelling of affected areas. (page 550)
2. Hypothermia: A condition in which the internal body temperature falls below 95°F/35°C, usually as a result of prolonged exposure to cool or freezing temperatures. (page 551)
3. Shock: A condition in which the circulatory system fails to provide sufficient circulation to enable every body part to perform its function. (page 548)
4. Autonomic nervous system: The part of the nervous system that regulates involuntary functions, such as digestion and sweating. (page 548)
5. Cyanosis: Bluish color of the skin resulting from poor oxygenation of the circulating blood. (page 552)
6. Dehydration: Loss of water from the tissues of the body. (page 551)
7. Sensitization: Developing a sensitivity to a substance that initially caused no allergic reaction. (page 552)

Fill-in

1. Hypoperfusion (page 548)
2. contraction (page 548)
3. nonessential, essential (page 548)
4. perfusion (page 548)
5. heart, vessels, blood (page 548)
6. shock (hypoperfusion) (page 548)
7. Sphincters, contract, dilate (page 548)
8. Diastolic, systolic (page 548)
9. Blood (page 548)
10. involuntary (page 548)

True/False

1. T (page 552)
2. T (page 551)
3. T (page 548)
4. F (page 548)
5. F (page 553)
6. T (page 558)
7. T (page 551)

Short Answer

1. Causes: allergic reaction (most severe form)

   Signs/Symptoms: Can develop within seconds; mild itching/rash; burning skin; vascular dilation; generalized edema; profound coma; rapid death

   Treatment: Manage airway. Assist ventilations. Administer high-flow oxygen. Determine cause. Assist with administration of epinephrine. Transport promptly.
   (pages 552, 556, 557)

2. Causes: Inadequate heart function; disease of muscle tissue; impaired electrical system; disease or injury

   Signs/Symptoms: Chest pains; irregular pulse; weak pulse; low blood pressure; cyanosis (lips, under nails); anxiety

   Treatment: Position comfortably. Administer oxygen. Assist ventilations. Transport promptly.
   (pages 550, 556, 557)

3. Causes: Loss of blood or fluid

   Signs/Symptoms: Rapid, weak pulse; low blood pressure; change in mental status; cyanosis (lips, under nails); cool, clammy skin; increased respiratory rate

   Treatment: Secure airway. Assist ventilations. Administer high-flow oxygen. Control external bleeding. Elevate legs. Keep warm. Transport promptly.
   (pages 551, 556, 557)

4. Causes: Excessive loss of fluid and electrolytes due to vomiting, urination, or diarrhea

   Signs/Symptoms: Rapid, weak pulse; low blood pressure; change in mental status; cyanosis (lips, under nails); cool, clammy skin; increased respiratory rate

   Treatment: Secure airway. Assist ventilations. Administer high-flow oxygen. Determine illness. Transport promptly.
   (pages 556, 557)

5. Causes: Damaged cervical spine, which causes widespread blood vessel dilation

   Signs/Symptoms: Bradycardia (slow pulse); low blood pressure; signs of neck injury

   Treatment: Secure airway. Spinal immobilization. Assist ventilations. Administer high-flow oxygen. Transport promptly. (pages 551, 556, 557)

6. Causes: Temporary, generalized vascular dilation; anxiety; bad news; sight of injury/blood; prospect of medical treatment; severe pain; illness; tiredness

   Signs/Symptoms: Rapid pulse; normal or low blood pressure

   Treatment: Determine duration of unconsciousness. Record initial vital signs and mental status. Suspect head injury if patient is confused or slow to regain consciousness. Transport promptly.

   (pages 552, 556, 557)

7. Causes: Severe bacterial infection

   Signs/Symptoms: Warm skin; tachycardia; low blood pressure

   Treatment: Transport promptly. Administer oxygen en route. Provide full ventilatory support. Elevate legs. Keep patient warm.

   (pages 551, 556, 557)

8. 1. Poor pump function

   2. Blood or fluid loss from blood vessels

   3. Poor vessel function (blood vessels dilate) (page 550)

Word Fun

Ambulance Calls

1. -Treat for shock (neurogenic).

-High-flow oxygen.

-Place patient in Trendelenburg's position.

-Keep patient warm.

-Rapid transport.

-Monitor vital signs continually en route.

-Consider PASG.

2. -Treat for shock (hypovolemic).

-High-flow oxygen via BVM.

-Place patient in Trendelenburg's position.

-Keep patient warm with consideration for burned areas.

-Consider PASG.

-Rapid transport.

-Monitor vital signs continually en route.

3. -Treat for anaphylactic shock.

-Apply high-flow oxygen while inquiring if the patient has an EpiPen.

-Obtain orders and administer EpiPen.

-Monitor vital signs.

-Rapid transport.

Skill Drills

1. **Skill Drill 23-1: Treating Shock (page 554)**

1. Keep the patient supine, open the airway, and check breathing and pulse.
2. Control obvious external bleeding.
3. Splint any broken bones or joint injuries.
4. Give high-flow oxygen if you have not already done so, and place blankets under and over the patient.
5. If no fractures are suspected, elevate the legs 6 to 12 inches.

Chapter 24: Soft-Tissue Injuries

Matching

1. G (page 564)

2. B (page 564)

3. D (page 564)

4. F (page 564)

5. I (page 564)

6. L (page 567)

7. E (page 568)

8. J (page 571)

9. A (page 569)

10. C (page 568)

11. K (page 572)

12. H (page 571)

Multiple Choice

1. C (page 564)
2. B (page 564)
3. B (page 564)
4. A (page 564)
5. B (page 565)
6. C (page 566)
7. B (page 566)
8. B (page 566)
9. D (page 566)
10. C (page 566)
11. C (page 567)
12. C (page 567)
13. B (page 567)
14. A (page 567)
15. D (page 568)
16. B (page 568)
17. D (page 569)
18. D (page 569)
19. A (page 570)
20. D (page 570)
21. D (page 570)
22. B (page 571)
23. C – Never touch exposed organs (A). Use moist sterile dressings (B). Never use adherent dressings (C). (page 572)
24. C – Hypovolemic shock (A) does not result from air being sucked in. Tracheal deviation (B) results from a tension pneumothorax. Subcutaneous emphysema (C) results from air outside the vessels. (page 574)
25. D (page 574)
26. D – Factors in helping to determine the severity of a burn also include whether or not there are any preexisting medical conditions or other injuries, and if the patient is younger than 5 years of age or older than 55 years of age. (page 575)
27. C (page 576)
28. B (page 576)
29. C (page 576)
30. D (page 576)
31. D – Significant airway burns may also be associated with soot around the nose and mouth. (page 577)
32. B – Personal safety always comes first. (page 581)
33. D – Apply high-flow oxygen (A). Be prepared to defibrillate (B). (page 582)
34. A – An occlusive dressing must be airtight; gauze pads (A) are not. (page 583)
35. D (page 583)

Labeling

1. The Skin (page 565)

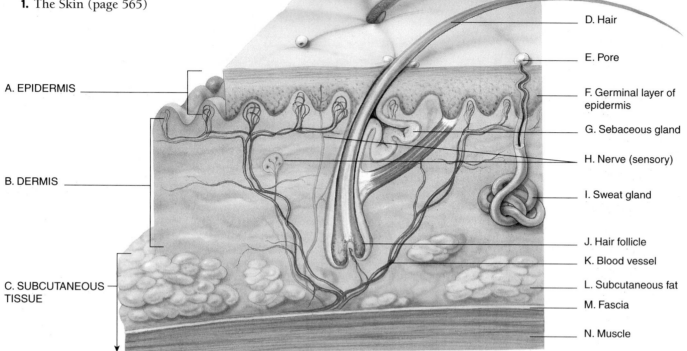

A. EPIDERMIS

B. DERMIS

C. SUBCUTANEOUS TISSUE

D. Hair

E. Pore

F. Germinal layer of epidermis

G. Sebaceous gland

H. Nerve (sensory)

I. Sweat gland

J. Hair follicle

K. Blood vessel

L. Subcutaneous fat

M. Fascia

N. Muscle

2. The Rule of Nines (page 578)

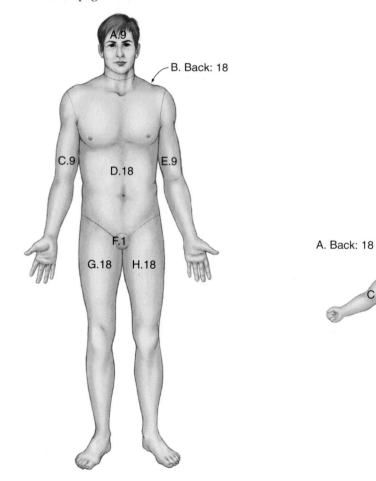

A.9

B. Back: 18

C.9

D.18

E.9

F.1

G.18 H.18

A. Back: 18

B.18

C.9

D.18

E.9

F.1

G.13.5 H.13.5

Vocabulary

1. Partial-thickness burn: A burn affecting the epidermis and some portion of the dermis; characterized by red, moist, mottled skin with blisters present. (page 576)
2. Closed injury: Injury in which damage occurs beneath the skin or mucous membrane, but the surface remains intact. (page 565)
3. Evisceration: An open wound in the abdominal cavity in which organs protrude through the wound. (page 572)
4. Compartment syndrome: Swelling in a confined space that produces dangerous pressure. (page 566)
5. Contamination: The presence of infective organisms or foreign bodies such as dirt, gravel, or metal. (page 567)

Fill-in

1. moist (page 565)
2. cool (page 565)
3. dermis (page 564)
4. subcutaneous (page 565)
5. constrict (page 565)
6. bacteria, water (page 565)
7. dermis (page 564)
8. temperature (page 565)
9. epidermis, dermis (page 564)
10. radiated (page 565)

True/False

1. T   (page 576)
2. T   (page 576)
3. T   (page 576)
4. T   (page 577)
5. T   (page 575)
6. T   (pages 577)
7. F   (page 577)
8. T   (page 581)
9. T   (page 582)
10. T   (page 583)
11. F   (page 583)
12. F   (page 583)
13. T   (page 583)
14. F   (page 566)
15. F   (page 568)

Short Answer

1. 1. superficial
   2. partial-thickness
   3. full-thickness (page 576)
2. 1. closed
   2. open
   3. burns (page 565)
3. I=ice
   C=compression
   E=elevation
   S=splinting (page 567)
4. -Full- or partial-thickness burns covering more than 20% of total body surface area
   -Burns involving the hands, feet, face, airway, or genitalia (page 577)
5. Brush off dry chemicals and/or remove clothing, then flush the burned area with large amounts of water. (page 580)
6. First, there may be deep tissue injury not visible on the outside. Second, there is a danger of cardiac arrest from the electrical shock. (pages 581 to 582)
7. A wound caused by a penetrating object into the chest that causes air to enter the chest. The air enters the chest area through the wound, but remains in the pleural space and the lung does not expand. With exhalation, air passes back through the wound, making a "sucking" sound. (page 571)
8. 1. Control bleeding
   2. Protect from further damage
   3. Prevent further contamination and infection (page 582)

9. 1. Abrasions

 2. Lacerations

 3. Avulsions

 4. Penetrating (pages 567 to 569)

10. 1. Depth (superficial/partial/full)

 2. Extent (% of body burned)

 3. Involvement of critical areas (face, upper airway, hands, feet, genitalia)

 4. Pre-existing medical conditions or other injuries

 5. Age of younger than 5 years or older than 55 years (page 575)

Word Fun

Ambulance Calls

1. -BSI precautions

 -Apply direct pressure.

 -Elevate the extremity and apply a pressure dressing.

 -Try a pressure point if bleeding is not controlled (tourniquet as a last resort).

 -Once bleeding is controlled, splint the arm to decrease movement.

 -Apply high-flow oxygen.

 -Transport in position of comfort, normal response.

 -Monitor vital signs.

2. -BSI precautions

    -Quickly assess ABCs.

    -Wrap foot with a pressure dressing to control bleeding.

    -Transport in position of comfort.

    -Standard transport.

    -Monitor vital signs en route.

    -Apply oxygen as needed.

    -Be alert for signs of hypovolemic shock.

3. -Apply high-flow oxygen, preferably humidified (dry oxygen may irritate the tissues further).

    -Cover burns with sterile dressings loosely.

    -Rapid transport in position of comfort.

    -Monitor airway and vital signs continuously.

    -Protect from hypothermia.

Skill Drills

1. Skill Drill 24-1: Controlling Bleeding from a Soft-Tissue Injury (page 571)

1. Apply **direct pressure** with a **sterile** bandage.
2. Maintain pressure with a **roller** bandage.
3. If bleeding continues, apply a second **dressing** and **roller** bandage over the first.
4. **Splint** the extremity.

2. Skill Drill 24-2: Sealing a Sucking Chest Wound (page 572)

1. Keep the patient **supine** and give **oxygen**.
2. Seal the wound with an **occlusive** dressing.
3. Follow **local protocol** regarding sealing or **leaving open** the dressing's fourth side

3. Skill Drill 24-3: Stabilizing an Impaled Object (page 573)

1. Do not attempt to **move** or **remove** the object.
2. Control **bleeding** and **stabilize** the object in place using **soft dressings**, **gauze**, and/or **tape**.
3. Tape a **rigid** item over the stabilized object to protect it from **movement** during transport.

4. Skill Drill 24-4: Caring for Burns (page 579)

1. Follow BSI precautions to help prevent infection.

    Remove the patient from the burning area; extinguish or remove hot clothing and jewelry as needed.

    If the wound(s) is still burning or hot, immerse the hot area in cool, sterile water, or cover with a wet, cool dressing.

2. Give supplemental oxygen and continue to assess the airway.
3. Estimate the severity of the burn, then cover the area with a dry, sterile dressing or clean sheet.

    Assess and treat the patient for any other injuries.

4. Prepare for transport.

    Treat for shock if needed.

5. Cover the patient with blankets to prevent loss of body heat.

    Transport promptly.

Chapter 25: Eye Injuries

Matching

1. D	(page 589)		**6.** G	(page 588)
2. C	(page 589)		**7.** A	(page 588)
3. B	(page 589)		**8.** F	(page 588)
4. E	(page 589)		**9.** I	(page 588)
5. J	(page 588)		**10.** H	(page 589)

Multiple Choice

1. B (page 588)

2. D (page 588)

3. A (page 589)

4. A (page 589)

5. D (page 589, 590)

6. C – Also record the severity and duration of signs and symptoms and any history of previous eye surgery. (page 590)

7. D (page 592)

8. D (page 594)

9. B (page 594)

10. D (page 594)

11. B (page 594)

12. C – Compression can interfere with the blood supply to the back of the eye and result in loss of vision from damage to the retina. (page 594)

13. D – Do not attempt to reposition an eyeball in its socket. Simply cover the eye, and stabilize it with a moist, sterile dressing. Have the patient lie supine while en route to the hospital. (page 595)

14. A (page 595)

15. B (page 595)

16. A (page 596)

17. D – Signs of a possible head injury also include the eyes not moving together or pointing in different directions, and failure of the eyes to follow the movement of your finger as instructed. (page 596)

18. C (page 597)

Vocabulary

1. Blowout fracture: Fracture of the orbit or of the bones that support the floor of the orbit. (page 596)

2. Hyphema: Bleeding into the anterior chamber of the eye. (page 595)

3. Conjunctivitis: Inflammation of the conjunctiva. (page 590)

Fill-in

1. lacrimal glands (page 588)

2. optic nerve (page 589)

3. camera (page 589)

4. chemical burn (page 597)

5. orbit (page 588)

6. abnormalities (page 589)

7. orbit (page 590)

8. Conjunctivitis (page 590)

True/False

1. F (page 591)
2. F (page 588)
3. T (page 588)
4. F (page 588)
5. F (page 597)

6. F (page 591)
7. F (page 589)
8. T (page 589)
9. F (page 590)

Short Answer

1. A condition in which the retina is separated from its attachments at the back of the eye. Common findings include flashing lights, floaters, or a cloud or shade over the patient's vision. However, pain is not a common complaint. (page 589)

2. 1. Never exert pressure on the eye.

   2. If part of the eyeball is exposed, gently apply a moist, sterile dressing to prevent drying.

   3. Cover the eye with a protective shield or sterile dressing. (page 595)

3. 1. One pupil larger than the other

   2. Eyes not moving together or pointing in different directions

   3. Failure of the eyes to follow movement when instructed

   4. Bleeding under the conjunctiva

   5. Protrusion or bulging of the eye (page 596)

Word Fun

Ambulance Calls

1. Stabilize object with bulky dressings.

   Cover both eyes.

   Place patient on stretcher in position of comfort.

   Reassure patient.

   Rapid transport

2. Position patient supine on stretcher.

   Set up IV bags to use as irrigation hoses.

   Connect IV line to a nasal cannula and place the prongs over the nose so that the fluid runs into the eyes.

   Monitor patient and flush eyes continuously en route to the hospital.

   Rapid transport

3. Position patient supine on stretcher.

   Cover both eyes to minimize movement.

   Monitor patient and vital signs.

   Rapid transport

Skill Drills

1. **Skill Drill 25-1: Removing a Foreign Object from Under the Upper Eyelid (page 591)**

   1. Have the patient look **down**, grasp the **upper lashes**, and gently pull the lid away from the eye.
   2. Place a **cotton-tipped applicator** on the upper lid.
   3. Pull the lid **forward** and **up**, folding it back over the applicator.
   4. Gently remove the foreign object with a **moistened**, **sterile** applicator.

2. **Skill Drill 25-2: Stabilizing a Foreign Object Impaled in the Eye (page 592)**

   1. To prepare a doughnut ring, wrap a 2" roll around your fingers and thumb **seven or eight** times. Adjust the diameter by **spreading** your fingers.
   2. Wrap the remainder of the roll...
   3. ...working around the ring.
   4. Place the dressing over the **eye** to hold the impaled object in place, then **secure** it with a **gauze** dressing.

Chapter 26: Face and Throat Injuries

Matching

1. C   (page 602)
2. B   (page 602)
3. E   (page 602)
4. F   (page 602)
5. G   (page 602)
6. D   (page 604)
7. A   (page 602)

Multiple Choice

1. D (page 602)
2. B (page 602)
3. C (page 602)
4. D (page 602)
5. A (page 602)

6. D (pages 602 to 603)

7. C (page 604)

8. D (page 604)

9. A – Always place amputated or avulsed parts in a plastic bag and in ice water. Placing the part directly on ice can cause frostbite. (page 606)

10. D (page 606)

11. B (page 606)

12. C (page 607)

13. B (page 606)

14. D – Also, assume a facial fracture in patients who have sustained a direct blow to the mouth or nose. A collision with a steering wheel or windshield, or a hit with a baseball bat or pipe are examples of blunt impact that cause facial fractures. (page 608)

15. D (page 608)

16. A (page 609)

Labeling

Face/Skull

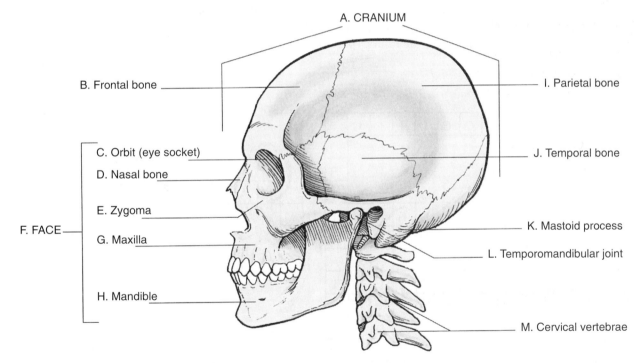

Vocabulary

1. Air embolism: The presence of air in the veins, which can lead to cardiac arrest if it enters the heart. (page 609)

2. Hematoma: The collection of blood in a space, tissue, or organ due to a break in the wall of a blood vessel. (page 604)

3. Sternocleidomastoid muscle: Muscles on either side of the neck that allow movement of the head. (page 604)

4. Subcutaneous emphysema: A characteristic crackling sensation on palpation, caused by the presence of air in soft tissues. (page 608)

5. Temporomandibular joint (TMJ): The joint that is formed where the mandible and the cranium meet, just in front of the ear. (page 602)

Fill-in

1. carotid (page 604)
2. cervical (page 602)
3. temporal (page 602)
4. trachea (page 603)
5. cartilage (page 604)

6. men, women (page 603)
7. foramen magnum (page 602)
8. maxilla (page 602)
9. parietal (page 602)
10. trachea (pages 604)

True/False

1. T   (page 604)
2. T   (page 605)
3. F   (page 605)
4. F   (page 607)
5. T   (page 608)
6. T   (page 602)

Short Answer

1. Direct manual pressure with a dry dressing. Use roller gauze around the circumference of the head to hold pressure dressing in place. (page 605)
2. Apply direct pressure above and below the injury. (page 609)

Word Fun

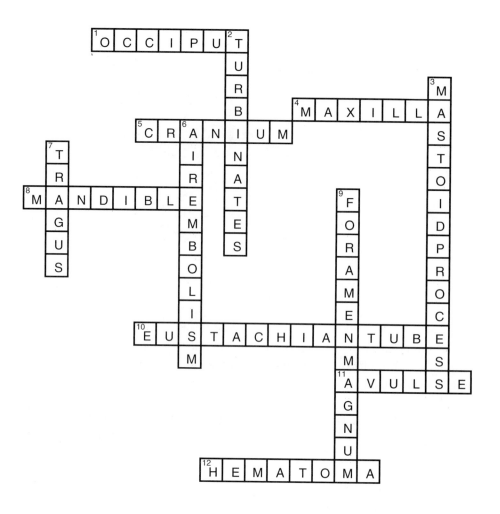

Ambulance Calls

1. -Have patient lean forward while pinching the nostrils together.

   -Assess the airway to assure patency.

2. -Apply direct pressure, being careful not to compress both sides of the neck.

   -Apply high-flow oxygen.

   -Transport in a reclining position.

   -Monitor vital signs.

3. -Transport in position of comfort.

   -Apply high-flow oxygen.

   -Bandage ear when bleeding is controlled.

   -Wrap avulsed part in moist, sterile dressing and place in plastic bag. Keep cool.

   -Monitor vital signs.

Skill Drills

1. **Skill Drill 26-1: Controlling Bleeding from a Neck Injury (page 609)**

1. Apply **direct pressure** to control bleeding.
2. Use a **roller gauze** to secure a dressing in place.
3. Wrap the bandage around and under the patient's **shoulder**.

Chapter 27: Chest Injuries

Matching

1. B   (page 614)
2. D   (page 614)
3. C   (page 614)
4. A   (page 614)
5. E   (page 614)

6. H   (page 615)
7. I   (page 617)
8. J   (page 621)
9. F   (page 615)
10. G   (page 616)

Multiple Choice

1. B (page 614)
2. A (page 614)
3. C (page 614)
4. D (page 615)
5. D – (A), (B), and (C) are signs. (page 616)
6. D (page 616)
7. B – Flail segment (A) is a common cause of paradoxical motion (pages 616, 620)
8. C (page 617)
9. D (page 617)
10. C (pages 617 to 618)
11. D (page 618)
12. D (page 619)
13. D – It also includes tachycardia, low blood pressure, cyanosis, and decreased breath sounds on the injured side. (page 619)
14. D (page 619)
15. D (page 619)

16. A (page 620)

17. C – Bruising of the lung (A) is called a pulmonary contusion. Broken ribs in two or more places (B) is called flail chest. (page 620)

18. D (page 620)

19. D – Signs and symptoms of pericardial tamponade also include jugular vein distention and a decrease in the difference between the systolic and diastolic blood pressure. (page 621)

20. B (page 621)

Labeling

1. Anterior Aspect of the Chest (page 615)

2. The Ribs (page 615)

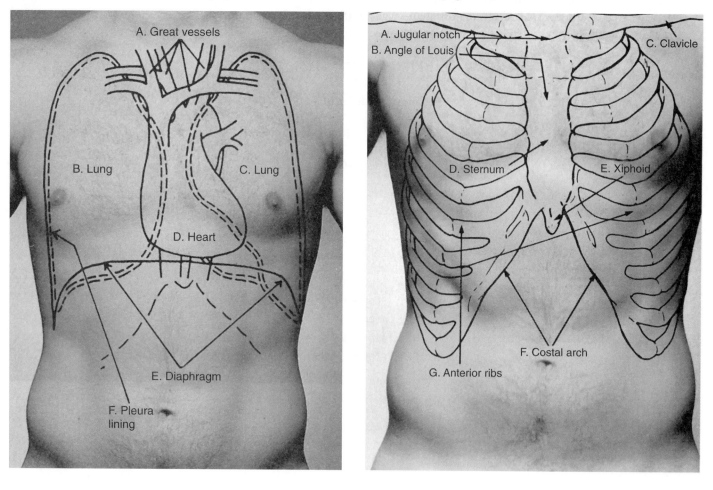

Vocabulary

1. Flail chest: A condition in which three or more ribs are fractured in two or more places, or in association with a fracture of the sternum, so that a segment of chest wall is effectively detached from the rest of the thoracic cage. (page 620)

2. Paradoxical motion: The motion of a portion of the chest wall that is detached in a flail chest; the motion—in during inhalation, out during exhalation—is exactly the opposite of normal motion during breathing. (page 620)

3. Pericardial tamponade: Compression of the heart due to a buildup of blood or other fluid in pericardial sac. (page 621)

4. Spontaneous pneumothorax: Pneumothorax that occurs when a weak area on the lung ruptures in the absence of major injury, allowing air to leak into the pleural space. (page 618)

5. Sucking chest wound: An opening or penetrating chest wall wound through which air passes during inspiration and expiration, creating a sucking sound. (page 617)

6. Tension pneumothorax: An accumulation of air or gas in the pleural cavity that progressively increases the pressure in the chest with potentially fatal results. (page 618)

Fill-in

1. back (page 614)
2. decreases (page 614)
3. sternum (page 614)
4. bronchi (page 614)
5. phrenic (page 614)
6. ribs (page 614)
7. diaphragm (page 614)
8. Pleura (page 617)
9. aorta (page 614)
10. contracts (page 614)

True/False

1. T   (page 616)
2. F   (page 616)
3. T   (page 619)
4. F   (page 619)
5. F   (page 618)
6. F   (page 621)
7. F   (page 614)
8. T   (page 614)
9. F   (page 615)

Short Answer

1. -Pain at the site of injury
   -Pain localized at the site of injury that is aggravated by or increased with breathing
   -Dyspnea (difficulty breathing, shortness of breath)
   -Hemoptysis (coughing up blood)
   -Failure of one or both sides of the chest to expand normally with inspiration
   -Rapid, weak pulse and low blood pressure
   -Cyanosis around the lips or fingernails. (page 616)

2. 1. Seal the wound with a large airtight dressing that seals all four sides.
   2. Seal the wound with a dressing that seals three sides with the fourth side as a flutter valve.
   Your local protocol will dictate the way you are to care for this injury. (page 618)

3. Tape a bulky pad against the segment of the chest. (page 620)

4. Sudden severe compression of the chest, causing a rapid increase of pressure within the chest. Characteristic signs include distended neck veins, facial and neck cyanosis, and hemorrhage in the sclera of the eye. (page 620)

Word Fun

Crossword answers:

1. HEMOTHORAX
2. HEMOPTYSIS
3. TENSIONPNEUMOTHORAX
4. PULMONARYCONTUSION
5. PARADOXICALMOTION
6. FLUTTERVALVE
7. TACHYPNEA
8. FLAILCHEST
9. PERICARDIALTAMPONADE
10. DYSPNEA
11. PNEUMOTHORAX

Ambulance Calls

1. -Immediately begin assisting ventilations with a BVM attached to 100% oxygen.
   -Begin chest compressions.
   -Rapid transport
   -Apply an AED as soon as possible.
   -Continue CPR and continuously monitor the patient.

2. -Apply high-flow oxygen.
   -Monitor vital signs.
   -Rapid transport in position of comfort
   -Continue to monitor patient for signs of distress en route to the hospital.

3. -Apply high-flow oxygen via nonrebreathing mask or BVM.
   -Fully c-spine immobilized patient.
   -Rapid transport
   -Monitor vital signs en route.

Chapter 28: Abdomen and Genitalia Injuries

Matching

1. G   (page 626)
2. D   (page 626)
3. C   (page 626)
4. H   (page 632)

5. E   (page 632)
6. B   (page 631)
7. A   (page 634)
8. F   (page 626)

Multiple Choice

1. D (page 626)
2. D – The intestines are also hollow. (page 626)
3. C (page 626)
4. D (page 626)
5. B – The abdomen becomes distended and firm to touch (A), normal bowel sounds diminish or disappear (C). (page 626)
6. D (page 626)
7. A (page 627)
8. C (page 627)
9. B (page 627)
10. A (page 627)
11. D (page 627)
12. C (page 628)
13. B – (A) and (C) are types of open injuries (page 627)
14. A (page 628)
15. B – (A) is a symptom (page 628)
16. B (page 628)
17. B (page 628)
18. D (page 629)
19. D (pages 629 to 630)
20. D – Using a diagonal shoulder safety belts alone can also cause rib fractures and other types of fractures. Far fewer head and neck injuries are seen when this belt is used in combination with a lap belt and a headrest. (page 630)
21. C (page 630)
22. D (page 630)
23. D – Only a surgeon can accurately assess the damage. (page 631)
24. D – Never attempt to replace abdominal contents (A), always keep organs moist (B), and never use any type of adherent dressings (C). (page 631)
25. A – (B), (C), and (D) are all hollow organs (page 632)
26. C (page 632)
27. D (page 633)
28. D (page 633)
29. A (page 634)
30. B (page 634)
31. D (page 635)
32. D (page 635)

Labeling

1. Hollow organs (page 626)

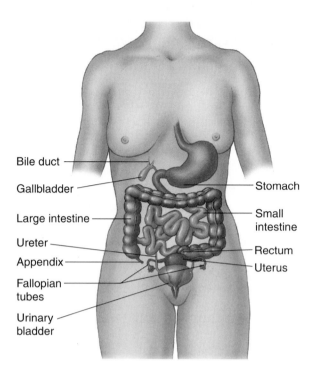

Bile duct
Gallbladder
Large intestine
Ureter
Appendix
Fallopian tubes
Urinary bladder
Stomach
Small intestine
Rectum
Uterus

2. Solid organs (page 627)

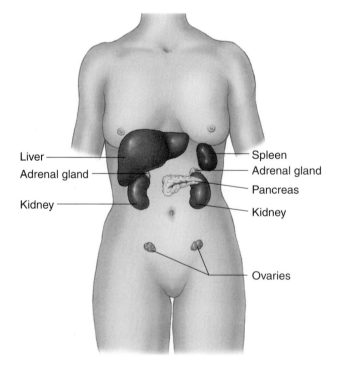

Liver
Adrenal gland
Kidney
Spleen
Adrenal gland
Pancreas
Kidney
Ovaries

Vocabulary

1. Closed abdominal injury: Any injury of the abdomen caused by a nonpenetrating instrument or force in which the skin remains intact; also called blunt abdominal injury. (page 627)
2. Open abdominal injury: Any injury of the abdomen caused by a penetrating or piercing instrument or force, in which the skin is lacerated or perforated and the cavity is opened to the atmosphere; also called penetrating injury. (pages 627 to 628)
3. Guarding: Contracting the stomach muscles to minimize the pain of abdominal movement; a sign of peritonitis. (page 629)

Fill-in

1. solid (page 627)
2. urinary (page 632)
3. retroperitoneal (page 633)
4. External signs (page 633)
5. Peritonitis (page 626)
6. Inflammatory response (page 627)
7. Blunt injuries (page 627)

True/False

1. F   (page 626)
2. T   (page 628)
3. F   (page 629)
4. T   (page 626)
5. F   (page 631)
6. F   (page 632)

Short Answer

1. Stomach, intestines, ureters, bladder, gallbladder, and rectum (page 626)
2. Liver, spleen, pancreas, and kidneys (page 626)
3. Pain, shock signs, bruises, lacerations, bleeding, tenderness, guarding, and difficulty with movement because of pain (page 628)
4. -Inspect the patient's back and sides for exit wounds.

   -Apply a dry, sterile dressing to all open wounds.

   -If the penetrating object is still in place, apply a stabilizing bandage around it to control external bleeding and to minimize movement of the object. (page 631)
5. -Cover with moistened sterile dressing.

   -Secure dressing with bandage.

   -Secure bandage with tape. (page 631)
6. -An abrasion, laceration, or contusion in the flank

   -A penetrating wound in the region of the lower rib cage (the flank) or the upper abdomen

   -Fractures on either side of the lower rib cage or of the lower thoracic or upper lumbar vertebrae (page 633)

Word Fun

Ambulance Calls

1. Assess the ABCs.

Apply high-flow oxygen.

Control any bleeding.

Stabilize the knife in place with bulky dressings – DO NOT REMOVE.

Keep movement of patient to the bare minimum, so as not to create further injury. (Sliding the patient very carefully onto a backboard may help to minimize movement.)

Monitor vital signs.

Rapid transport

Bandage minor lacerations en route.

2. C-spine immobilization

Apply high-flow oxygen.

Normal transport unless patient's condition changes

Continue your assessment en route.

Monitor vital signs and airway continuously.

3. Apply high-flow oxygen via nonrebreathing mask.

Cover the abdominal contents with moist, sterile gauze and an occlusive dressing.

Transport in the position of comfort (Probably supine with knees bent).

Rapid transport

Monitor vital signs and airway continuously.

Chapter 29: Musculoskeletal Care

Matching

1. G   (page 640)

2. J   (page 640)

3. D   (page 640)

4. I   (page 642)

5. F   (page 642)

6. B   (page 644)

7. K   (page 645)

8. A   (page 644)

9. C   (page 642)

10. E   (page 644)

11. H   (page 653)

Multiple Choice

1. A (page 668)

2. B (page 640)

3. A – Minerals (B) and electrolytes (C) are stored in the bone marrow, which serves as a reservoir. (page 641)

4. B (page 642)

5. C (page 642)

6. B (page 642)

7. C – With a strain (B), no ligament or joint damage occurs. (page 643)

8. A (page 643)

9. D – The size of the zone of injury depends on the amount of kinetic energy the tissues absorb from forces acting in the body. (page 643)

10. A (page 644)

11. B (page 644)

12. C (page 644)

13. D (page 644)

14. B (page 644)

15. C (page 644)

16. D (pages 644 to 645)

17. A (page 645)

18. C (page 645 to 646)

19. D (page 645)

20. A (page 645)

21. C (page 645)

22. B (page 645)

23. A (page 646)

24. C (page 646)

25. D (page 647)

26. C – Marked deformity occurs with dislocations and fractures. (page 647)

27. D (page 648)

28. D (page 648)

29. D (page 649)

30. D (page 652)

31. C (page 653)

32. D (page 654)

33. B (page 657)

34. D – Hazards of improper splinting also include aggravation of the injury, and injury to tissue, nerves, blood vessels, or muscles as a result of excessive movement of the bone or joint. (page 659)

35. B (page 661)

36. D (page 666)

37. D (page 670)

38. C – With open fractures, the amount of blood loss may be even greater. (page 671)

39. B (page 672)

40. D (page 672)

41. C (page 672)

42. B (page 673)

43. A (page 674)

44. C (page 674)

Labeling

1. The Human Skeleton

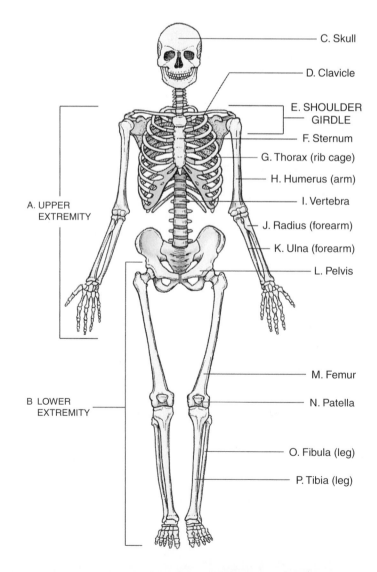

Vocabulary

1. Acromioclavicular (A/C) joint: A simple joint where the bony projections of the scapula and the clavicle meet at the top of the shoulder. (page 661)

2. Compartment syndrome: An elevation of pressure within the fascial compartment, characterized by extreme pain, decreased sensation, pain on stretching of affected muscles, and decreased power; most frequently seen below the elbow or knee in children. (page 648)

3. Dislocation: Disruption of a joint in which ligaments are damaged and the bone ends are completely displaced. (page 642)

4. Nondisplaced fracture: A simple crack in the bone that has not caused the bone to move from its normal anatomic position; also called a hairline fracture. (page 644)

5. Position of function: A hand position in which the wrist is slightly dorsiflexed and all finger joints are moderately flexed. (page 667)

6. Sling: A bandage or material that helps to support the weight of an injured upper extremity. (page 662)

7. Swathe: A bandage that passes around the chest to secure an injured arm to the chest. (page 662)

Fill-in

1. wasting (page 640)

2. red (page 641)

3. hinge (page 642)

4. clavicle (page 661)

5. mechanism of injury (page 648)

6. open fracture (page 648)

7. sciatic nerve (page 669)

8. femur (page 667)

9. crepitus (page 646)

10. reduce (page 647)

11. neurovascular status (page 648)

True/False

1. T (pages 651 to 652)

2. T (page 652)

3. F (page 653)

4. T (page 653)

5. T (page 653)

6. F (page 653)

7. T (page 649)

8. T (page 649)

9. T (page 649)

Short Answer

1. 1. Direct

2. Indirect

3. Twisting

4. High-energy (page 644)

2. Deformity

Tenderness (point)

Guarding

Swelling

Bruising

Crepitus

False motion

Exposed fragments

Pain

Locked joint (pages 645 to 646)

3. 1. Pulse

2. Capillary refill

3. Sensation

4. Motor function (page 649)

4. 1. Remove clothing from the area.

2. Note and record the patient's neurovascular status distal to the site of the injury.

3. Cover all wounds with a dry, sterile dressing before splinting.

4. Do not move the patient before splinting.

5. For a suspected fracture, immobilize the joints above and below the fracture.

6. For a joint injury, immobilize the bones above and below the injured joint.

7. Pad all rigid splints.

8. Maintain manual immobilization to minimize movement of the limb and to support the injury site.

9. Use a constant, gentle manual traction to align the limb.

10. If you encounter resistance to limb alignment, splint the limb in its deformed position.

11. Immobilize all suspected spinal injuries in a neutral in-line position on a backboard.

12. If the patient has signs of shock, align the limb in the normal anatomic position and provide transport.

13. When in doubt, splint. (page 653)

5. 1. Stabilize the fracture.

2. Align the limb.

3. Avoid potential neurovascular compromise. (page 653)

Word Fun

Ambulance Calls

1. -Evaluate ABCs and pulse/motor/sensation in all extremities.

-Apply high-flow oxygen.

-Splint fractured scapula using a sling and swathe.

-Transport patient in position of comfort.

-Consider c-spine immobilization if patient fell, was thrown, or complains of any neck pain.

-Upgrade to rapid transport if patient shows any signs of altered mental status, respiratory distress/compromise, or circulatory compromise.

2. -Splint the arm in position found, since circulation is intact.

-Use a board splint for support with a sling and swathe.

-Immobilize hand in the position of function.

-Apply oxygen as needed.

-Transport in the position of comfort.

-Normal transport.

-Monitor vital signs.

3. -Treat for shock (hypovolemic).

-Apply high-flow oxygen.

-Apply PASG with c-spine immobilization.

-Place patient in Trendelenburg's position.

-Keep patient warm.

-Rapid transport.

-Continue to reassess constantly en route to the hospital.

Skill Drills

1. Skill Drill 29-1: Assessing Neurovascular Pulse (page 650)

1. Palpate the **radial** pulse in the upper extremity.
2. Palpate the **posterior tibial** pulse in the lower extremity.
3. Assess capillary refill by blanching a fingernail or **toenail**.
4. Assess sensation on the flesh near the **tip** of the **index** finger.
5. Also check hand sensation on the **little** finger.
6. On the foot, first check sensation on the flesh near the **tip** of the **great toe**.
7. Also check foot sensation on the **lateral side**.
8. Evaluate motor function by asking the patient to **open** the hand. (Perform motor tests only if the hand or foot is not **injured**. **Stop** a test if it causes pain.)
9. Also ask the patient to **make a fist**.
10. To evaluate motor function in the foot, ask the patient to **extend** the foot.
11. Also have the patient **flex** the foot and **wiggle** the toes.

2. Skill Drill 29-2: Caring for Musculoskeletal Injuries (page 652)

1. Cover open wounds with a **dry**, **sterile** dressing, and **apply pressure** to control bleeding.
2. Apply a splint and elevate the extremity about 6" (slightly above the level of the **heart**).
3. Apply **cold packs** if there is swelling, but do not place them **directly** on the skin.
4. **Position** the patient for transport and **secure** the injured area.

3. Skill Drill 29-3: Applying a Rigid Splint (page 654)

1. Provide gentle **support** and **in-line traction** of the limb.
2. Second EMT places the splint **alongside** or **under** the limb.
   Pad between the limb and the splint as needed to ensure even pressure and contact.
3. Secure the splint to the limb with **bindings**.
4. Assess and record **distal neurovascular** functions.

4. Skill Drill 29-4: Applying a Zippered Air Splint (page 655)

1. Support the injured limb and apply gentle **traction** as your partner applies the open, deflated splint.
2. Zip up the splint, inflate it by **pump** or by **mouth**, and test the **pressure**. Check and record **distal neurovascular** function.

5. Skill Drill 29-5: Applying an Unzipped Air Splint (page 656)

1. **Support** the injured limb. Have your partner place his or her arm through the splint to grasp the patient's **hand** or **foot**.
2. Apply gentle **traction** while sliding the splint onto the injured limb.
3. **Inflate** the splint.

6. Skill Drill 29-6: Applying a Vacuum Splint (page 657)

1. **Stabilize** and **support** the injury.
2. Place the splint and **wrap** it around the limb.
3. **Draw** the air **out of** the splint and **seal** the valve.

7. Skill Drill 29-7: Applying a Hare Traction Splint (page 658)

1. Expose the injured limb and check pulse, motor, and sensory function.
   Place the splint beside the uninjured limb, adjust the splint to proper length, and prepare the straps.
2. Support the injured limb as your partner fastens the ankle hitch about the foot and ankle.
3. Continue to support the limb as your partner applies gentle in-line traction to the ankle hitch and foot.
4. Slide the splint into position under the injured limb.
5. Pad the groin and fasten the ischial strap.
6. Connect the loops of the ankle hitch to the end of the splint as your partner continues to maintain traction. Carefully tighten the ratchet to the point that the splint holds adequate traction.
7. Secure and check support straps.
   Assess distal neurovascular functions.

8. Skill Drill 29-8: Applying a Sager Traction Splint (page 660)

1. After exposing the injured area, check the patient's pulse and motor and sensory function.
   Adjust the thigh strap so that it lies anteriorly when secured.
2. Estimate the proper length of the splint by placing it next to the injured limb.
   Fit the ankle pads to the ankle.
3. Place the splint at the inner thigh, apply the thigh strap at the upper thigh, and secure snugly.
4. Tighten the ankle harness just above the malleoli.
   Snug the cable ring against the bottom of the foot.
5. Extend the splint's inner shaft to apply traction of about 10% of body weight.
6. Secure the splint with elasticized cravats.
7. Secure the patient to a long spine board.
   Check pulse, motor and sensory functions.

9. Skill Drill 29-9: Splinting the Hand and Wrist (page 668)

1. Move the hand into the **position of function**. Place a soft **roller bandage** in the palm.

2. Apply a **padded board** splint on the **palmar** side with fingers **exposed**.

3. Secure the splint with a **roller bandage**.

Chapter 30: Head and Spine Injuries

Matching

1. D   (page 683)   **6.** C   (page 682)

2. E   (page 683)   **7.** B   (page 698)

3. H   (pages 684 to 685)   **8.** I   (page 684)

4. G   (page 685)   **9.** J   (page 686)

5. A   (page 686)   **10.** F   (page 683)

Multiple Choice

1. D (page 682)

2. B (page 682)

3. C (page 683)

4. C (page 688)

5. A (page 683)

6. B (page 685)

7. D (page 684)

8. D (page 684)

9. C (page 685)

10. A (page 686)

11. D (page 687)

12. D (page 688)

13. B (page 686)

14. D – Assess pulse and motor and sensory function in all extremities (page 690)

15. B (page 689)

16. A (page 690)

17. A – Remember BSI. (page 693)

18. B (page 692)

19. D (page 696)

20. B (page 696)

21. D (page 697)

22. D – An intracerebral hemorrhage (B) is within the substance of the brain tissue itself. A subdural hematoma (C) is below the dura but outside the brain. (page 697)

23. B (page 697)

24. B (page 698)

25. D – Cyanosis (A) and hypoxia (B) are results of cerebral edema. Vomiting (C) results from increased intracranial pressure. (page 698)

26. C (page 698)

27. C (page 699)

28. D – An example of a congenital problem (B) that could cause unequal pupils is anisocoria. (page 700)

29. C (page 700)

30. B – BSI and scene safety always come first, but (B) is the best answer of the choices given. (pages 700 to 701)

31. D (page 701)

32. C (page 704)

33. A (page 707)

Labeling

1. The Brain (page 683)

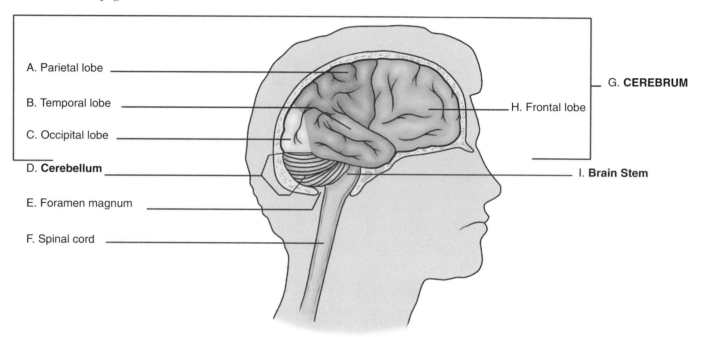

A. Parietal lobe

B. Temporal lobe

C. Occipital lobe

D. **Cerebellum**

E. Foramen magnum

F. Spinal cord

G. **CEREBRUM**

H. Frontal lobe

I. **Brain Stem**

2. The Connecting Nerves in the Spinal Cord (page 685)

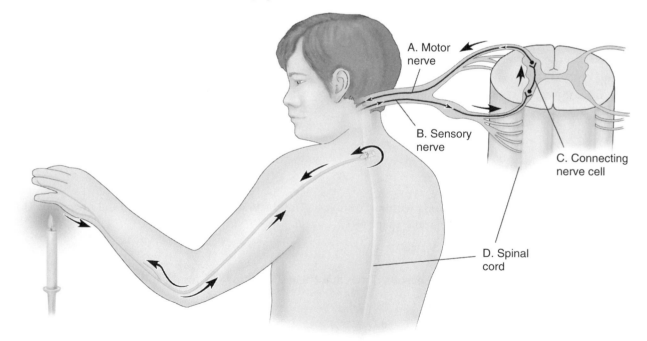

A. Motor nerve

B. Sensory nerve

C. Connecting nerve cell

D. Spinal cord

3. The Spinal Column (page 686)

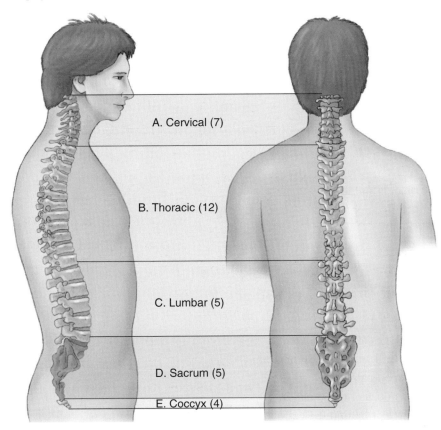

A. Cervical (7)

B. Thoracic (12)

C. Lumbar (5)

D. Sacrum (5)

E. Coccyx (4)

Vocabulary

1. Retrograde amnesia: The inability to remember events leading up to a head injury. (page 697)

2. Anterograde (posttraumatic) amnesia: Inability to remember events after an injury. (page 697)

3. Closed head injury: Injury usually associated with trauma in which the brain has been injured but the skin has not been broken and there is no obvious bleeding. (page 698)

4. Eyes-forward position: A position in which the head is gently lifted until the patient's eyes are looking straight ahead and the head and torso are in line. (page 689)

5. Open head injury: Injury to the head often caused by a penetrating object in which there may be bleeding and exposed brain tissue. (page 699)

Fill-in

1. motor (page 684)

2. meninges (page 683)

3. central (page 682)

4. 31 (page 684)

5. cranial (page 684)

6. intervertebral discs (page 686)

7. cranium, face. (page 685)

8. arachnoid, pia mater (pages 683 to 684)

9. sympathetic (page 685)

10. parasympathetic (page 685)

True/False

1. T   (page 686)
2. F   (page 684)
3. F   (page 684)
4. T   (page 685)

5. F   (page 685)
6. F   (page 690)
7. F   (page 687)

Short Answer

1. 1. Does your neck or back hurt?

2. What happened?

3. Where does it hurt?

4. Can you move your hands and feet?

5. Can you feel me touching your fingers? Your toes? (page 687)

2. -Muscle spasms in the neck

-Increased pain

-Numbness, tingling, or weakness

-Compromised airway or ventilations (page 690)

3. 1. Concussion

2. Contusion

3. Intracranial bleeding (pages 696 to 697)

4. -Lacerations, contusions, or hematomas to the scalp

-Soft area or depression upon palpation

-Visible fractures or deformities of the skull

-Ecchymosis about the eyes or behind the ear over the mastoid process

-Clear or pink cerebrospinal fluid leakage from a scalp wound, the nose, or the ear

-Failure of the pupils to respond to light

-Unequal pupil size

-Loss of sensation and/or motor function

-A period of unconsciousness

-Amnesia

-Seizures

-Numbness or tingling in the extremities

-Irregular respirations

-Dizziness

-Visual complaints

-Combative or other abnormal behavior

-Nausea or vomiting (page 699)

5. 1. Establish an adequate airway.

2. Control bleeding.

3. Assess the patient's baseline level of consciousness. (page 700)

6. 1. Is the patient's airway clear?

2. Is the patient breathing adequately?

3. Can you maintain the airway and assist ventilations if the helmet remains in place?

4. How well does the helmet fit?

5. Can the patient move within the helmet?

6. Can the spine be immobilized in a neutral position with the helmet on?
(page 704)

Word Fun

Ambulance Calls

1. -Apply a cervical collar and hold manual stabilization.

-Use a short spinal extrication device.

-Secure patient to long backboard.

-Apply high-flow oxygen.

-Monitor vital signs and continue with assessment.

-Normal transport

2. -Cervical spine immobilization

-Maintain the airway/apply high-flow oxygen—suction as needed.

-Cover the open laceration with gauze, being careful to gently press over the area to avoid further damage.

-Put patient on long backboard and rapid transport.

-Monitor vital signs en route.

3. -Leave the patient in his car seat.

   -Pad appropriately to immobilize the patient.

   -Use blow-by oxygen if the patient will tolerate it.

   -Monitor vital signs.

   -Continue assessment.

   -Rapid transport due to mechanism of injury and death in vehicle

Skill Drills

1. Skill Drill 30-1: Performing Manual In-Line Stabilization (page 689)

1. Kneel behind the patient and place your hands firmly around the **base** of the **skull** on either **side**.
2. Support the lower jaw with your **index** and **long** fingers, and the head with your **palms**.
3. Gently **lift** the head into a **neutral**, **eyes-forward** position, aligned with the torso. Do not **move** the head or neck excessively.
4. Continue to **support** the head manually while your partner places a rigid **cervical collar** around the neck. Maintain **manual support** until you have the patient secured to a backboard.

2. Skill Drill 30-2: Immobilizing a Patient to a Long Backboard (pages 691 to 692)

1. Apply and maintain **cervical stabilization**.
   Assess **distal functions** in all extremities.
2. Apply a **cervical collar**.
3. Rescuers **kneel** on one side of the patient and place **hands** on the far side of the patient.
4. On command, rescuers **roll** the patient toward themselves, quickly examine the **back**, slide the backboard under the patient, and roll the patient onto the board.
5. **Center** the patient on the board.
6. Secure the **upper torso** first.
7. Secure the **chest**, **pelvis**, and **upper legs**.
8. Begin to secure the patient's head using a commercial immobilization device or **rolled towels**.
9. Place **tape** across the patient's forehead.
10. Check all **straps** and readjust as needed.
    Reassess **distal functions** in all extremities.

3. Skill Drill 30-3: Immobilizing a Patient Found in a Sitting Position (page 694)

1. Stabilize the head and neck in a neutral, in-line position.
   Assess pulse, motor, and sensory function in each extremity.
   Apply a cervical collar.
2. Insert a short spine immobilization device between the patient's upper back and the seat.
3. Open the side flaps, and position them around the patient's torso, snug around the armpits.
4. Secure the upper torso flaps, then the midtorso flaps.
5. Secure the groin (leg) straps. Check and adjust torso straps.
6. Pad between the head and the device as needed.
   Secure the forehead strap and fasten the lower head strap around the collar.
7. Wedge a long backboard next to the patient's buttocks.
8. Turn and lower the patient onto the long board.
   Lift the patient, and slip the long board under the spine device.
9. Secure the immobilization devices to each other.
   Reassess pulse, motor, and sensory functions in each extremity.

4. Skill Drill 30-4: Immobilizing a Patient found in a Standing Position (page 695)

1. After **manually** stabilizing the head and neck, apply a **cervical collar**. Position the board **behind** the patient.
2. Position EMT-Bs at **sides** and **behind** the patient.
   Side EMT-Bs reach under patient's **arms** and grasp **handholds** at or slightly above **shoulder** level.
3. Prepare to lower the patient. EMT-Bs on the sides should be **facing** the EMT-B at the head and **wait** for his or her **direction**.
4. On command, **lower** the backboard to the ground.

5. Skill Drill 30-5: Application of a Cervical Collar (page 703)

1. Apply **in-line** stabilization.
2. Measure the proper **collar size**.
3. Place the **chin support** first.
4. **Wrap** the collar around the neck and **secure** the collar.
5. Assure proper **fit** and maintain **neutral**, **in-line** stabilization.

6. Skill Drill 30-6: Removing a Helmet (page 706)

1. Kneel down at the patient's head with your **partner** at one side.
   Open the face shield to assess **airway** and **breathing**. Remove **eyeglasses** if present.
2. Prevent head movement by placing your **hands** on either side of the helmet and fingers on the **lower jaw**. Have your partner **loosen** the strap.
3. Have your partner place one hand at the **angle** of the **lower jaw** and the other at the **occiput**.
4. Gently slip the helmet about **halfway** off, then stop.
5. Have your partner slide the hand from the **occiput** to the **back** of the head to prevent it from snapping back.
6. Remove the helmet and **stabilize** the cervical spine.
   Apply a **cervical collar** and secure the patient to a **long backboard**.
   Pad as needed to prevent neck flexion or extension.

Section 6 Special Populations

Chapter 31: Pediatric Assessment

Matching

1. I   (page 727)
2. F   (page 728)
3. E   (page 726)
4. J   (page 719)
5. B   (page 720)

6. A   (page 722)
7. H   (page 718)
8. G   (page 718)
9. C   (page 717)
10. D   (page 716)

Multiple Choice

1. D (page 716)
2. D (page 717)
3. A (page 717)
4. C – Infants have very little use of their chest muscles to make their chest expand during inspiration, so they depend on the diaphragm. (page 717)
5. B (page 717)
6. D (page 717)
7. C – Physical stimuli include light, warmth, cold, hunger, sound, and taste. (page 718)
8. D (page 718)
9. B (page 719)
10. C (page 720)
11. D (page 722)
12. C (page 722)
13. D (page 724)
14. B (page 726)
15. C (page 726)
16. D (page 726)
17. A (page 728)

Vocabulary

1. Central IV lines: Indwelling IV catheters placed near the heart for long-term use. (page 727)
2. Fluid reservoir: A device on the side of the head, behind the ear, beneath the skin, that collects fluid from a shunt. (page 728)
3. Work-of-breathing (WOB): Often seen as faster breathing, retractions along the chest wall, or the way the child sits and positions himself or herself. (page 722)
4. Mucous plugs: Secretions built up by the body that can plug the tracheostomy tube. (page 726)

Fill-in

1. pediatrics (page 716)
2. tongue (page 716)
3. trachea (page 717)
4. 200 (page 717)
5. Pale skin (page 717)
6. fontanels (page 717)
7. adolescents (page 722)

True/False

1. T (page 719)
2. T (page 717)
3. F (page 718)
4. F (page 718)
5. T (page 720)
6. F (page 724)
7. T (page 725)

8. F (page 725)
9. T (page 726)
10. F (page 728)
11. T (page 728)
12. F (page 729)
13. F (page 729)
14. T (page 729)

Short Answer

1. At first, infants respond mainly to physical stimuli. Crying is the main avenue of expression. Later, infants learn to coo, smile, roll over, and recognize caregivers. Usually, they demonstrate no stranger anxiety. Observe infants from a distance first and allow the caregiver to hold the baby as you perform the examination. (pages 718-719)

2. - Children who were born prematurely and have associated lung disease problems.
   - Small children or infants with congenital heart disease.
   - Children with neurologic disease, such as cerebral palsy.
   - Children with chronic disease or with functions that have been altered since birth.
   (page 726)

3. Toddlers begin to walk and explore the environment. Because they are able to open doors, boxes, etc., injuries in this age group are more frequent. Stranger anxiety develops early in this period. They are also developing their own ideas about almost everything. Make as many observations as you can before touching the child. When appropriate, examine the child on the caregiver's lap, and use a distraction. Allowing the child to examine your equipment may be enough distraction to allow you to complete an assessment. (pages 719 to 720)

4. These children are beginning to act more like adults and can think in concrete terms, respond sensibly to direct questions, and help take care of themselves. Talk to the child, not just the parent when obtaining a history. Give the child choices and explain any procedures. Give them simple explanations and carry on a conversation to distract them. Reward the child after a procedure. (page 721)

5. Adolescents are able to think abstractly, can participate in decision-making, and are more susceptible to peer pressure. They are very concerned about body image and how they appear to peers and others. Respect their privacy at all times. They can understand complex concepts and treatment options, and should be included on decision-making concerning their care. Advise them in advance of painful procedures and talk with them about their interests to distract them. (page 722)

Word Fun

Ambulance Calls

1. - Talk to the child directly.
- Explain that you need to check for a pedal pulse before touching the child
- Once equipment is prepared, show it to the child and explain how it will work.
- Explain that moving the injured leg will be painful and ask the mother to help by allowing the child to squeeze her hand, etc.
- Manipulate the leg as little as possible, but be expedient.
- Transport in the position of comfort, allowing the child to dictate movement within reason.

2. - Immediately open and suction the airway.
- Assess breathing and apply high-flow oxygen by nonrebreathing mask or BVM device.
- Assess patient further en route.
- Rapid transport
- Obtain history from grandmother en route.
- Reassess patient's airway and vital signs en route.

3. - High-flow oxygen—assist ventilations as needed via BVM device.
- Suction if needed.
- Transport child sitting up or with head elevated.
- Rapid transport
- Continually monitor airway and vital signs en route.

Chapter 32: Pediatric Airway and Medical Emergencies

Matching

1. I   (page 742)
2. K   (page 742)
3. B   (page 748)
4. M   (page 742)
5. O   (page 742)
6. E   (page 748)
7. F   (page 752)
8. Q   (page 742)
9. N   (page 738)
10. G   (page 763)
11. P   (page 752)
12. H   (page 746)
13. J   (page 737)
14. D   (page 751)
15. L   (page 762)
16. C   (page 748)
17. A   (page 746)

Multiple Choice

1. C (page 734)
2. D (page 735)
3. A – The ribs (B) are lifted by the intercostal muscles. (page 735)
4. D (page 735)
5. B – Using a nasal airway with a possible basilar skull fracture may result in the adjunct being pushed into the brain. It may also increase intracranial pressure in patients with head trauma. (page 737)
6. D (page 739)
7. D (page 740)
8. A (page 742)
9. D – Signs of complete airway obstruction also include an ineffective cough (no sound) and/or loss of consciousness. (page 743)
10. C – Cyanosis (C) is a late sign if seen at all. (page 746)
11. D – Signs of increased work of breathing also include grunting respirations, retractions, and the tripod position. (page 746)
12. A (page 747)
13. D (page 747)
14. B (page 748)
15. D – Signs that a patient is not breathing adequately also include snoring respirations, caused by the tongue blocking the airway. (page 748)
16. C – Never use alcohol or cold water to cool a patient. (page 749)
17. D (page 749)
18. D (page 749)
19. D (page 750)
20. B (page 750)
21. C (page 751)
22. B (page 751)
23. D (page 751)
24. B – This is a symptom—it is a sign if you can feel nuchal rigidity. (page 752)
25. D (page 752)
26. C (page 752)
27. A – In children, shock is rarely due to a cardiac event. Other common causes of shock in children include severe infection, traumatic injury with blood loss, severe allergic reaction, and blood or fluid around the heart. (page 752)

28. B (page 753)

29. B (page 754)

30. C (page 755)

31. D (page 756)

32. D – Respiratory problems leading to cardiopulmonary arrest also include injury, infections of the respiratory tract or other organ system, electrocution, and poisoning or drug overdose. (page 757)

33. C – Unlike an adult, an unconscious child may respond quickly to ventilation and oxygenation. (page 758)

34. B (page 760)

35. D (page 760)

36. A (page 763)

37. D – When dealing with the death of an infant, your assessment of the scene should also include any signs of illness, including medications, humidifiers, thermometers, etc. (page 763)

38. D – A classic apparent life-threatening event (ALTE) is also characterized by cyanosis and apnea. (page 764)

39. D – Signs of posttraumatic stress also include restlessness, a constant need for food, etc. (page 765)

40. D – In dealing with the family after the death of a child, also:

-Speak to the family members at eye level

-Use the word "dead" or "died" instead of euphemisms

-Offer to call other family members or clergy

-Ask each adult family member individually whether or not he or she wants to hold the child

-Wrap the dead child in a blanket

-Stay with the family while they hold the child

-Ask them not to remove equipment that was used in attempted resuscitation
(page 765)

Vocabulary

1. Meningeal irritation: Pain that accompanies movement of the head in patients with meningitis. Bending the neck forward or back increases the tension within the spinal canal and stretches the meninges. Patients refuse to move neck, lift legs, or curl into a "C" position. (page 752)

2. Sudden infant death syndrome (SIDS): Death of an infant or young child that remains unexplained after a complete autopsy. (page 763)

3. Sellick maneuver: Application of pressure over the trachea just below the Adam's apple to decrease gastric distention and aspiration of vomitus by pushing the larynx back to compress and close off the esophagus. (page 760)

4. Pediatric resuscitation tape measure: A tape that estimates an infant or child's weight on the basis of length and lists appropriate drug doses and equipment sizes on the tape. (page 736)

5. Altered level of consciousness: A mental state in which infants and children may be unresponsive, combative, or confused, may thrash about, or may drift into and out of an alert state. (page 749)

6. Apparent life-threatening event (ALTE): An event that causes unresponsiveness, cyanosis, and apnea in an infant, who then resumes breathing with stimulation. (page 764)

7. Dependent lividity: Pooling of the blood in the lower parts of the body after death. (page 763)

8. Epiglottitis: An infection of the soft tissue in the area above the vocal chords. (page 742)

9. Meconium: A dark green material in the amniotic fluid that can cause lung disease in the newborn. (page 757)

Fill-in

1. status epilepticus (page 747)

2. see, hear (page 734)

3. tongue (page 735)

4. Gastric distention (page 735)

5. Dehydration (page 754)

6. Airway adjuncts (page 735)
7. gag reflex (page 735)
8. Skin condition (page 738)
9. tidal volume (page 739)
10. Infection (page 742)
11. seizure (page 747)
12. Fever (page 751)

True/False

1. F (page 746)
2. F (page 748)
3. T (page 747)
4. T (page 748)
5. F (page 749)
6. T (page 752)

Short Answer

1. By the number of wet diapers; 6 to 10 per day is normal. (page 753)
2. -Alcohol
   -Epilepsy, endocrine, or electrolyte abnormalities
   -Insulin or low blood glucose levels
   -Opiates or other drugs
   -Uremia
   -Trauma or temperature
   -Infection
   -Psychogenic or poison
   -Shock, stroke, or shunt obstruction (page 749)
3. -Nasal flaring
   -Grunting
   -Wheezing, stridor, or other abnormal airway sounds
   -Use of accessory muscles
   -Retractions
   -Tripod position (page 746)
4. -Signs of illness
   -General condition of house
   -Family interaction
   -Site where the infant was discovered (page 763)
5. Things you can do for a family of a SIDS baby:
   -Tell the family how sorry you are.
   -Tell the family whom they can call if they have questions later.
   -Give written instructions and referrals.
   Things you should not say to the family of a SIDS baby:
   -"I know how you feel."
   -"You have other children," or "You can have other children."
   -Don't try to explain why it happened.
   -Don't tell the family they will feel better in time. (page 765)

Word Fun

Ambulance Calls

1. - Position the child with the airway in a neutral sniffing position.
   - Insert a nasopharyngeal airway.
   - Assist ventilations with a BVM and 100% oxygen.
   - Rapid transport, continuing assessment en route.
   - Obtain SAMPLE history from staff en route.

2. - Allow the child to remain in the mother's arms to decrease her anxiety.
   - Offer oxygen via nonrebreathing mask with the mother holding it.
   - If she will not tolerate the nonrebreathing mask, use blow-by oxygen.
   - Allow the mother to ride in the patient compartment to comfort the child.
   - Rapid transport in position of comfort with as much oxygen as she will tolerate.
   - Continually assess patient for signs of altered mental status and decreasing tidal volume; be prepared to assist ventilations.
   - Obtain further history en route.

3. - Assess ABCs, apply high-flow oxygen if patient will tolerate.
   - Place patient in ambulance and begin cooling him with tepid water.
   - Prompt transport without lights and siren unless patient starts seizing again.
   - Obtain further history en route.
   - Be alert to the possibility of another seizure.

Skill Drills

1. **Skill Drill 32-1: Positioning the Airway in a Child (page 736)**

1. Position the child on a **firm** surface.
2. Place a **folded** towel about **1** inch thick under the **shoulders** and **back**.
3. **Immobilize** the forehead to limit **movement**.

2. **Skill Drill 32-2: Inserting an Oropharyngeal Airway in a Child (page 737)**

1. Determine the **appropriately sized** airway.
   Confirm the correct size **visually**, next to the patient's **face**.
2. Position the patient's **airway** with the appropriate method.
3. Open the mouth.
   Insert the airway until the **flange** rests against the **lips**.
   Reassess the airway.

3. **Skill Drill 32-3: Inserting a Nasopharyngeal Airway in a Child (page 738)**

1. Determine the correct airway size by comparing its **diameter** to the opening of the **nostrils** (naris).
   Place the airway next to the patient's **face** to confirm correct **length**.
   Position the airway.
2. **Lubricate** the airway.
   Insert the **tip** into the right naris with the bevel pointing toward the **septum**.
3. Carefully move the tip forward until the **flange** rests against the **outside** of the nostril.
   Reassess the **airway**.

4. **Skill Drill 32-4: One-Person BVM Ventilation on a Child (page 741)**

1. Open the airway and insert the appropriate airway adjunct.
2. Hold the mask on the patient's face with a one-handed head tilt-chin lift technique ("E-C grip").
   Ensure a good mask-face seal while maintaining the airway.
3. Squeeze the bag 20 times/min for a child, or 30 times/min for an infant. Allow adequate time for exhalation.
4. Assess effectiveness of ventilation by watching bilateral rise and fall of the chest.

5. **Skill Drill 32-5: Removing a Foreign Body Airway Obstruction in an Unconscious Child (page 744)**

1. Position the child on a firm, flat surface.
2. Inspect the airway. Remove any visible foreign object, if you can see it.
3. Attempt rescue breathing. If unsuccessful, reposition the head and try again.
4. If ventilation is unsuccessful, position your hands on the abdomen above the navel and well below the chest cage.
   Give five abdominal thrusts.
5. Open the airway again to try to see the object.
6. Only try to remove the obstruction if you can see it.
   Attempt rescue breathing. If unsuccessful, reposition the head and try again.
   Repeat abdominal thrusts if obstruction persists.

6. Skill Drill 32-6: Caring for a Child in Shock (page 754)

1. Open the airway.
   Be prepared to ventilate.
   Control any bleeding.
   Begin supplemental oxygen.

2. Position the patient with the head lower than the feet.
3. Keep the patient warm with blankets.
   Transport immediately.
   Continue to monitor vital signs.
   Consider ALS backup.
4. Allow a caregiver to accompany the child if possible.

7. Skill Drill 32-7: Performing Infant Chest Compressions (page 761)

1. Position the infant on a **firm** surface while **maintaining** the airway.
   Place two **fingers** in the **middle** of the sternum just below a line between the **nipples**.
2. Use two fingers to **compress** the chest about ½ inch to **1** inch at a rate of **100** times/min.
   Allow the sternum to return **briefly** to its **normal** position between compressions.
3. Coordinate rapid **compressions** and **ventilations** in a **5:1** ratio.
   Check for return of **breathing** and **pulse** after **1** minute, then every **few** minutes.

8. Skill Drill 32-8: Performing CPR on a Child (page 762)

1. Place the child on a **firm** surface and maintain the airway with one **hand**.
2. Place the **heel** of your other hand over the lower half of the **sternum**, avoiding the **xiphoid**.
3. Compress the chest about **1** inch to 1½ inches at a rate of **100**/min (about **80**/min with pauses for ventilations).
   Coordinate compressions with ventilations in a **5:1** ratio, pausing for **ventilations**.
4. Reassess for **breathing** and **pulse** after about **1** minute, and then every **few** minutes.
   If the child resumes **effective** breathing, place him or her in the **recovery** position.

Chapter 33: Pediatric Trauma

Matching

1. H	(page 779)		**7.** D	(page 775)
2. K	(page 775)		**8.** B	(page 778)
3. F	(page 772)		**9.** C	(page 778)
4. E	(page 773)		**10.** G	(page 778)
5. I	(page 778)		**11.** A	(page 782)
6. J	(page 773)			

Multiple Choice

1. D (page 772)
2. B – Pulse is determined through palpation, not visualization. (page 773)
3. A – (B) and (C) deal with injury patterns as opposed to psychological maturity. (page 773)

4. D (page 773)

5. C – Elevate (A) only injured extremities if needed. Only assist ventilations (B) if needed. Remove helmets (D) as the situation or local protocols dictate. (page 774)

6. C –Airway is top priority in ANY patient. (page 774)

7. D (page 775)

8. B – Hypotension is a sign of shock in an adult. (page 775)

9. C (page 775)

10. B (page 776)

11. D (page 776)

12. C (page 777)

13. A (page 777)

14. C (page 778)

15. B (page 778)

16. D (page 778)

17. C (page 778)

18. D (page 779)

19. D – Child abuse also includes sexual abuse and any improper or excessive action that injures or otherwise harms a child or infant. (page 779)

20. B (page 782)

21. D – Falls from a bed are not usually associated with fractures. (page 782)

22. A (page 782)

23. D (page 783)

24. D (page 783)

25. C (page 783)

Vocabulary

1. Shaken baby syndrome: Bleeding within the head and damage to the cervical spine of an infant who has been intentionally and forcibly shaken; a form of child abuse. (pages 782 to 783)

2. Child neglect: Children who are often dirty, too thin, and appear developmentally delayed because of lack of stimulation. (page 779)

3. Decorticate posturing: Pulling in; abnormal flexion usually resulting from head trauma. (page 778)

4. Decerebrate posturing: Pushing away; abnormal extension usually resulting from head trauma. (page 778)

Fill-in

1. Trauma (page 772)

2. ribs (page 772)

3. head (page 773)

4. abdominal (page 773)

5. bumper (page 774)

6. head injury (page 774)

7. shock (page 775)

8. Growth plates (page 775)

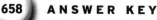

True/False

1. F   (page 772)
2. T   (page 772)
3. T   (page 773)
4. F   (page 774)
5. T   (page 775)
6. T   (page 772)
7. F   (page 779)
8. T   (page 783)
9. T   (page 783)

Short Answer

1. Minor: Partial-thickness burns involving less than 10% of the body surface.

Moderate: Partial-thickness burns involving 10% to 20% of the body surface

Critical: Any full-thickness burn; any partial-thickness burn involving more than 20% of the body surface; any burn involving the hands, feet, airway, or genitalia

(page 777)

2.   1. Is the injury typical for the developmental level of the child?

2. Is the method of injury reported by the parent or caregiver consistent with the child's injuries?

3. Is the caregiver behaving appropriately (concerned about the child's well-being)?

4. Is there evidence of drinking or drug use at the scene?

5. Was there a delay in seeking care for the child?

6. Is there a good relationship between the child and the caregiver?

7. Does the child have multiple injuries at different stages of healing?

8. Does the child have any unusual marks or bruises that may have been caused by cigarettes, grids, or branding injuries?

9. Does the child have several types of injuries, such as burns, fractures, and bruises?

10. Does the child have any burns on the hands and feet that involve a glove distribution (marks that encircle a hand or foot in a pattern that looks like a glove)?

11. Is there an unexplained decreased level of consciousness?

12. Is the child clean and an appropriate weight for his or her age?

13. Is there any rectal or vaginal bleeding?

14. What does the home look like? Clean or dirty? Is it warm or cold? Is there food? (pages 782)

3. -Hot, scalding water

-Hot items on the stove

-Caustic substances (page 776)

Word Fun

Ambulance Calls

1. Suction the airway if needed.

   Immediately begin assisting ventilations with a BVM and 100% oxygen.

   Determine if c-spine control is needed.

   Rapid transport and continue assessment en route.

   Continue to monitor the airway and be alert for vomiting.

   Monitor vital signs and keep patient warm to prevent hypothermia.

2. Immediately begin hyperventilating the patient with a BVM device and 100% oxygen.

   Provide c-spine immobilization and check for a pulse—begin chest compressions if no pulse.

   Be alert for possible vomiting.

   Rapid transport

   Monitor patient closely and continuously en route while continuing to ventilate.

   Alert the ER staff to the possibility of child abuse.

   Report the incident to the proper authorities as set forth by your EMS service.

3. If you are not female, have your female partner talk with the patient.

Do not allow her to bathe, urinate, change clothes, etc.

Take her out to the ambulance (preferably walking to prevent further humiliation).

Close the doors to protect her privacy and question her about where she may hurt or be injured.

Ask her if you may look at her vaginal area to assess the extent of injury and place a trauma dressing in her panties to help with bleeding control.

Cover her with a sheet and assess any other injuries she may have.

Take baseline vital signs and give oxygen if needed.

Talk with patient en route to the hospital about whatever she wants to talk about.

Monitor patient en route and give a thorough report to ER staff.

Skill Drills

1. Skill Drill 33-1: Immobilizing a Child (page 780)

1. Use a towel under the **shoulders** to maintain the head in a **neutral** position.
2. Apply an appropriately sized **cervical collar**.
3. **Log roll** the child onto the **immobilization** device.
4. Secure the **torso** first.
5. Secure the **head**.
6. Ensure that the child is **strapped** properly.

2. Skill Drill 33-2: Immobilization of an Infant (page 781)

1. Stabilize the head in neutral position.
2. Place an immobilization device between the patient and the surface he or she is resting on.
3. Slide the infant onto the board.
4. Place a towel under the shoulders to ensure neutral head position.
5. Secure the torso first; pad any voids.
6. Secure the head.

Chapter 34: Geriatric Assessment

Matching

1. E	(page 790)	**7.** D	(page 795)
2. C	(page 789)	**8.** H	(page 795)
3. G	(page 796)	**9.** A	(page 789)
4. I	(page 797)	**10.** J	(page 796)
5. K	(page 795)	**11.** B	(page 790)
6. F	(page 796)		

Multiple Choice

1. B – Leading causes of death in the elderly also include stroke and trauma. (page 788)

2. D (page 788)

3. D (page 788)

4. A (page 788)

5. C (page 789)

6. D (page 789)

7. B (page 789)

8. C (page 789)

9. B (page 789)

10. D (page 790)

11. A (page 790)

12. C (page 790)

13. A – Kidney function in the elderly declines and they are unable to effectively eliminate medications, which causes a buildup that equates to a drug overdose. (page 790)

14. D – The 10% reduction in brain weight can also result in a decrease in the ability to perform psychomotor skills. (page 790)

15. B – This decrease in bone density is also known as osteoporosis. (page 790)

16. D (page 791)

17. B (page 791)

18. A (page 791)

19. D (page 791)

20. C (page 792)

21. B (page 792)

22. D (page 793)

23. A (page 793)

24. D – Although an injury may be considered isolated and not alarming in most adults, an elderly patient's overall physical condition may lessen the body's ability to compensate for the affects of even simple injuries. (pages 793 to 794)

25. D (page 794)

26. A (page 794)

27. D (page 795)

28. C – Delirium has an acute or recent onset. (page 796)

29. D (page 797)

30. B (page 797)

31. B (page 798)

32. D – Poor temperature regulation is also a sign of neglect. (page 799)

Vocabulary

1. **Advance directives:** Written document that specifies medical treatment for a competent patient should he or she become unable to make decisions. (page 797)

2. **Atherosclerosis:** The most common form of arteriosclerosis in which fatty material is deposited and accumulates in the innermost layer of medium- and large-sized arteries. (page 790)

3. **Arteriosclerosis:** A disease that is characterized by hardening, thickening, and calcification of the arterial walls. (page 790)

4. **Elder abuse:** Any action on the part of an elderly individual's family member, caretaker, or other associated person that takes advantage of the elderly individual's person, property, or emotional state; also called granny battering or parent battering. (page 798)

5. **Osteoporosis:** A generalized bone disease, commonly associated with postmenopausal women, in which there is a reduction in the amount of bone mass leading to fractures after minimal trauma in either sex. (page 794)

Fill-in

1. 65 (page 788)

2. mask (page 788)

3. stereotypes (page 789)

4. sweat glands (page 789)

5. Cardiac output (page 790)

6. aneurysm (page 790)

7. kyphosis (page 790)

True/False

1. F   (page 796)

2. T   (page 788)

3. T   (page 789)

4. F   (page 790)

5. T   (page 792)

6. T   (page 792)

7. T   (page 793)

8. T   (page 795)

9. F   (page 798)

Short Answer

1. -Physical

-Psychological

-Financial (page 799)

2. 1. Cardiac dysrhythmias/dysrhythmias/heart attack: The heart is beating too fast or too slowly, the cardiac output drops, and blood flow to the brain is interrupted. A heart attack can also cause syncope.

2. Vascular and volume: Medication interactions can cause venous pooling, and vasodilation, widening of the blood vessel, results in a drop in blood pressure and inadequate blood flow to the brain. Another cause of syncope can be a drop in blood volume because of hidden bleeding from a condition such as an aneurysm.

3. Neurologic: A transient ischemic attack (TIA) or "brain attack" can sometimes mimic syncope. (page 796)

3. -Repeated visits to the emergency department or clinic

-A history of being "accident prone"

-Soft-tissue injuries

-Unbelievable or vague explanations of injuries

-Psychosomatic complaints

-Chronic pain

-Self-destructive behavior

-Eating and sleep disorders

-Depression or lack of energy

-Substance and/or sexual abuse (page 798)

4. -Dyspnea

-Weak feeling

-Syncope/confusion/altered mental status (page 796)

Word Fun

Ambulance Calls

1. -Survey the scene for any signs or clues to the patient's care.

   -Ask about patient's normal mental status.

   -Apply high-flow oxygen.

   -Obtain patient's medications to take to the hospital.

   -Place patient supine on stretcher with legs elevated.

   -Keep patient warm.

   -Rapid transport

   -Monitor vital signs en route.

2. -C-spine immobilization

   -Apply high-flow oxygen (especially important due to altered mental status that may have caused crash).

   -Control bleeding with direct pressure.

   -Rapid transport due to altered mental status—keep patient warm

   -Question patient about previous medical history and medications.

   -Monitor vital signs and continue the secondary assessment en route.

3. -Fully immobilize patient.

   -Apply high-flow oxygen and keep the patient warm.

   -Treat for possible shock in case of bleeding from possible hip fracture.

   -Rapid transport and continue a complete assessment en route.

   -Monitor vital signs.

   -Accurately document all findings and alert the proper authorities at the receiving facility or your department (according to local protocols) to the possibility of elder abuse.

Section 7 Operations

Chapter 35: Ambulance Operations

Matching

1. B   (page 818)
2. E   (page 818)
3. G   (page 818)
4. I   (page 825)
5. A   (page 818)

6. H   (page 806)
7. D   (page 825)
8. F   (page 818)
9. C   (page 825)

Multiple Choice

1. C (page 806)
2. A (page 806)
3. A (page 807)
4. B (page 807)
5. D (page 808)
6. D (page 808)
7. A (page 809)
8. D (page 809)
9. D (page 810)
10. B (page 811)
11. D (page 812)
12. B (page 812)
13. A (page 813)
14. D (page 814)
15. D (page 815)
16. C (page 816)
17. C (page 816)
18. D – You should not be driving if you are taking medications that may cause drowsiness or slow reaction times. Emotional stability is closely related to the ability to operate under stress. (page 816)
19. B (page 816)
20. D (page 817)
21. B (page 817)
22. A (page 817)
23. A (page 818)
24. D (page 818)
25. A (page 818)
26. C (page 818)
27. D (page 820)
28. B (page 821)
29. C (page 821)
30. C (page 821)
31. B (page 821)

32. D – Guidelines for sizing up the scene also include determining the nature of the illness for a medical patient, determining the mechanism of injury in a trauma patient, and taking BSI precautions. (page 822)

33. D (page 823)

34. C (page 824)

35. D (page 825)

36. D – Air medical unit crews can also include flight nurses. (page 826)

37. D (page 826)

38. D (page 827)

Vocabulary

1. Air ambulances: Fixed-wing aircraft and helicopters that have been modified for medical care; used to evacuate and transport patients with life-threatening injuries to treatment facilities. (page 825)

2. Coefficient of friction: A measure of the grip of the tire on the road surface. (page 818)

3. Decontaminate: To remove or neutralize radiation, chemical, or other hazardous material from clothing, equipment, vehicles, and personnel. (page 809)

4. Hydroplaning: A condition in which the tires of a vehicle may be lifted off the road surface as water "piles up" under them, making the vehicle feel as though it is floating. (page 820)

Fill-in

1. jump kit (page 812)

2. Star of Life (page 807)

3. hearse (page 806)

4. First responder vehicles (page 806)

5. nine (page 807)

6. decontaminate (page 809)

7. airway (page 810)

8. CPR board (page 810)

9. Friction (page 818)

True/False

1. T (page 808)

2. F (page 810)

3. T (page 808)

4. T (page 817)

5. F (page 816)

6. T (page 819)

7. F (page 817)

8. T (page 822)

9. T (page 826)

10. F (page 826)

11. F (page 827)

Short Answer

1. Type I: Conventional, truck cab-chassis with modular ambulance body that can be transferred to a newer chassis as needed.

Type II: Standard van, forward-control integral cab-body ambulance

Type III: Specialty van, forward-control integral cab-body ambulance (page 807)

2.   1. Preparation for the call

2. Dispatch

3. En route to scene

4. Arrival at scene

5. Transfer of patient to the ambulance

6. En route to the receiving facility (transport)

7. At the receiving facility (delivery)

8. En route to station

9. Postrun (page 807)

3. 1. Lack of experience of the dispatcher

2. Inadequate equipment in the ambulance

3. Inadequate training of the EMT

4. Inadequate driving ability

5. Siren syndrome (page 817)

4. 1. The unit must be on a true emergency call to the best of your knowledge.

2. Both audible and visual warning devices must be used simultaneously.

3. The unit must be operated with due regard for the safety of all others, on and off the roadway. (page 821)

5. 1. Select the shortest and least congested route to the scene at the time of dispatch.

2. Avoid routes with heavy traffic congestion.

3. Avoid one-way streets.

4. Watch carefully for bystanders as you approach the scene.

5. Park the ambulance in a safe place once you arrive at the scene.

6. Drive within the speed limit while transporting patients, except in the rare extreme emergency.

7. Go with the flow of the traffic.

8. Use the siren as little as possible en route.

9. Always drive defensively.

10. Always maintain a safe following distance.

11. Try to maintain an open space in the lane next to you as an escape route in case the vehicle in front of you stops suddenly.

12. Use your siren if you turn on the emergency lights, except when you are on a freeway.

13. Always assume that other drivers will not hear the siren or see your emergency lights. (page 822)

6. Approach from the front of the aircraft using an approach area of between nine o'clock and three o'clock as the pilot faces forward. (page 826)

7. -A clear site that is free of loose debris, electric or telephone poles and wires, or any other hazards that might interfere with the safe operation of the helicopter.

-A minimum of 100' by 100' is recommended for the landing zone. (page 827)

Word Fun

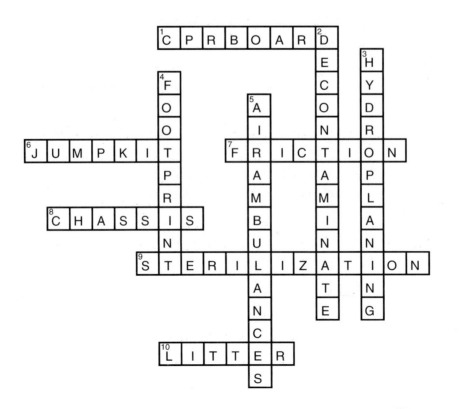

Ambulance Calls

1. -Consult your street and area maps.

   -Ask the dispatcher for a cross street or reference point.

   -During down time in between calls, familiarize yourself with streets and alternative routes.

2. -Take your foot off the gas.

   -Continue to steer, possibly shimmying the steering wheel.

   -Apply a light pressure to the brakes to gradually slow down.

   -Once clear of the water, lightly tap the brakes to dry them.

3. -Park your ambulance in a safe area.

   -Ensure personal safety.

   -Either you or your partner should handle traffic control until the police arrive.

   -The person not handling traffic control should assess the patient.

   -Provide patient care while ensuring personal safety and patient safety.

Chapter 36: Gaining Access

Matching

1. E (page 832)
2. A (page 836)
3. D (page 836)
4. F (page 832)

5. B (page 838)
6. G (page 839)
7. C (page 839)

Multiple Choice

1. D (page 832)
2. B (page 832)
3. C (page 832)
4. A (page 832)
5. D (page 833)
6. A – You have no control over the patient's level of consciousness (B). The patient may be unconscious and the person holding the head is responsible for immobilization (C). (page 834)
7. B – You must triage when multiple patients are involved to ensure that as many patients as possible receive optimal care. (page 835)
8. D (page 836)
9. D (page 837)
10. C (page 837)
11. D (page 838)
12. B (page 839)

Vocabulary

1. Entrapment – To be caught (trapped) within a vehicle, room, or container with no way out, or to have a limb or other body part trapped. (page 832)
2. Technical rescue situation – A rescue that requires special technical skills and equipment in one of many specialized rescue areas. (page 837)
3. Danger zone – An area where individuals can be exposed to sharp metal edges, broken glass, toxic substances, lethal rays, or ignition or explosion of hazardous materials. (page 834)

4. Tactical situation – A hostage, robbery, or other situation in which armed conflict is threatened or shots have been and the threat of violence remains. (page 839)

5. Technical rescue group – A team of individuals from one or more departments in a region that is trained and on call for certain types of technical rescue. (page 837)

Fill-in

1. Removal (page 832)
2. safety (page 832)
3. communication (page 833)
4. stable (page 834)
5. safest (page 835)
6. harm (page 837)
7. self-contained breathing apparatus (page 839)

True/False

1. F (page 832)
2. T (page 833)
3. T (page 834)
4. F (page 835)
5. F (page 836)
6. T (page 835)

Short Answer

1. -Firefighters: responsible for extinguishing any fire, preventing additional ignition, ensuring that the scene is safe, and washing down spilled fuel.

   -Law enforcement: responsible for traffic control and direction, maintaining order at the scene, investigating the crash or crime scene, and establishing and maintaining lines so that bystanders are kept at a safe distance and out of the way of rescuers.

   -Rescue group: responsible for properly securing and stabilizing the vehicle, providing safe entrance and access to patients, extricating any patients, ensuring that patients are properly protected during extrication or other rescue activities, and providing adequate room so that patients can be removed properly.

   -EMS personnel: responsible for assessing and providing immediate medical care, triage and assigning priority to patients, packaging the patient, providing additional assessment and care as needed once the patient has been removed, and providing transport to the emergency department. (pages 832 to 833)

2. Is the patient in a vehicle or in some other structure?

   Is the vehicle or structure severely damaged?

   What hazards exist that pose risk to the patient and rescuers?

   In what position is the vehicle? On what type of surface? Is the vehicle stable or is it apt to roll or tip? (page 835)

3. 1. Provide manual stabilization to protect c-spine, as needed.

   2. Open the airway.

   3. Provide high-flow oxygen

   4. Assist or provide for adequate ventilation.

   5. Control any significant external bleeding.

   6. Treat all critical injuries. (page 836)

4. The extent of injury and whether there is a possibility of hidden bleeding

   Evaluate sensation in the trapped area to discover if an object is pressing on or impaled in the patient. (page 836)

5. Ensure that each EMT-B can be positioned so that he/she can lift and carry at all times.

   Move the patient in a series of smooth, slow, controlled steps, with stops designed between to allow for repositioning and adjustments.

   Plan the exact steps and pathway that you will follow.

   Choose a path that requires the least manipulation of the patient or equipment.

   Make sure that sufficient personnel are available.

   Make sure that you move the patient as a unit.

   While moving the patient, continue to protect him/her from any hazards. (page 837)

Word Fun

Ambulance Calls

1. Prepare all equipment that may be required to immobilize the patient if he has fallen as well as any other equipment you may need.

   Leave all equipment in the back of the unit in case you need to drive to the site where the patient is found.

   Question family members about further history.

   Advise medical control as soon as possible and be prepared for a patient that may be hypothermic.

2. Have someone assume c-spine control.

   Apply high-flow oxygen via nonrebreathing mask.

   Examine as much of the patient as is physically possible and control any bleeding.

   Have someone prepare the stretcher with the backboard and PASG to use as soon as the patient is free.

   Continue to talk to the patient and try to obtain any history.

   Protect patient from debris as extrication is carried out.

   Advise medical control as soon as possible.

   Rapid transport with treatment for shock as soon as patient is disentangled.

3. Try to learn as much about the chemical as possible by having the dispatcher contact CHEMTREC or another agency to find out about possible effects to the patient.

Prepare necessary equipment to manage the airway and ventilation.

Be prepared to do CPR if necessary.

Have all equipment within reach.

Once patient is brought to you, rapidly begin to manage the ABCs and prepare for rapid transport.

Chapter 37: Special Operations

Matching

1. J   (page 853)
2. K   (page 844)
3. L   (page 850)
4. D   (page 853)
5. H   (page 853)
6. E   (page 851)

7. B   (page 858)
8. I   (page 858)
9. A   (page 846)
10. F   (page 857)
11. G   (page 857)
12. C   (page 856)

Multiple Choice

1. D (page 845)
2. D – Extended operations also have finance and administrative sectors as part of their typical incident command structure. (page 845)
3. C (page 846)
4. A (page 856)
5. D (page 857)
6. C (page 857)
7. D (page 857)
8. B (page 856)
9. D – Bus crashes are also an example of a mass-casualty incident. (page 853)
10. D (page 852)
11. B (page 852)
12. D (pages 850 to 851)
13. A (page 850)
14. D (page 849)
15. B (page 848)
16. D (page 848)
17. D (page 848)
18. D (page 847)
19. D (page 847)
20. B (page 847)
21. D (page 847)
22. A (page 854)
23. B (page 854)
24. D (page 854)
25. C (page 854)

26. A (page 854)

27. D (page 854)

28. A (page 854)

29. B (page 854)

30. D (page 854)

31. C (page 854)

Vocabulary

1. Command post: The designated field command center where the incident commander and support personnel are located. (page 845)

2. Danger zone: An area where individuals can be exposed to toxic substances, lethal rays, or ignition or explosion of hazardous materials. (page 848)

3. Hazardous material: Any substance that is toxic, poisonous, radioactive, flammable, or explosive and causes injury or death with exposure. (page 846)

4. Incident commander: The individual who has overall command of the scene in the field. (page 845)

Fill-in

1. incident command system (page 844)

2. command post (page 845)

3. safety officer (page 845)

4. hazardous materials incident (page 846)

5. extrication (page 856)

6. disaster (page 858)

7. Triage (pages 853 to 854)

8. contaminated (page 852)

9. airway, breathing (page 851)

10. Decontamination (page 849)

11. relocate (page 848)

12. toxic (page 847)

13. placard (page 847)

True/False

1. T   (page 846)

2. F   (page 851)

3. F   (page 850)

4. F   (page 851)

5. T   (page 854)

6. F   (page 854)

7. T   (page 847)

8. F   (page 849)

9. F   (page 851)

10. T   (page 858)

11. F   (page 849)

Short Answer

1. Level 0: Materials that would cause little, if any, health hazard if you came into contact with them

Level 1: Materials that would cause irritation on contact, but only mild residual injury, even without treatment

Level 2: Materials that could cause temporary damage or residual injury unless prompt medical treatment is provided

Level 3: Materials that are extremely hazardous to health

Level 4: Materials that are so hazardous that minimal contact will cause death
(page 850)

2. Level A: Fully encapsulated, chemical-resistant protective clothing; SCBA; special, sealed equipment

Level B: Nonencapsulated protective clothing, or clothing designed to protect against a particular hazard; SCBA; eye protection

Level C: Nonpermeable clothing; eye protection; facemasks

Level D: Work uniform (page 851)

3. 1. Command

2. Staging

3. Extrication

4. Triage

5. Treatment

6. Supply

7. Transportation

8. Rehabilitation (pages 856 to 857)

4. First priority (red): Patients who need immediate care and transport

Second priority (yellow): Patients whose treatment and transportation can be temporarily delayed

Third priority (green): Patients whose treatment and transportation can be delayed until last

Fourth priority (black): Patients who are already dead or have little chance for survival (page 854)

5. (page 848)

Class	Type
Class 1	Explosives
Class 2	Gases
Class 3	Flammable liquids
Class 4	Flammable solids
Class 5	Oxidizers
Class 6	Poisons
Class 7	Radioactive
Class 8	Corrosives
Class 9	Miscellaneous

6. The process of removing or neutralizing and properly disposing of hazardous materials from equipment, patients, and rescue personnel. The decontamination area is the designated area where contaminants are removed before an individual can go to another area. (page 849)

Word Fun

Ambulance Calls

1. -The 4-year-old should be triaged as first priority (red).
   -The 27-year-old should be triaged as third priority (green).
   -The 42-year-old should be triaged as fourth priority (black).

2. -Move upwind at least 100 feet.
   -Call for a HazMat response.
   -Try to keep others out of the scene.
   -Call for help from law enforcement.
   -Have the tag number of the vehicle traced to possibly call the company for identification.
   -Do not provide any patient care until the scene is safe.

3. -The EMT riding in the patient compartment should wear goggles, a mask, a gown, two pairs of gloves, a respirator, a HazMat suit, etc.
   -Cabinet doors should be taped shut.
   -All unneeded equipment should be stored in the front of the vehicle or in outside compartments.
   -Turn on the power vent ceiling fan and the patient compartment air-conditioning unit fan.
   -Open the windows in the driver's area and sliding side windows in the patient compartment.

Section 8 Enrichment

Chapter 38: Advanced Airway Management

Matching

1. K (page 867)
2. G (page 867)
3. L (page 867)
4. E (page 867)
5. J (page 867)
6. C (page 867)
7. B (page 867)

8. M (page 880)
9. H (page 869)
10. A (page 874)
11. I (page 869)
12. D (page 871)
13. F (page 872)

Multiple Choice

1. D (page 866)
2. B (pages 866 to 867)
3. C – The larynx (C) is part of the lower airway. (page 867)
4. C (page 867)
5. A (page 867)
6. B (page 867)
7. B (page 867)
8. A (page 867)
9. D (page 867)
10. C (page 868)
11. D (page 868)
12. B (page 868)
13. A (page 868)
14. B (page 869)
15. D (page 869 to 870)
16. B (page 870)
17. D – Intubation is done after defibrillation and one minute of CPR. (page 871)
18. A (page 871)
19. B (page 872)
20. C (page 873)
21. A (page 873)
22. C (page 873)
23. B (page 874)
24. B (page 874)
25. A (page 875)
26. D (page 876)
27. C (page 877)
28. A (page 879)
29. C (page 880)
30. D – BSI is the first step in all procedures. (page 881)

31. D – Other complications of endotracheal intubation include intubating the right mainstem bronchus, aggravating spinal injury, taking too long to intubate, patient vomiting, patient intolerance of the ETT, and a decrease in heart rate. (page 881 to 882)

32. B (page 881)

33. D (page 882)

34. B – Other benefits of using a multi-lumen airway include that it requires minimal skill and practice to maintain, it is easily used in spinal injury patients, it protects the airway from upper airway secretions, it stays in place well, and the balloon is sturdy. (page 883)

35. D – Other contraindications of the ETC include adults who are shorter than 5 feet tall and patients who have a known esophageal disease. (page 883)

36. D (page 885)

37. C – Other contraindications of the PtL include adults shorter than 5 feet tall and patients who have ingested a caustic substance. (page 886)

38. D (page 887)

Labeling

The Upper and Lower Airways (page 866)

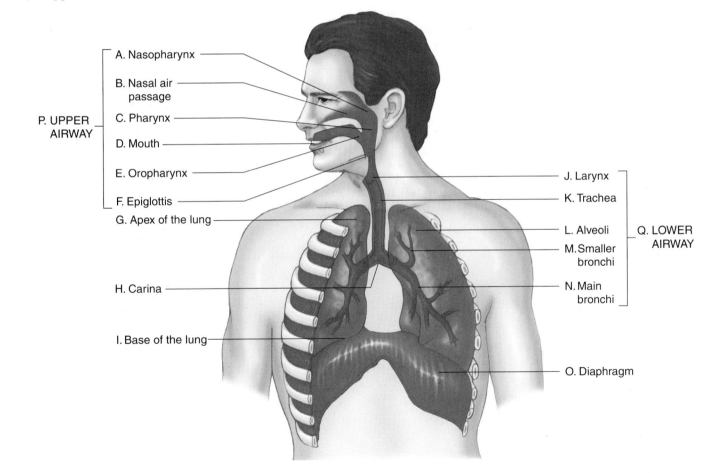

A. Nasopharynx
B. Nasal air passage
C. Pharynx
D. Mouth
E. Oropharynx
F. Epiglottis
G. Apex of the lung
H. Carina
I. Base of the lung
P. UPPER AIRWAY

J. Larynx
K. Trachea
L. Alveoli
M. Smaller bronchi
N. Main bronchi
O. Diaphragm
Q. LOWER AIRWAY

Vocabulary

1. Endotracheal intubation: Insertion of an endotracheal tube (ETT) directly through the larynx between the vocal cords, and into the trachea, to open and maintain an airway. (page 869)

2. Gastric tube: An advanced airway adjunct that provides a channel directly into a patient's stomach, allowing you to remove gas, blood, and toxins, or insert medications and nutrition. (page 868)

3. Nasotracheal intubation: Endotracheal intubation through the nose. (page 869)

4. Orotracheal intubation: Endotracheal intubation through the mouth. (page 869)

5. Sellick maneuver: A technique used with intubation, in which pressure is applied on either side of the cricoid cartilage to prevent gastric distention and allow better visualization of vocal cords; also called cricoid pressure. (page 869)

Fill-in

1. inhalation (page 867)

2. bronchi (page 867)

3. passive (page 867)

4. oxygen (page 867)

5. carbon dioxide (page 867)

6. 4 to 6 (page 868)

7. bronchioles (page 867)

8. capillaries (page 867)

True/False

1. T   (page 872)

2. T   (page 869)

3. F   (page 873)

4. F   (page 874)

5. T   (page 874)

6. F   (page 875)

Short Answer

1. To perform the Sellick maneuver, place a thumb and index finger on either side of the midline of the cricoid cartilage. Apply firm—but not excessive—pressure, as too much pressure could collapse the larynx. Maintain this pressure until the patient is intubated. (page 869)

2. 1. Completely controls and protects the airway

2. Delivers better minute volume without the difficulty of maintaining an adequate mask seal

3. If prolonged ventilation is required, it may be left in place for a long time.

4. Prevents gastric distention

5. Minimizes risk of aspiration of stomach contents into the respiratory system

6. Allows for direct access to the trachea for suctioning

7. Allows for delivery of high volumes of oxygen at higher than normal pressures

8. Provides a route for administration of certain medications (page 870)

3. 1. Intubating the right mainstem bronchus

2. Intubating the esophagus

3. Aggravating a spinal injury

4. Taking too long to intubate

5. Patient vomiting

6. Soft-tissue trauma

7. Mechanical failure

8. Patient intolerance of ETT

9. Decrease in heart rate (pages 881 to 882)

4. 1. Conscious or semi-conscious patients with a gag reflex

2. Children younger than age 16 years

3. Adults shorter than 5 feet tall

4. Patients who have ingested a caustic substance

5. Patients who have a known esophageal disease (page 883)

5. Benefits: Ease of proper placement; no mask seal necessary; requires minimal skill and practice to maintain; easily used in spinal injury patients; may be inserted blindly; protects the airway from upper airway secretions; stays in place well (ETC); sturdy balloon (ETC).

Complications: Loses effectiveness (cuff malfunction); requires deeply comatose patient; requires constant balloon observation; cannot be used on patients shorter than 5' tall; requires great care in listening to breath sounds; large balloon is easily broken and tends to push the PtL out of the mouth when inflated. (page 883)

Word Fun

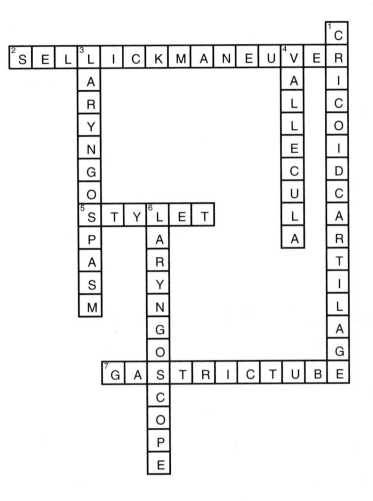

Ambulance Calls

1. -Initiate ventilations by opening the airway, inserting an adjunct, and ventilating with a BVM and 100% O_2.

-Begin chest compressions while your partner readies the AED.

-Defibrillate if the patient is in a shockable rhythm.

-Perform one minute of CPR.

-Prepare for endotracheal intubation.

-Rapid transport

2. -Open the airway and suction if necessary.

-Insert an oral adjunct.

-Ventilate patient with a BVM and 100% oxygen.

-Intubate with an endotracheal tube because the combitube is contraindicated.

-Rapid transport

3. -Maintain c-spine control.

-Use portable suction if needed.

-Ventilate with BVM and 100% oxygen while c-spine and jaw thrust is maintained.

-Insert a combitube to protect the airway (great because it is a blind insertion).

-Continue to ventilate through combitube until patient is extricated.

Skill Drills

1. Skill Drill 38-1: Performing the Sellick Maneuver (page 870)

1. Visualize the **cricoid** cartilage.
2. **Palpate** to confirm its location.
3. Apply **firm** pressure on the cricoid **ring** with your thumb and index finger on either side of **midline**. Maintain pressure until **intubated**.

2. Skill Drill 38-2: Performing Orotracheal Intubation

1. Open and clear the airway.
   Place an oropharyngeal airway and hyperventilate with a BVM.
2. Assemble and test intubation equipment as your partner continues to ventilate.
3. Confirm adequate hyperventilation and remove the oral airway.
   If available, have another rescuer perform the Sellick manuever to improve visualization of the cords.
   Use the head tilt–chin lift maneuver to position the nontrauma patient for insertion of the laryngoscope.
4. In a trauma patient, maintain the cervical spine in-line and neutral as your partner lies down or straddles the patient's head to visualize the vocal cords.
5. Insert the laryngoscope from the right side of mouth and move the tongue to the left. Lift the laryngoscope away from the posterior pharynx to visualize the vocal cords.
   Insert the ETT from the right side. Remove laryngoscope and stylet. Hold the tube carefully until secured.
6. Inflate the balloon cuff and remove the syringe as your partner prepares to ventilate.
7. Begin ventilating and confirm placement of the ETT by listening over both lungs and the stomach. Also confirm placement with an end tidal carbon dioxide detector, if available.
8. Secure the tube and continue to ventilate.
   Note and record depth of insertion, and reconfirm position after each time you move the patient.

Chapter 39: BLS Review

Matching

1. E   (pages 895)
2. D   (page 913)
3. B   (page 894)
4. C   (page 896)
5. A   (page 895)

Multiple Choice

1. D (page 894)
2. C (page 896)
3. B (page 896)
4. D (page 896)
5. A (page 896)
6. D (page 897)
7. D – Causes of respiratory arrest in infants and children also include near-drowning incidents or electrocution. (page 897)
8. D – Signs of irreversible death also include putrefaction or decomposition of the body. (page 897)

9. C (page 898)

10. D (page 899)

11. A (page 900)

12. B – Use the jaw-thrust maneuver with head tilt if there is no suspected spinal injury. If spinal injury is suspected, use the jaw-thrust maneuver without the head tilt. (page 901)

13. C (page 902)

14. B (page 902)

15. A (page 903)

16. D (page 904)

17. A (page 905)

18. D (page 905)

19. B (page 909)

20. D (page 911)

21. D (page 911)

22. B (page 912)

23. D (page 912)

24. D (page 913)

25. B – The abdominal-thrust maneuver (A), also called the Heimlich maneuver (C) may injure the liver or other abdominal organs in an infant, and may not be effective for an obese patient. It also may injure the fetus in a pregnant woman. (page 913)

26. C – You should stay with the patient, give 100% oxygen via nonrebreathing mask, and provide prompt transport. (page 915)

27. D – Transmission of HBV and meningitis can also occur while performing CPR. (page 917)

Vocabulary

1. Gastric distention: A condition in which air fills the stomach as a result of high volume and pressure during artificial ventilation. (page 902)

2. Head tilt–chin lift maneuver: A technique to open the airway that combines tilting back the forehead and lifting the chin. (page 900)

3. Jaw-thrust maneuver: A technique to open the airway by placing the fingers behind the angles of the patient's lower jaw and forcefully moving the jaw forward; can be performed without head tilt. (page 901)

4. Recovery position: A position that helps to maintain a clear airway in a patient with a decreased level of consciousness who has not had traumatic injuries and is breathing on his or her own. (page 903)

Fill-in

1. 4 to 6 (page 895)

2. barrier (page 896)

3. ABCs (page 896)

4. opening (page 897)

5. biological (page 897)

6. Clinical death (page 897)

7. Advance directives (page 898)

8. firm (page 898)

9. airway (page 899)

True/False

1. T (page 896)

2. F (page 896)

3. T (page 896)

4. F (page 903)

5. F (page 900)

6. T (page 897)

7. T (page 905)

8. F (page 907)

9. F (page 904)

10. T (page 904)

11. F (page 904)

Short Answer

1. 1. Rigor mortis, or stiffening of the body after death

2. Dependent lividity (livor mortis), a discoloration of the skin due to pooling of blood

3. Putrefaction or decomposition of the body

4. Evidence of nonsurvivable injury, such as decapitation (page 897)

2. (page 895)

Procedure	Infants (younger than age 1 year)	Children (age 1 to 8 years)
Airway	Head tilt–chin lift; jaw-thrust if spinal injury is suspected	Head tilt–chin lift; jaw thrust if spinal injury is suspected

Breathing
  Initial breaths
 	
  Subsequent breaths
	2 breaths with duration of 1 to 1½ seconds each
	1 breath every 3 seconds; 20 breaths/min
		2 breaths with duration of 1 to 1½ seconds each
		1 breath every 3 seconds; 20 breaths/min

Circulation
  Pulse check
  Compression area
  Compression width
  Compression depth
  Compression rate
  Ratio of Compressions to Ventilations
  Foreign Body Obstruction
	Brachial/femoral arteries
	Lower half of sternum
	2 or 3 fingers
	½ to 1"
	At least 100/min
	5:1 (pause for ventilation)
	
	Back blows and chest thrusts
		Carotid arteries
		Lower half of sternum
		Heel of hand
		1" to 1½"
		100/min
		5:1 (pause for ventilation)
		
		Abdominal thrusts

3. S – The patient starts breathing and has a pulse.

T – The patient is transferred to another person who is trained in BLS or ALS, or another emergency medical responder.

O – You are out of strength or too tired to continue.

P – A physician who is present assumes responsibility for the patient. (page 898)

4. To perform the head tilt–chin lift maneuver, make sure the patient is supine. Place one hand on the patient's forehead, and apply firm backward pressure with your palm to tilt the head back. Next, place the tips of your fingers of your other hand under the lower jaw near the bony part of the chin. Lift the chin forward, bringing the entire lower jaw with it, helping to tilt the head back. (page 900)

5. To perform the jaw-thrust maneuver, kneel above the patient's head. Place your index or middle finger behind the angle of the lower jaw on both sides. Forcefully move the jaw forward, and tilt the head back. Use your thumbs to pull the patient's lower jaw down to allow breathing through the mouth and nose. (page 901)

6. Ensure that the patient is on a firm, flat surface. Place your hands in the proper position. Lock your elbows with your arms straight and your shoulders directly over your hands. Give 15 compressions at a rate of about 100 beats/minute for an adult. Using a rocking motion, apply pressure vertically from your shoulders down through both arms to depress the sternum 1½" to 2" in the adult, then rise up gently. Count the compressions aloud. The ratio of compressions and relaxation should be 1:1. (pages 905 to 907)

7. First EMT-B moves into position to begin chest compressions after giving a breath. Second EMT-B gives the 15th compression, then moves to the patient's head. Second EMT-B checks the carotid pulse for 5 to 10 seconds. If the patient has no pulse, say, "No pulse, continue CPR." (page 909)

8. -Standing: Stand behind the patient and wrap your arms around his or her waist. Press your fist into the patient's abdomen in a series of five quick inward and upward thrusts.

   -Supine: Straddle the hips or legs. Place the heel of one hand against the patient's abdomen and the other hand on top of the first. Press your hands into the patient's abdomen in a series of five quick inward and upward thrusts. (page 913)

9. -Standing: Stand behind the patient and wrap your arms under the armpits and around the patient's chest. Press your fist into the patient's chest and perform backwards thrusts until the object is expelled or the patient becomes unconscious.

   -Supine: Kneel next to the patient. Place your hands as you would to deliver chest compressions. Deliver slow chest thrusts until the object is expelled.(page 914)

10. 1. "Sandwich" the infant between your hands and arms.

    2. Deliver five quick back blows between the shoulder blades, using the heel of your hand.

    3. Turn the infant face up.

    4. Give five quick chest thrusts on the sternum at a slightly slower rate than you would give for CPR. (pages 916 to 917)

Word Fun

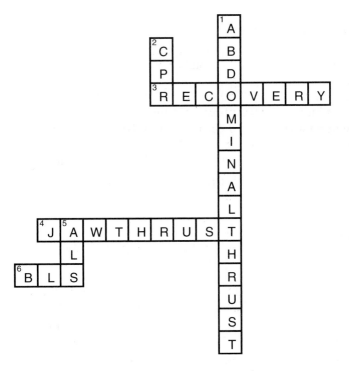

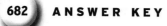
Ambulance Calls

1. -Question the family about the last time they spoke with her.

   -Explain that she has been down too long for CPR to be effective.

   -Comfort family members.

   -Notify your dispatcher to alert the supervisor and either law enforcement, the coroner, or a funeral home according to local protocols.

2. -Immediately assess the airway and apply high-flow oxygen, assisting ventilations as needed.

   -Prepare for rapid transport.

   -Obtain a SAMPLE history from coworkers if possible.

   -Continue your assessment and monitor vital signs en route.

3. -BSI precautions.

   -Position the patient and begin rescue breathing with a BVM and 100% oxygen.

   -Assess circulation at the carotid artery.

   -Find proper hand positioning for chest compressions.

   -Depress sternum 1½ to 2 inches.

   -Move patient onto stretcher on a backboard after delivering 2 breaths via BVM device.

   -Begin CPR again with 2 breaths and assessing pulse.

   -Rapid transport, if no pulse/respirations, continue CPR en route.

Skill Drills

1. Skill Drill 39-1: Positioning the Patient (page 899)

1. Kneel beside the patient, leaving room to roll the patient toward you.
2. Grasp the patient, stabilizing the cervical spine if needed.
3. Move the head and neck as a unit with the torso as your partner pulls on the distant shoulder and hip.
4. Move the patient to a supine position with legs straight and arms at the sides.

2. Skill Drill 39-2: Performing Chest Compressions (page 906)

1. Slide your **index** and **middle** fingers along the rib cage to the **notch** in the **center** of the chest.
2. Push the middle finger high into the notch, and lay the index finger on the **lower portion** of the sternum.
3. Place the **heel** of the second hand on the lower half of the sternum, touching the **index finger** of your first hand.
4. Remove your first hand from the notch, and place it over the **hand** on the sternum.
5. With your arms straight, **lock** your elbows, and position your shoulders directly **over** your hands. Depress the sternum 1½ inches to 2 inches using a rhythmic motion.

3. Skill Drill 39-3: Performing One-Rescuer Adult CPR (page 908)

1. Establish unresponsiveness and call for help.
2. Open the airway.
3. Look, listen, and feel for breathing. If breathing, place in the recovery position and monitor.
4. If not breathing, give two breaths of 2 seconds each.
5. Check for carotid pulse.
6. If no pulse is found, place your hands in the proper position for chest compressions.

   Give 15 compressions at about 100/min.

   Open the airway and give two ventilations of 2 seconds each.

   Perform four cycles of compressions.

   Stop CPR and check for return of the carotid pulse. Depending on patient condition, continue CPR, continue rescue breathing only, or place in the recovery position and monitor.

4. Skill Drill 39-4: Performing Two-Rescuer Adult CPR (page 910)

1. Establish **unresponsiveness** and take positions.
2. **Open** the airway.
3. Look, listen, and feel for breathing. If breathing, place in the **recovery** position and **monitor**.
4. If not breathing, give **two** breaths of **2** seconds each.
5. Check for **carotid** pulse.
6. If no pulse, begin **chest compressions** at about 100/min (**15** compressions to **two** ventilations).

   After **1 minute**, check for **carotid** pulse. Check every few **minutes** thereafter. Depending on patient **condition**, continue CPR, continue rescue breathing only, or place in the recovery position and monitor.

Notes